LOCALISATION AND INTERACTION

in disordered metals and doped semiconductors

Proceedings of the Thirty First Scottish
Universities Summer School in Physics
ST. ANDREWS August 1986

A NATO Advanced Study Institute

edited by

D M FINLAYSON

Published by the
Scottish Universities Summer School in Physics

SUSSP PUBLICATIONS

Edinburgh University Physics Department
King's Buildings, Mayfield Road
Edinburgh

Further copies of this book may be obtained
directly from the above address

ISBN 0 905945 14 X

Produced by Edinburgh University Press
and printed in Great Britain by
Redwood Burn Limited, Trowbridge

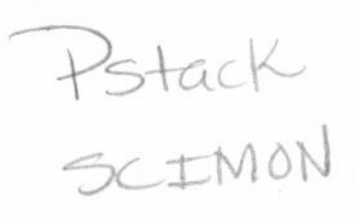

SUSSP PROCEEDINGS

1	1960	Dispersion Relations
2	1961	Fluctuation, Relaxation and Resonance in Magnetic Systems
3	1962	Polarons and Excitons
4	1963	Strong Interactions and High Energy Physics
5	1964	Nuclear Structure and Electromagnetic Interactions
6	1965	Phonons in Perfect and Imperfect Lattices
7	1966	Particle Interactions at High Energies
8	1967	Mathematical Methods in Solid State and Superfluid Theory
9	1968	Physics of Hot Plasma
10	1969	Quantum Optics
11	1970	Hadronic Interactions of Electrons and Photons
12	1971	Atoms and Molecules in Astrophysics
13	1972	Electronic/Structural Properties of Amorphous Semiconductors
14	1973	Phenomenology of Particles at High Energies
15	1974	The Helium Liquids
16	1975	Non-Linear Optics
17	1976	Fundamentals of Quark Models
18	1977	Nuclear Structure Physics
19	1978	The Metal Non-Metal Transition in Disordered Systems
20	1979	Laser-Plasma Interactions
21	1980	Gauge Theories and Experiments at High Energies
22	1981	Magnetism in Solids
23	1982	Lasers: Physics, Systems and Techniques
24	1982	Laser-Plasma Interactions 2
25	1983	Quantitative Electron Microscopy
26	1983	Statistical and Particle Physics
27	1984	Fundamental Forces
28	1985	Superstrings and Supergravity
29	1985	Laser-Plasma Interactions 3
30	1985	Synchrotron Radiation
31	1986	Localisation and Interaction in Disordered Systems.
32	1987	Computational Physics
33	1987	Laboratory and Astrophysical Plasma Spectroscopy

1. S. Sorella
2. P.J. Stiles
3. H. Fukuyama
4. N.C. McGill
5. D.J. Holcomb
6. N.F. Mott
7. D.P. Tunstall
8. D.M. Finlayson
9. J. Kubrycht
10. K. Lumsden
11. A. MacKinnon
12. J. Piquet
13. A. Poustie
14. M. Maliepaard
15. P. Barnsley
16. G. Strinati
17. R. Wehrhahn
18. J. Vogel
19. D. Reich
20. F. Komori
21. V. Matijasevic
22. D.V. Baxter
23. D. Gauthier
24. A. Kent
25. W.P. Beyermann
26. H.P. Loebl
27. N.A. Missert
28. W.F. Smith
29. M. Esguerra
30. R. Chicon
31. P.C. Holdworth
32. M. Migliuolo
33. A. Gorski
34. M. Trudeau
35. T. Capra
36. V.R. Vieira
37. O. Adigüzel
38. C. Shearwood
39. B. Sas
40. E. Castaño
41. J.H. Burnett
42. U. Sivan
43. S. Abboudy
44. Y. Meir
45. H. Yurtseven
46. M.A.K. Mohamed
47. N. Hess
48. R. Germain
49. J. Leo
50. L.R. Tessler
51. G.P. Whittington
52. T. Ohtsuki
53. C.W.J. Beenakker
54. R.P. Taylor
55. J.L. Pichard
56. K.D. Mackay
57. J.C. Licini
58. D.A. Wharam
59. M. Leadbeater
60. C. Gros
61. C. Klingshirn
62. M. Severin
63. R. Richter
64. R. Sollie
65. T. Meisenheimer
66. M. Saint Jean
67. P.Lindqvist
68. P.F. Hopkins
69. V. Chandrasekhar
70. M. Lakrimi
71. H.R. Krauss
72. P.D.S. Sacramento
73. H. White
74. M.N. Ozer
75. S. Yalcin
76. J.C. Villagran
77. F.I.B. Williams
78. G. Dousselin
79. D. El-Khatouri
80. S. Hershfield
81. P. Dellouve
82. N. de Courtenay
83. H. Trzeciakowski
84. T. Wojtowicz
85. T. Giamarchi
86. Y. Nishio

EXECUTIVE COMMITTEE

Tunstall, Dr. D.P.	*St. Andrews (Director)*
Finlayson, Dr. D.M.	*St. Andrews (Secretary and Editor)*
McGill, Dr. N.C.	*St. Andrews (Treasurer)*
Tunstall, Mrs Rosemarie	*St. Andrews (Social Secretary)*
Stradling, Professor R.A.	*Imperial College*
LeComber, Professor P.G.	*Dundee*
Poustie, A.	*St. Andrews (Steward)*
Barnsley, P.	*St. Andrews (Steward)*

LECTURERS

Bergmann, Professor G.	*University of Southern California, Los Angeles, USA.*
Fukuyama, Professor H.	*Tokyo University, Japan.*
Hajdu, Professor J.	*Koln University, West Germany.*
Holcomb, Professor D.F.	*Cornell University, USA.*
MacKinnon, Dr. A.	*Imperial College, UK.*
Mott, Professor N.F.	*Cambridge University, UK.*
Nicholas, Dr. R.	*Oxford University, UK.*
Pepper, Dr. M.	*GEC, Wembley & Cambridge University, UK.*
Stiles, Professor P.	*Brown University, USA.*
Thomas, Dr. G.A.	*AT&T Bell Laboratories, USA.*
von Klitzing, Professor K.	*Stuttgart University, West Germany.*
Wagner, Dr. J.	*Fraunhofer-Institute, Freiburg, West Germany.*

DIRECTOR'S PREFACE

The 31st Scottish Universities Summer School in Physics was held at the University of St. Andrews from the 22nd July to the 7th August, 1986. The title of the School was 'Localization and Interaction in Disordered Metals and Doped Semiconductors', and in many ways it formed a sequel and an up-date of the 19th School, also in St. Andrews, in 1978, on the Metal Non-Metal Transition in Disordered Systems. The interval between these two Schools had seen much progress, both on the experimental front in the millikelvin data in Si:P and in elegant research into the magnetoresistance of thin metal films, and on the theoretical front with the diagrammatic understanding of the phenomenon of weak localization and with the advent of scaling theories of localization in 1-, 2- and 3-dimensional systems. The 31st School presented a synthesis of this new material, and of other exciting recent developments in closely related areas of solid-state physics, such as the Quantum Hall Effect, and its anomalous variant.

The twelve lecturers formed just about the most distinguished group that could be assembled around the topic, and their lectures were enthusiastically received by the 76 post-graduate and post-doctoral student participants. Particular evidence for this latter point was provided by the lively and determined questioning that accompanied the lectures. Overall, there was, I believe, a genuine sense, throughout the body of lecturers and students alike, of being participants in a School of a genuinely tutorial ambiance. This ambiance was further fostered by an extensive programme of short colloquia presented by the students themselves, and by a substantial series of student posters.

These Proceedings represent one of the lasting fruits of our three week School, and constitute an important contribution to the literature surrounding the topic. They will convey to people who did not come to the School at least some of the flavour of the ideas, experiments and concepts that were discussed during those three weeks, and hence maybe lead on to new theories and new experiments. Furthermore, many of the individual articles constitute 'state-of-the-art' reviews of sub-topics within the general area of interest of the School.

The School was predominantly financed by NATO as an Advanced Study Institute, but other substantial contributions came from the Scottish Universities, the Science and Engineering Research Council in the U.K. and the National Science Foundation in the U.S.A., and from the S.U.S.S.P. parent organization itself. This funding allowed us to be generous with student bursaries and hence to attract participation from a cosmopolitan group of bright students, many of whom would have been unable to come otherwise.

I would like to thank for their help in organizing the School my fellow members of the Executive and Bursaries Committees, in particular David Finlayson (Secretary and Editor) and Neil McGill (Treasurer). Special thanks are also due to Jan Kubrycht and Karen Lumsden, for their efficient running of the School office, to Peter Barnsley and Alistair Poustie, the Junior Stewards, and to my wife, Rosemarie, who organized the excellent social programme.

EDITOR'S NOTE

The articles in this volume have been prepared by the authors and closely follow the lectures presented at the School. These were provided in camera-ready form and, because of the excellent co-operation of the lecturers, required minimal editing.

Special thanks are due to Jan Kubrycht for her patience and secretarial skill. Thanks are also due to Karen Lumsden, Brian McAndie, Tom McQueen and Derek Bayne who assisted in various ways towards the production of the School Proceedings.

CONTENTS

THE SCALING THEORY OF LOCALISATION

A.MacKinnon

Physics Dept. Imperial College, London SW7 2BZ

1.
OPENING REMARKS

In these lectures I shall be giving an introduction to the scaling theory of localisation. This approach to the metal insulator transition in disordered systems, which has been generally, but not universally, accepted in recent years, is usually said to have started with the work of Abrahams, Anderson, Licciardello and Ramakrishnan in 1979[1] although it leans heavily on the work of Thouless[2] and Wegner[3] from the mid '70s and indeed on the much earlier perturbation theory of Langer and Neal[4].

I want to provide a more physical than mathematical approach to the subject. In doing so I shall try to use only undergraduate mathematics and quantum mechanics. Occasionally I shall have to sacrifice mathematical rigour for the sake of clarity. This is a subject which has tended to spawn some very difficult mathematics, such as to represent a barrier to communication not only between theory and experiment but also between theorists of different schools. Partly for this reason I propose to completely ignore many body effects, although they certainly play an important role.

I certainly make no apology for trying to avoid these difficulties. I will apologise however if the resulting presentation should turn out to be rather too idiosyncratic.

2.
THE BAND EDGE IN ONE DIMENSIONAL SOLIDS

2.1 Free Electrons

I wish to start by considering the 1-dimensional Schrödinger equation:

$$\frac{d^2\Phi}{dx^2} = \frac{2m}{\hbar^2}(V-E)\Phi \tag{2.1}$$

where V is a constant potential energy and the other symbols have their usual meanings. It is useful at this point to write everything in terms of dimensionless units. To do this let us choose a length scale $\underline{b}$, use the substitution $x'=x/\underline{b}$, and rewrite (2.1) to obtain

$$\frac{d^2\Phi'}{dx'^2} = \underline{b}^2\frac{2m}{\hbar^2}(V-E)\Phi' = -E'\Phi'. \tag{2.2}$$

Since the scale factor, $\underline{b}$,is completely arbitrary it may be chosen to give $E' = \pm 1$. Thus we can write

$$\frac{d^2\Phi'}{dx'^2} = \pm\Phi' \qquad \underline{b} = \left|\frac{2m}{\hbar^2}(E-V)\right|^{-\frac{1}{2}} \tag{2.3}$$

where the sign in (2.3) is positive when $E<V$ and negative when $E>V$. A length which behaves in this way is often called a <u>correlation length</u>. The index $\frac{1}{2}$ which describes the divergence is called a <u>critical index</u> or <u>critical exponent</u>. In the present case the correlation length is, of course, the wavelength or the decay length.

2.2 Tight-Binding Model

Let us now consider a more complicated example. Instead of free electrons we use a tight-binding model for a single band, represented by the Hamiltonian equation

$$E\Phi_n = V\Phi_{n+1} + V\Phi_{n-1}. \tag{2.4}$$

The full solution of this is given by

$$E = 2V\cos \underline{ka} \qquad \Phi_n = \Phi_0 \exp(i\underline{k}n\underline{a}) \tag{2.5}$$

where $\underline{a}$ is the lattice constant.

We can approach this problem in a different way which is more closely related to the analysis used for free electrons. For odd values of n in (2.4) we may write

$$\Phi_{2m+1} = \frac{V}{E}\Phi_{2m+2} + \frac{V}{E}\Phi_{2m}. \tag{2.6}$$

Substituting this form in the equations for even n gives

$$E\Phi_{2m} = \frac{V^2}{E}(\Phi_{2m+2} + \Phi_{2m}) + \frac{V^2}{E}(\Phi_{2m} + \Phi_{2m-2}) \tag{2.7a}$$

$$(E^2 - 2V^2)\Phi_{2m} = V^2(\Phi_{2m+2} + \Phi_{2m-2}). \tag{2.7b}$$

Equation (2.7c) has the same form as (2.4) with the substitutions

$$E' = E^2 - 2V^2 \qquad V' = V^2. \tag{2.8}$$

Actually it is simpler than this. The units can always be chosen to make V=1. We thus have the single equation

$$\varepsilon' = \varepsilon^2 - 2. \tag{2.9}$$

This is an example of a decimation group transformation. The new Hamiltonian, (2.7b), describes a system of half the number of atoms or lattice sites of the original (2.4) but with a lattice constant, $\underline{a}$, which has been doubled. If the solution of (2.4) contains a length, $\underline{b}$, the corresponding solution of (2.7b) will contain $\underline{b}' = \underline{b}/2$, when $\underline{b}$ is expressed in units of $\underline{a}$.

Equation (2.9) is an example of a non-linear mapping very similar to that associated with Feigenbaum[5]. Of particular interest are the fixed points, where $\varepsilon' = \varepsilon$. In this case these occur when $\varepsilon = 2$ or $\varepsilon = -1$. Let us consider now a value of ε very close to one of these fixed points (eg $\varepsilon=2$ $\Rightarrow \varepsilon = 2 + \delta$). Substituting this in (2.9) and ignoring terms quadratic in δ gives

$$\delta' = 4\delta. \tag{2.10}$$

In both cases δ tends to get larger as (2.9) is iterated. The fixed points are said to be unstable.

As in the case of completely free electrons we should like to establish a relationship between the energy ε and the correlation length $\underline{b}$. Close to the fixed points, (2.10) tells us that when $\underline{b}$ is halved δ is multiplied by 4, or $\delta' \sim \underline{b}^2$. In other words

$$\underline{b} \sim (\varepsilon - \varepsilon_c)^{-\frac{1}{2}} \tag{2.11}$$

just as in the first case. This is of course the result we would expect since we know that $\varepsilon = 2\cos \underline{ka}$ is the solution of the problem and $\varepsilon = 2 - (\underline{ka})^2$ close to the band edge.

What happens away from the band edge? When $|\varepsilon| > 2$, ε simply diverges to infinity. This corresponds to an energy outside the band. On the other hand, when $|\varepsilon| < 2$, the mapping, (2.9), has some very rich structure: when the length $\underline{b} = 2\pi/\underline{k}$ is a rational multiple of the lattice constant the mapping has limit cycles. It repeats itself after several iterations. When b is irrational the repeat time becomes infinite and the mapping may be described as chaotic.

This complicated behaviour occurs because, as $\underline{b}$ is repeatedly halved $\underline{k}$ is doubled and eventually moves into the next Brillouin zone. There it becomes equivalent to a smaller $\underline{k}$ in the 1st zone. What is it about the tight-binding case that makes it so much more complicated than free electrons? Whereas the free electrons are characterised by a single length scale, the tight binding model has two such scales: the wavelength or decay length, $\underline{b}$, of the wavefunctions and the lattice constant, $\underline{a}$. When $\underline{b}$ is much larger than $\underline{a}$, the lattice may be ignored. When $\underline{b}$ becomes comparable with $\underline{a}$, the complicated structure appears, and both scales must be taken into account. Close to the band edge, $\underline{b}=\infty$, $\underline{b}$ is often called the relevant length scale and $\underline{a}$ the irrelevant length scale.

2.3 Another Scaling Argument

An alternative approach to either of the above problems is to consider the behaviour of particles in a box of length L. The Heisenberg uncertainty principle tells us that the momentum of the particle must be at least, $p=\hbar/L$. Therefore the energy, $E = p^2/2m$, must be at least

$$E = \hbar^2/(2mL^2) \qquad (2.12)$$

Such a relationship can always be rewritten in the form

$$E = E_0(\underline{b}/L)^2 \qquad (2.13)$$

where E_0 is some energy scale and $\underline{b}$ a length scale. Thus we automatically retrieve the result that $E \sim \underline{b}^2$ as before.

This argument is an example of finite size scaling. Arguments of this type are particularly suited to computer simulation. Whereas it is generally impossible to simulate an infinite system, it is relatively easy to do several calculations for systems of different sizes.

2.4 The Renormalisation Group

The arguments used in the last three sections are all examples of a Renormalisation Group. They each derive information about critical behaviour at the band edge by studying the dependence of the energy on some length scale.

In general we write an equation of the form

$$\underline{A} = \underline{F}(L/\underline{b}) \tag{2.14}$$

where $\underline{A}$ is some property of the system, $\underline{b}$ is a length scale characteristic of the energy (or disorder, temperature, etc.) and L is a length scale imposed on the system. The decimation and finite size scaling arguments are examples of this.

We can also write a differential equation for A

$$d\underline{A}/d\ln L = \beta(\underline{A}). \tag{2.15}$$

Strictly speaking (2.15) does not follow from (2.14), but (2.14) does represent the general solution of (2.15), where $\underline{F}$ is related to β and $\underline{b}$ appears as a constant of integration.

3.
NON-LINEAR DYNAMICS

3.1 Fixed Points and Attractors

The general 1st order differential equation

$$d\underline{x}/dt = \underline{f}(\underline{x}) \tag{3.1}$$

has been the subject of considerable study in recent years[5]. The equations describing most dynamical processes can be written in this general form. Most sets of equations of this type which are of physical interest have the property

$$\text{div}\underline{f}(\underline{x}) < 0 \tag{3.2}$$

Consider a small hypercube in the phase space described by the vectors $\underline{x}$. If all the points in the cube move according to (3.1) and (3.2) is true then the cube gets smaller as time progresses.

If, for example, div$\underline{f}$ is constant then the volume of any hypercube falls exponentially to zero. The simplest possibility is the case of a simple fixed point, $\underline{x}_0$. In this case

$$d\underline{x}/dt) = 0 \quad @ \ \underline{x} = \underline{x}_0 \tag{3.3}$$

and every starting point $\underline{x}$ eventually converges to $\underline{x}_0$.

This is not the only possibility, however. In general the D-dimensional space must condense to a subspace of less than D dimensions. For D=3 the subspace may be a surface (D=2), a line (D=1) or a point (D=0). Such a subspace is called an attractor, as it "attracts" all neighbouring points.

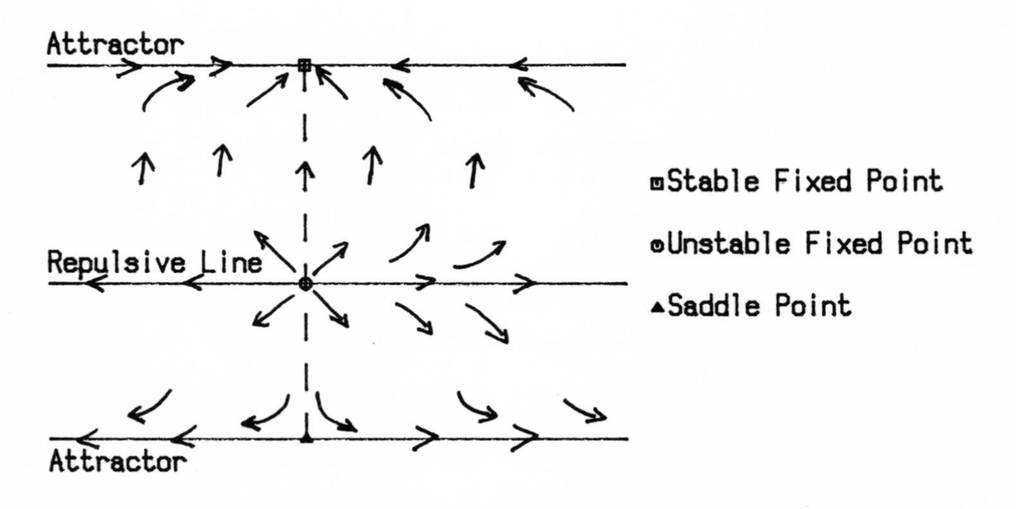

Figure 1.

Figure 1. shows the various possibilities starting from D=2. The diagram illustrates a number of important cases:

a) Between attractors there can be repulsive lines or points.
b) There can be both stable and unstable fixed points.
c) It is possible to have an unstable fixed point on an attractive line. Such a point is stable in directions perpendicular to the line and unstable along the line. This is usually called a saddle point.
d) Only the stable fixed point can actually be reached. Unstable and saddle points are approached and then left again.

3.2 Fixed Point Analysis.

Obviously the general solution of such a system depends on the details of the equations. However, a number of general principles relating to the long time behaviour can be derived by considering the regions close to the fixed points.

Using a Taylor expansion around $\underline{x} = \underline{x}_0$ (3.1) can be written

$$\dot{\underline{x}}_0 + \delta\dot{\underline{x}} = \underline{f}(\underline{x}_0) + [\underline{Df}]\delta\underline{x} \qquad [\underline{Df}] \equiv df_i/dx_j \tag{3.4}$$

Thus $\delta\underline{x}$ obeys the linear equation

$$\delta\dot{\underline{x}} = [\underline{Df}]\delta\underline{x} \tag{3.5}$$

Such an equation is easily solved by looking at the eigenvalues of $[\underline{Df}]$. This gives us a system of scalar equations

$$\delta x_n = \alpha_n \delta x_n \tag{3.6}$$

for each eigenvalue, α_n. The α_n are often called Lyapunov exponents. The solution of (3.6) has the form

$$\delta x_n(t) = \delta x_n(0)\exp(\alpha_n t) \tag{3.7}$$

Two cases must be distinguished:

a) $Re\cdot\alpha_n < 0$: $\delta x_n \to 0$ as $t \to \infty$.

b) $Re\cdot\alpha_n > 0$: $\delta x_n \to \infty$ as $t \to \infty$.

In addition, if $Im\cdot\alpha_n \neq 0$ δx_n oscillates as it converges to or diverges from zero.

The general solution is a linear combination of the eigenfunctions. This leads to 3 distinct situations:

a) all $Re\cdot\alpha_n < 0$: the fixed point is stable.

b) all $Re\cdot\alpha_n > 0$: the fixed point is unstable.

c) some $Re\cdot\alpha_n > 0$ some $Re\cdot\alpha_n < 0$: the fixed point is a saddle point.

3.3 Relationship with Scaling.

By comparing (2.15) with (3.1) we observe that they have the same form apart from the substitution $t \to \ln L$. We thus have the general renormalisation group equation

$$d\underline{x}/d\ln L = \underline{f}(\underline{x}). \tag{3.8}$$

The whole of the above discussion on non-linear dynamics can be directly applied to this case. In particular (3.7) becomes

$$\underline{\delta x_n} = \underline{\delta x}_n^0 L^{\alpha_n} \tag{3.9}$$

and the general solution close to a fixed point takes the form

$$\underline{x} = \underline{x_0} + \Sigma_n a_n \underline{\delta x}_n^0 L^{\alpha_n}. \tag{3.10}$$

The starting point for the dynamics described here is some point $\underline{x}$ in phase space which is determined by the parameters (Energy, disorder, temperature, etc) of the problem in which we are interested. Then the coefficients a_n in (3.10) are (analytical) functions of these parameters.

If we rewrite (3.10) in a dimensionless form we obtain

$$\underline{x}' = \underline{x}'_0 + \Sigma_n c_n \underline{\delta x}_n^0 (L/b_n)^{\alpha n} \tag{3.11}$$

where b_n is a function of the parameters.

Among the sets of parameters {P} there will be some P_n for which the coefficient $c_n(a_n)$ of a particular term in (3.10) or (3.11) is zero. If we now write (3.10) as a Taylor expansion in P around this point we obtain the equation

$$\underline{x}' = \underline{x}'_0 + \Sigma_n c_n (P - P_n) \underline{\delta x}_n^0 L^{\alpha n} \tag{3.12}$$

Comparison of (3.11) and (3.12) leads to the conclusion that

$$b_n = |P - P_n|^{-1/\alpha n} \tag{3.13}$$

This is the general form of the procedure used earlier to derive the behaviour at the band edge of a 1-D system. In that case there was only one exponent $\alpha = 2$.

In general, if we can find the Lyapunov exponents we also have the critical indices or exponents.

3.4 Relevant and Irrelevant Variables.

Let us look more carefully at the limit of large L. Eventually that term will dominate for which $\mathrm{Re}\cdot\alpha_n$ is largest. This term may be referred to as dominant. Any terms for which $\mathrm{Re}\cdot\alpha_n < 0$ can be ignored in this limit. They are irrelevant.

If there is only a single term or at least only one with $\mathrm{Re}\cdot\alpha_n > 0$, the system is said to obey one parameter scaling. This property makes it easier to derive useful results. It will therefore often be assumed or the problem defined to make sure that it is true.

3.5 Universality Classes

Often there will be a whole class of problems, which although differing significantly, have the same type of fixed points, or at least as far as the relevant variables and critical exponents are concerned. The differences are either in the irrelevant variables, the starting point of the integration, or in the behaviour far from the fixed point. Systems which behave similarly in this way are said to belong to the same universality class.

In determining the universality class a particular problem belongs to, symmetry tends to be important rather than the details of interactions or potentials. Thus, in the study of phase transitions, there is a whole class of rather diverse problems which involve variables that can only take one of 2 values and belong to the same universality class as the Ising model of magnetism. In the problem of localisation in disordered systems, as we shall see, all systems belong to the same universality class as long as they have time-reversal symmetry (ie no magnetic fields) and do not involve spin-orbit coupling.

The concept of universality classes provides some justification for the use of mathematical models which bear little apparent relationship to reality. In particular, it is sometimes possible to define a model such that only the relevant variables appear and the irrelevant ones are eliminated.

4. SCALING AND DISORDER

4.1 Length Scales

Before discussing the application of the above ideas to electrons in disordered solids it is worth while to consider the different length scales in the problem. This will be of considerable assistance when we come to try to justify the one parameter scaling of the system. These length scales are of two types: those associated with the random potential and those associated with the wave functions. The following is a (not necessarily complete) list.

a) Potential Length Scales
- i) Lattice constant
- ii) Correlation length
- iii) Average distance between scatterers
- iv) Average size of scatterers

b) Wave Function Length Scales
- i) Wavelength
- ii) Decay length
- iii) Localisation length
- iv) Phase correlation length
- v) Amplitude correlation length
- vi) Mean free path

c) Other Important Length Scales
- i) Inelastic scattering length (mean free path)
- ii) Cyclotron Radius
- iii) Spin-Orbit scattering length
- iv) Magnetic scattering length.

Although some of these are equivalent it is still clear that there are a large number of such scales which could be expected to appear in any application of (3.11) or (3.19) to our problem.

4.2 General Scaling Equation

In fact the situation is even worse than this: we have not yet taken the randomness into account.

Whereas the equations previously discussed were completely deterministic (albeit including the possibility of chaotic behaviour), in the case of a disordered system only the probability distribution of a particular quantity is defined at any time/length scale. Thus the general form of the dynamical equation becomes

$$dP(\underline{x})/d\ln L = \beta[P(\underline{x}),\underline{x}] \tag{4.1}$$

where $P(\underline{x})$ is the probability distribution of $\underline{x}$ and $\beta[\,]$ represents a functional of the distribution. In other words, every possible value of the $\underline{x}$ is represented by an orthogonal direction and $P(\underline{x})$ is a vector in this infinite dimensional space.

The general form of the condition (3.2) that the system condenses onto a subspace in the limit of large system size (the <u>thermodynamic limit</u>) becomes

$$\int D\beta/DP(\underline{x})d\underline{x} < 0 \tag{4.2}$$

where $D/DP(\underline{x})$ represents the functional derivative with respect to the distribution $P(\underline{x})$.

For 1-dimensional systems it can be shown[6] that the distribution is <u>asymptotically log-normal</u>. This condition is sufficient to guarantee the existence of an attractor. It is not necessary to demand that the distribution becomes narrow. In more than 1-D the situation is not so clear.

4.3 One Parameter Scaling

So far what we have derived does not represent a particularly useful approach to the problem. What is required is a proof of the existence of an attractive line or a fixed point. Such an argument was provided by Thouless[2].

Consider a cube of disordered material (as in Fig 2a). This has a spectrum like that in Fig.2b. The eigenenergies are distributed randomly (assumption?) consistent with the density of states. The time taken to cross the cube is

$$t=L^2/D \tag{4.3}$$

where L is the (linear) size of the cube and D is the diffusion constant (assumption?).

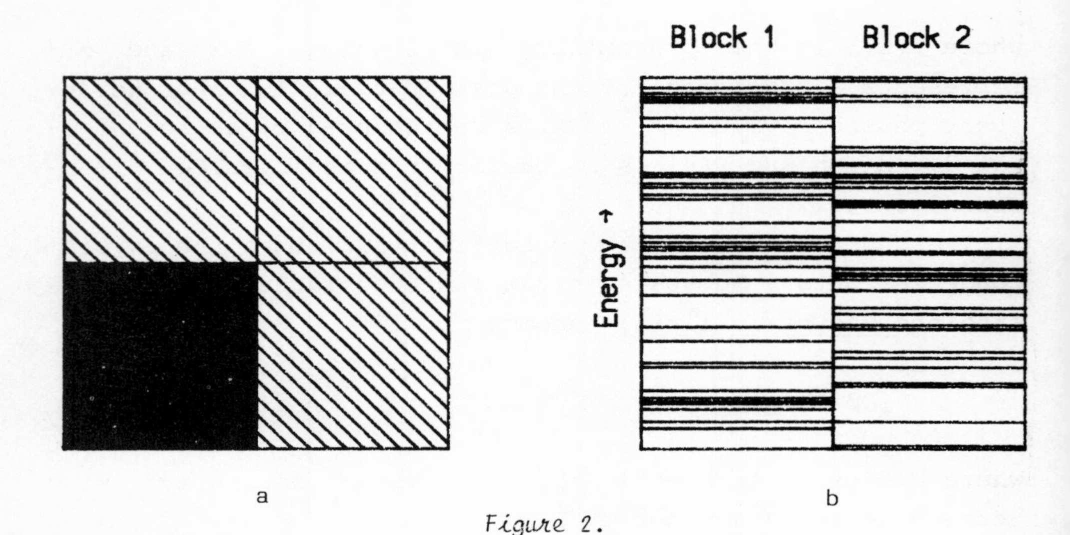

Figure 2.

What happens when several such cubes are joined together to produce a larger cube? The spectra of the cubes are similar but distinct. It is very unlikely that the energies of different cubes match up. However, since the electron only spends a finite time in any cube, there is an uncertainty in its energy given by

$$\delta E = \hbar/t \tag{4.4}$$

If δE is small (t is large), the electron will not be able to get into the next cube and will be localised. If δE is large (t is small) the electron will easily find a compatible state in the next cube. Therefore the important quantity is the ratio of δE to the average separation of the eigenenergies of a cube.

The diffusion constant, D, is related to the conductivity, σ, by the Einstein relation

$$\sigma = e^2 D \frac{dn}{dE} \tag{4.5}$$

where e is the electronic charge and $\frac{dn}{dE}$ is the density of states. Substituting (4.4) and (4.5) in (4.3) gives

$$\delta E = \frac{\hbar}{e^2}\sigma\frac{dE}{dn}L^{-2} = \frac{\hbar}{e^2}\sigma\frac{dE}{dN}L^{d-2} = g\frac{dE}{dN} \qquad (4.6)$$

where $\frac{dE}{dN}$ is the average splitting between eigenenergies, and g is the conductance (apart from a constant factor $e^2/\hbar$). Thus, if the physical picture is correct, that only $\delta E/\frac{dE}{dN}$ is relevant, then it should be possible to express the scaling behaviour of all transport quantities in terms of the dimensionless conductance, g. In particular it should be possible to write down a single parameter scaling function

$$\frac{dg}{d\ln L} = \beta(g) \quad \text{or} \quad \frac{d\ln g}{d\ln L} = \beta(\ln g) \qquad (4.7)$$

It should be noted at this point that this argument is cast in terms of averages. It is not clear however which averages are being discussed: arithmetic, geometric, harmonic or some other mean.

4.4 Diffusion and the Mean Free Path.

The crucial assumption is that the electrons diffuse. What does this mean? The behaviour is diffusive if the motion after some time is uncorrelated with that at an earlier time. In other words, if we observe the electrons only every t seconds, then their behaviour at each observation appears to be unrelated to that at any earlier time. There will be some time scale, t_D, below which the electrons behave ballistically (ie highly correlated) and above which they move diffusively (ie randomly). Associated with this time scale, t_D, will be a distance, l_D, the distance the electrons can travel in time, t_D. This distance is the mean free path.

A more precise treatment of this process, shows that the phases of the wave functions are correlated over a distance, l_D, and uncorrelated beyond this distance. If we choose two points a distance l apart, then, if $l > l_D$ the phase difference between the wave functions at the two points will be a completely random function of which two points are chosen. On the other hand, if $l < l_D$ the phase differences will tend to cluster around a particular value.

All this allows us to identify one of the irrelevant length scales in our problem: the mean free path or <u>phase correlation length</u>. This led Wegner[7] to define a Hamiltonian for which this length scale is zero:

$$H = \Sigma_{ij} V_{ij} |i\rangle\langle j| \tag{4.8}$$

where $\langle V_{ij}\rangle = 0$ is the ensemble average value of the matrix element V_{ij} and the different V_{ij} are statistically independent. This Hamiltonian has the property that any change of phase in V does not change the ensemble. The model is said to be <u>local gauge invariant</u>. Using such a model it is possible to study the one parameter scaling behaviour without the complications of irrelevant variables.

4.5 Anderson Hamiltonian.

For the purpose of the rest of these lectures, I shall concentrate on the Anderson[8] Hamiltonian

$$H = \Sigma_i \varepsilon_i |i\rangle\langle i| + \Sigma_{ij} V_{ij} |i\rangle\langle j| \tag{4.9}$$

where the ε's are independent random variables with mean zero and the V's are zero except between nearest neighbours on a square or cubic lattice. Usually V is chosen to be a constant equal to unity (<u>diagonal disorder</u>), but this is not necessary and probably does not affect the physical results.

Often it is useful to express the Hamiltonian in a basis of plane waves

$$H = \Sigma_k (2V\cos \underline{k}\cdot\underline{a}) |\underline{k}\rangle\langle\underline{k}| + \Sigma_{k,k'} \varepsilon'(\underline{k}-\underline{k}') |\underline{k}\rangle\langle\underline{k}'| \tag{4.10}$$

where the first term is the dispersion of the ordered band and the $\varepsilon'(\underline{k}-\underline{k}')$ is a random function of $\underline{k}-\underline{k}'$ alone [except for the hermiticity condition $\varepsilon'(\underline{k}-\underline{k}')=\varepsilon'^{*}(\underline{k}'-\underline{k})$]. The variance of the distribution of ε' is $\langle\varepsilon^2\rangle/N$, where $\langle\varepsilon^2\rangle$ is the variance of the distribution of ε and N is the number of lattice sites in the system (and the number of distinct $\underline{k}$'s). Note that the ensemble averaged value of H is given by H_0, the first term in (4.10) representing the ordered system, whereas the average of H^2 is

$$\langle H^2\rangle = \langle\varepsilon^2\rangle\Sigma_k|\underline{k}\rangle\langle\underline{k}| + H_0{}^2 \qquad (4.11)$$

This illustrates the important point that terms of order 1/N are not always negligible. There are sometimes N of them.

4.6 Diffusive Solutions

The general solution of any time independent Hamiltonian can be written as

$$\Phi(\underline{k},t) = \exp(iHt)\Phi(\underline{k}_0,0) \quad (\hbar = 1) \qquad (4.12)$$

The quantity we should like to know is the probability of finding the system in a state $|\underline{k}\rangle$ if it started in state $|\underline{k}_0\rangle$, ie.

$$P(\underline{k},t) = |\langle\underline{k}|\exp(iHt)|\underline{k}_0\rangle|^2. \qquad (4.13)$$

The exponentials may be expanded to give

$$P(\underline{k},t) = |\Sigma_{n=0}^{\infty}\frac{1}{n!}i^n t^n\langle\underline{k}|H^n|\underline{k}_0\rangle|^2. \qquad (4.14)$$

This has not been solved in general. It is possible however to identify the most important terms. Each term in (4.14) is composed of a product of steps over various $\underline{k}$ values from $\underline{k}_0$

to $\underline{k}$ and back. For each step $\underline{k} - \underline{k}'$ the complementary step $\underline{k}' - \underline{k}$ must appear. Otherwise the average is zero.

The most important group of terms are those which go from $\underline{k}_0$ to $\underline{k}$ and then return by exactly the same route. Including only such contributions, (4.14) can be rewritten

$$P(\underline{k},t) = \frac{1}{N}\Sigma_{n=0}^{\infty} f(n,t)t^{2n}\langle\varepsilon^2\rangle^n \tag{4.15}$$

where f(n,t) contains the effects of H_0 as well as various combinatorial factors. This equation contains a time scale of $\langle\varepsilon^2\rangle^{-\frac{1}{2}}$. Even without actually summing the series it is clear that every $\underline{k}$ value has equal weight (except $\underline{k} = \underline{k}_0$ which we excluded earlier). The equation therefore describes diffusion.

If this were the only term what would it tell us about the scaling behaviour? Simple diffusion implies that the conductivity is classical and independent of the size of the system. Therefore the dimensionless conductance, g, is given by

$$g = \sigma L^{d-2} \tag{4.16}$$

and the scaling equation (4.7) becomes

$$\frac{d\ln g}{d\ln L} = d-2. \tag{4.17}$$

Actually equation (4.15) must be modified to take account of the effects of Fermi-Dirac statistics. In reality only $\underline{k}$ vectors on the Fermi surface can take part in the scattering process. This does not effect our conclusions, however.

4.7 Correlated Back Scattering

As pointed out by Langer & Neal[4] and by AALR[1] there is an additional special case. When $\underline{k} = -\underline{k}_0$ it is not sufficient to consider those terms in (4.14) which contribute to diffusion but also a second class of terms: a path which goes from $\underline{k}_0$ to $\underline{k}$ and then returns by taking the reverse steps in the same order as before. In the diffusive case the reverse steps are taken in reverse order. That is

$$\text{Case 1:} \quad S_1S_2....S_{n-1}S_n\Sigma_n\Sigma_{n-1}....\Sigma_2\Sigma_1$$
$$\text{Case 2:} \quad S_1S_2....S_{n-1}S_n\Sigma_1\Sigma_2....\Sigma_{n-1}\Sigma_n \qquad (4.18)$$

The symmetry of these terms makes it clear that each contributes exactly the same weight. Thus the probability of back scattering is doubled compared with what it would otherwise have been.

Another way of interpreting this effect is to consider 2 paths from $\underline{k}_0$ to $\underline{k}$. In the second the same steps are taken but in the reverse order. When $\underline{k} = -\underline{k}_0$ the two paths are equivalent and the waves interfere constructively rather than randomly. Back-scattering is thus enhanced.

There is one important difference between this contribution and the diffusive one. There is only a single $\underline{k}$ value involved (for each $\underline{k}_0$), whereas the number of diffusive terms increases as the system size increases. The fact that there are more intermediate terms involved in the scattering is taken care of by the fact that at each stage there are N contributions with weight 1/N.

Whereas previously we wrote the conduction in the simple form $g = \sigma L^{d-2}$, (4.16) this must now be modified to read

$$g = \sigma_D L^{2-d} - a \qquad (4.19)$$

and the scaling equation (4.17) takes the form

$$\frac{d\ln g}{d\ln L} = d-2 - \frac{a}{g} \qquad (4.20)$$

where $\underline{a}$ is a positive constant calculated from the enhanced back scattering.

4.8 Strong Localisation

The last section derived the leading deviation from classical behaviour. We can also derive the form of the one parameter scaling function in the limit of large disorder, or strong localisation. In that case all states are expected to be exponentially localised. The conductance will therefore fall exponentially with system size, L. This gives a scaling function

$$\frac{d\ln g}{d\ln L} = \ln(g/g_0) \tag{4.21}$$

4.9 The AALR β-Function.

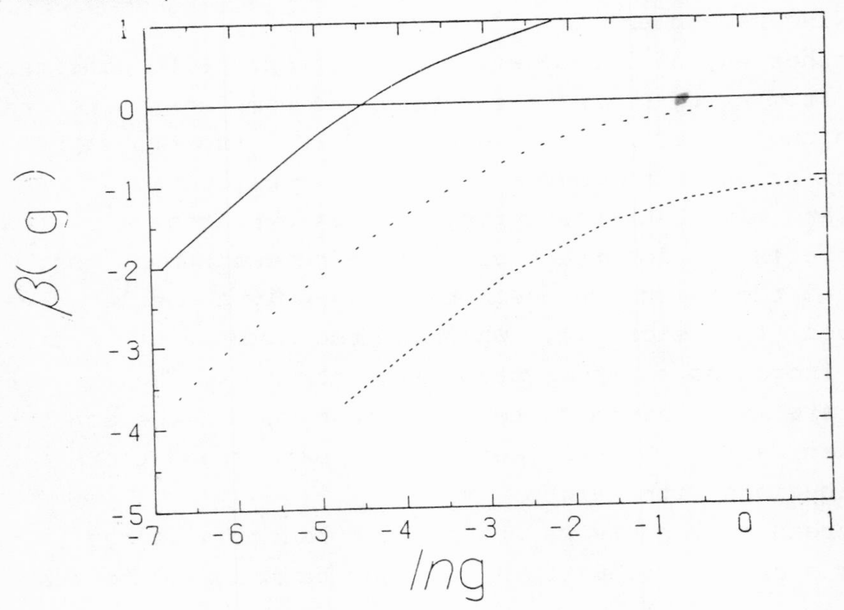

Calculated β-function[10]: ________ *3D*; . . . *2D*; ..... *1D*.

Figure 3.

The one parameter scaling function discussed here is often called the β-function. We now have all the ingredients required to derive the results of AALR[1].

We assume that the β-function is a smooth, monotonic function. Smoothness is a consequence of the fact that the system must behave analytically for all finite system sizes. Singularities can only emerge in the thermodynamic limit, $L\to\infty$. Monotonicity is an assumption which must be proved, but seems to be valid, at least in simple cases.

Consider therefore Figure 3. where the β-function has been sketched from its known properties, namely the limits for small and large conductance, g [(4.20) & (4.21)]. In one dimension, β is always negative: g always gets smaller as L increases.

In 3 dimensions β must cross zero. For small g, $g\to 0$ as $L\to\infty$, whereas for large g, $g\to\sigma L$ as $L\to\infty$.

2 dimensions is the most interesting case. $\beta\to 0$ as $g\to\infty$. Only the enhanced back scattering ensures that β is always negative and g always tends to zero. There are thus no extended states in 2-dimensions.

The β-function can be integrated in the form of (4.20) to give

$$g = g_0 - a\ln(L/L_0) \qquad (4.22)$$

The constants g_0 and L_0 may be combined into one. There are two common choices for these. When L is equal to the mean free path the simple diffusion equation is valid. Therefore choose $L_0 = l_{mfp}$ and $g_0 = \sigma_D$. Alternatively we can compare (4.22) with (3.11) and identify L_0 with the relevant length scale, namely the localisation length.

4.10 Critical Exponents (3D).

In the case of a system of 3 (or more) dimensions we can use the analysis derived earlier for fixed points. For this we require the gradient of β at the fixed point. As a first approximation let us use (4.20). This gives us a fixed point at $g^* = a/(d-2)$ and

$$\frac{d\beta}{d\ln g} = \frac{a}{g} = (d-2). \qquad (4.23)$$

Thus the critical exponent for a 3-dimensional system is unity. This value is preserved through more complicated analyses and seems to be the correct value for the simple universality class.

This is of course the exponent for the correlation length. How is this related to the exponents for the localisation length and the conductivity?

The general solution of an equation of the form of (4.7) is (2.14) $g = f(L/b)$. By comparing this with the forms for the fixed point and strong (4.19) and weak disorder limits, we observe that these can only be consistent with (2.14) and each other if

$$\sigma \sim b^{2-d} \qquad\qquad l_{loc} \sim b. \tag{4.24}$$

Thus in 3 dimensions the critical exponents for the localisation length and conductivity are equal and given by the reciprocal of the β-function at β=0.

4.11 Other Universality Classes

The analysis given above, and in particular the argument about back-scattering is changed in the presence of spin--orbit scattering and magnetic fields.

In the case of pure spin-orbit scattering the Hamiltonian has the property

$$H_{ij}^{++} = (H_{ij}^{--})^* \qquad\qquad H_{ij}^{+-} = -(H_{ij}^{-+})^* \tag{4.25}$$

Let us carry out the same analysis as before but this time in real space. The question to be answered is: what is the probability that an electron which starts at site $\underline{i}$ will have returned there after time $\underline{t}$, with either spin? This gives an equation of similar structure to (4.13) & (4.14). We can identify all the same terms as before plus a few extra ones.

Consider a path which starts at i=0 with spin up and arrives at i=0 but with spin down. The return journey can be accomplished by repeating each step, in the style of (4.18b), but with the opposite spin. (4.26) guarantees that the result will be finite and negative. There must be an odd number of pairs of spin-flipping terms.

This negative contribution tends to reduce the probability of returning to the origin. The constant a in (4.19) and (4.20) becomes negative and the β-function is positive for large g in 2d.

There are now two alternatives:

a) The β-function is always positive. There are no localised states.

b) The β-function becomes negative for small g, there is a fixed point and a transition. The critical exponent for the conductivity is zero, from (4.25), and there may be a minimum metallic conductivity.

The universality class for this case is often called symplectic.

The case of a system with magnetic fields is more complicated. The back-scattering terms no longer contribute and if the magnetic field is random, due to magnetic impurities, the classical result prevails (to leading order).

In a spatially constant magnetic field there are other contributions associated with boundary currents which do not vanish in the thermodynamic limit. The system follows a 2-parameter scaling law in σ_{xx} and σ_{xy}. This is the case of the quantum Hall effect.

4.12 Temperature

So far we have discussed only conductance as a function of length. While theoretically simple it is almost impossible to measure experimentally. What is required is to note that the time taken before an electron is scattered inelastically is usually proportional to a negative power of the temperature: $t_{in} \sim T^{-\alpha}$. Let us then use Thouless' argument again

(4.6). Setting the density of states as independent of length we obtain

$$\frac{d\ln t}{d\ln L} = d - \frac{d\ln g}{d\ln L}$$

$$\beta_t = \frac{d\ln g}{d\ln t} = \frac{d\ln g}{d\ln L}.\frac{d\ln L}{d\ln t} = \frac{\beta}{d-\beta} \tag{4.26}$$

$$\frac{d\ln \sigma}{d\ln T} = \alpha\{\frac{d-2-\beta}{d-\beta}\}$$

From this result a number of temperature regimes can be derived:

a) Weak Disorder: $\beta = d-2 - \frac{a}{g}$

$$\sigma \sim \tfrac{1}{2}\alpha a \cdot \ln T \qquad 2d$$

$$\sigma \sim \sigma_0 + AT^{\alpha/2} \qquad 3d \tag{4.27}$$

b) Near Critical Point: $\beta = 0$

$$\sigma \sim T^{\alpha/3} \tag{4.28}$$

c) Strong Disorder: $\beta \to -\infty$

$$\sigma \sim T^{\alpha} \tag{4.29}$$

It is important to remember that these formulae are only useful when the inelastic scattering length is larger than the mean free path. At smaller length scales the behaviour will be classical. Other corrections will occur if the density of states has structure on a scale smaller than kT or $\hbar/t_{in}$.

5. STATISTICS

5.1 Averages versus Distribution

In section 4.1 we derived a general scaling equation for a distribution. Thereafter we have discussed only averages. Is this consistent? In general one-parameter scaling of the whole distribution implies the existence of an attractive line in the phase space of the distribution. If this line is a monotonic function of a particular parameter of the distribution (e.g. the mean), this parameter will also obey 1-parameter scaling. There may however be several such parameters. Thus most analytical work averages the conductance whereas several numerical techniques calculate the geometric mean[9]. Can this be responsible for such differences as the critical index? Numerical methods currently stubbornly refuse to agree with the generally accepted value of unity.

If there exists a single attractive line in the phase space of the distribution of conductance it should be possible to rewrite (4.1) as a Taylor series around the fixed point:

$$\frac{dP(g)}{d\ln L} = B[P(g) - P^*(g)] \tag{5.1}$$

leading to the solution

$$P(g) = P^*(g) + A(g)L^B \tag{5.2}$$

where A(g) depends on energy, disorder, etc. and $\int_{-\infty}^{+\infty} A(g)dg = 0$. As before it must be possible to express A(g) as a Taylor expansion in the parameters (eg energy, disorder) and to extract a single correlation length _b_ with a critical exponent 1/B.

Any average calculated using appropriate integrals of (5.2) must have the same critical exponent, as long as the integral is defined.

This caveat may represent a serious restriction. Although the conductance of a finite system is extremely unlikely to be zero or infinite, it is not clear that this is necessarily applicable to the attractor which by definition describes asymptotic behaviour.

5.2 Crystal-like Model

It is also possible to construct a system where not all averages are defined. Instead of a finite size system, let us consider a block of disordered material which is repeated infinitely often. The model is then a crystal with a very large unit cell, and the associated bands and gaps in the spectrum.

In this case the conductivity is either zero or infinity, in a gap or a band respectively. Even if we consider the transmission coefficient, T, instead, it is then either zero or unity. The mean value of T is simply the probability of finding a band rather than a gap at the Fermi energy. The geometric and harmonic means are undefined.

This peculiar property does not vanish in the thermodynamic limit, so we are left with 2 alternatives:

a) The attractive line in phase space also has this property.

b) The $L \to \infty$ properties depend on the boundary conditions. In which case there is no single attractive line in phase space, and there is no single scaling parameter.

5.3 Distributions and Experiment.

Clearly the question which requires an answer: what is actually measured in an experiment? There is some evidence that the distribution of conductance becomes narrower at finite temperatures and the system <u>self-averages</u>, ie the distribution becomes narrower for large system sizes.

If $\delta E = kT$ or $\hbar/t_{in}$ is larger than the splitting between states in the system, $\frac{dE}{dN}$ in (4.6), then the binary structure (0,1 or 0,∞) is smeared out and replaced by an average over δE. In the $L \to \infty$ limit at finite temperatures this will always be the case, although it may be possible to construct samples which are not large enough. In such a case quantum size effects will be observed and these may be accompanied by statistical fluctuations which are reproducible within one sample but not between samples.

6.
CONCLUSIONS

In these lectures I have tried to provide a non-mathematical introduction to scaling theory. I have concentrated on the ideas behind the theory and the assumptions involved. I hope it will provide a firm foundation for understanding the more detailed approaches to be found elsewhere.

I have not discussed experimental results in detail, mainly because, in spite of some notable successes, the agreement between theory and experiment is still far from satisfactory.

A complete theory must include not only the effects of disorder but also interactions and temperature.

REFERENCES

1. E.Abrahams, PW.Anderson, DC.Licciardello and TV.Ramakrishnan, Phys.Rev.Letters 42, 673 (1979)
2. DJ.Thouless, Physics Reports C13, 93 (1974)
   Phys.Rev.Letters 39, 1167 (1977)
3. F.Wegner, Z.Physik B25, 327 (1976)
4. JS.Langer and T.Neal, Phys.Rev.Letters 16, 984 (1966)
5. MJ.Feigenbaum and RD.Kenway, "Statistical and Particle Physics", Ed. KC.Bowler and AJ.McKane (SUSSP 26, EUP, 1983)
6. PD.Kirkman and JB.Pendry, J.Phys. C17, 4327 (1984)
   J.Phys. C17, 5707 (1984)
7. F.Wegner, Z.Physik B35, 207 (1079)
8. PW.Anderson, Phys.Rev. 109, 1492 (1958)
9. A.MacKinnon and B.Kramer, Z.Physik B63, 1 (1983)

BIBLIOGRAPHY

Books:

Electronic Processes in Non-Crystalline Materials,
NF.Mott and EA.Davis (Clarendon, Oxford, 1979)

Reviews:

Disordered Electronic Systems,
PA.Lee and TV.Ramakrishnan, Rev.Mod.Phys. (1985)

Weak Localisation in Thin Films,
G.Bergmann, Phys.Reports 107 (1984)

Anderson Localisation
Ed. Y.Nagaoka, Prog.Theor.Phys. 84 (1985)

Conference Proceedings etc.

Electron-Electron Interactions in Disordered Systems,
Ed. M.Pollak and A.L.Efros (North Holland, 1984)

Disordered Systems and Localisation,
Ed. C.Castellani, C.DiCastro and L.Peliti
(Springer, 1981)

Localisation, Interaction and Transport Phenomena,
Ed. B.Kramer, G.Bergmann and Y.Bruynseraede
(Springer, 1984)

Localisation in Disordered Systems,
Ed. W.Weller and P.Ziesche (Teubner, 1984)

The Metal Non-Metal Transition in Disordered Systems,
Ed. LR.Friedman and DP.Tunstall (SUSSP 19, EUP, 1978)

METAL-INSULATOR TRANSITIONS IN DOPED SEMICONDUCTORS; A SURVEY

N.F. Mott

Cavendish Laboratory, Cambridge

1.
INTRODUCTION

It is just eight years since I lectured here on metal-insulator transitions in doped semiconductors. Now the organisers of this school have again given me the same subject. A lot has happened in the last eight years. The change in our understanding began in 1979 with the scaling theory of Abrahams et al[1)], about which you will hear from Dr. MacKinnon. As a consequence of this, and of the experimental work of Gordon Thomas and co-workers and of others, it is now clear that a "minimum metallic conductivity" does not exist, though the theoretical arguments against it based on scaling theory have recently been queried by Efetov[2]. One theme of these lectures will be, why is this quantity, apparently, so often observed? A further major advance is the realisation that in "dirty" metals long-range electron-electron interaction has a major effect on the density of states at the Fermi level, which is not present when the mean free path is long, and that this has a large effect also on the conductivity, introducing a term in the resistivity which varies as $T^{\frac{1}{2}}$, where T is the temperature. This was first shown by Altshuler and Aronov[3]. This is quite different from the effects of the Hubbard U, the short-range intra-atomic interaction, which is only meaningful in the tight-binding approximation and thus for electrons in an impurity band. The distinction between the two kinds of interaction will be another theme of these lectures. Long-range interaction is more important than in ordered systems, and indeed it is maintained by some that to treat it as a perturbation on the wave function calculated without interaction is incorrect[4]. Long-range interaction can also have a major effect on hopping conduction, as first pointed out by Efros and Shklovskii[5] (see § 11).

A problem much discussed in the literature is whether, in doped semi-conductors, the metal-insulator transition takes place in an impurity band or in the conduction band[6].

There is of course no doubt that for a low concentration of donors, for which in compensated materials conduction is by hopping, a band of localized levels exists which is separated from the conduction band. The question whether a metallic band exists just above the transition can be put in two ways:

(a) Is there a "metallic" band of levels separated from the conduction band?

(b) Is there a band of levels which may strongly overlap the conduction band, but for which, at the Fermi energy at any rate, the tight binding approximation is a good one? The famous paper of Anderson[7] on "Absence of diffusion in certain random lattices", which gave criteria for localization as a result of disorder, uses what we call a tight binding approximation. It is not certain that localization can occur in a broad conduction band as a result of the random field of the donors alone, and it seems to us likely that, in the sense (b), the transition is always in an impurity band*.

*Localisation can occur in a narrow conduction band which can be described by the tight-binding approximation. This seems to be the case for tungsten bronzes, cf. Davies and Franz[8].

Some of the evidence is as follows:

A donor level in Si:P is split (valley splitting) as in fig. 1, the impurity band being usually considered as formed from the lowest of these levels. Raman scattering (Jain et al[9]) shows transitions between these two levels, as illustrated schematically in fig. 2. In germanium this splitting is preserved well into the metallic regime, which suggests that an impurity band must be present for concentrations above that for the transition. In Si:P, on the other hand, the valley splitting disappears at $n \sim n_c$; this may mean that the upper impurity band merges with the conduction band. In Si:P it does not disappear. Berggren (priv. comm.) suggests that this is because the valley splitting is greater, so that an empty impurity band formed from the upper part of the split level remains, while perhaps for Si the two have merged.

If so, however, the uncompensated system will still behave as a Mott insulator, as pointed out to the author by Bhatt.

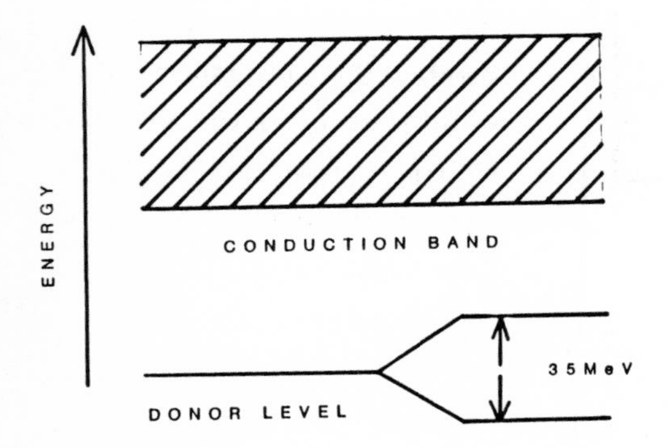

Fig. 1. Showing splitting of an impurity level in many-valley conduction band.

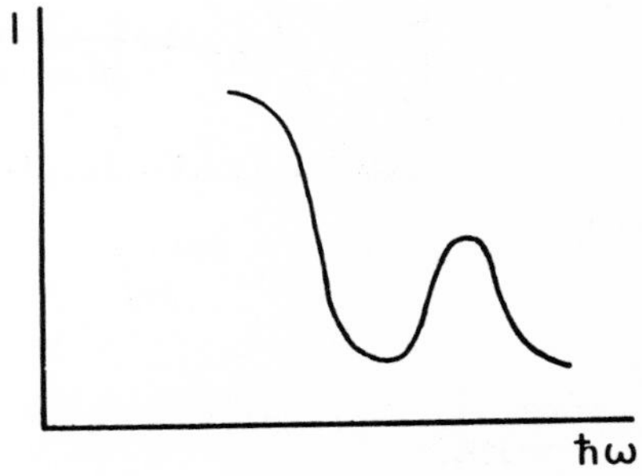

Fig. 2. Raman scattering from split level of fig. 1.

Further evidence on the existence of the impurity band at the transition is:-

(i) The work of Tunstall and Sohal[10] on the Knight shift K in germanium under stress. K is not changed by a stress which would lift all but one of the conduction band valleys above the Fermi energy E_F. These authors deduce that E_F cannot lie in the conduction band.

(ii) The results of Davis and Compton [11], who measured the quantities ε_1 (the ionisation energy for a donor) and ε_2 (interpreted as $E_c - E_F$) as a function of concentration for germanium (fig. 3). At the concentration for which ε_2 vanishes, ε_1 is still finite, indicating that E_F lies in an impurity band with a gap separating it from the conduction band.

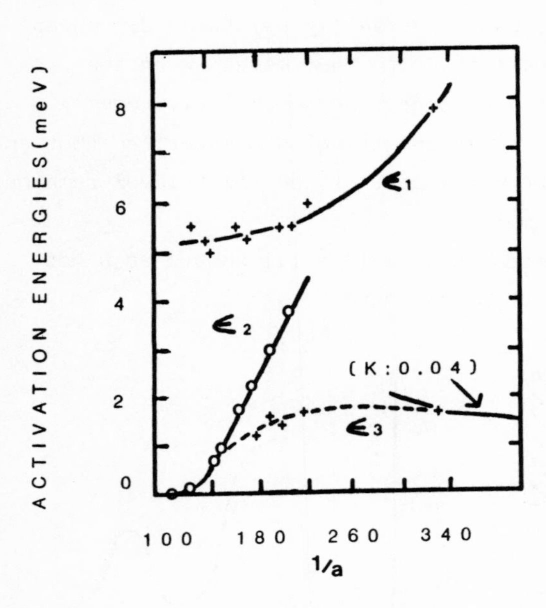

Fig. 3. ε_1, ε_2, and ε_3 for slightly compensated n-type Ge as a function of mean distance between centres (Davis and Compton 1965).

For silicon we do not have equivalent information. But in favour of the hypothesis that at the transition, E_F lies in the impurity band, we have the analysis of Jerome et al [12] on the phosphorus and silicon Knight shift of Si:P, which favours an impurity band. The published calculations of Meyer et al [13] on the Knight shift were comparable with a conduction band but more recent considerations (Meyer [14]) favour an impurity band. The chief evidence in favour of the conduction band is the value of the electronic specific heat C_v. Above ~ 1.5 K, C_v is of the form γT and γ is close to the value calculated for electrons in the many-valley conduction band. But according to calculations by Kaveh and Liebert [15] the assumption of an impurity band, without electron-electron interactions, fits the data equally well. We are inclined to believe then, that the transition *always* takes place in an impurity band in the sense (ii) above. In our subsequent discussion we shall make this assumption.

A further point made by Takemori and Kamimura [67] and Kamimura [68] is that, while the lower Hubbard band does not show valley splitting, the calculated upper Hubbard band does. So, while we adhere to the belief that in compensated semiconductors the transition takes place in an impurity band, in uncompensated semiconductors the situation is less clear. Our discussion will however be based on the concept of a tight binding treatment for electrons at the Fermi energy.

2.

THE THEORY OF LOCALIZATION, ACCORDING TO ANDERSON[7]

Anderson considered, within a tight binding approximation, an electron moving in the potential shown in fig. 4. In the absence of a random potential, the solutions of Schrödinger's equation are of the form

$$\sum_n \exp(ika_n)\, \psi_n(r) \tag{1}$$

where the a_n are lattice points and ψ_n is an s-wave function at the point a_n. The energy as a function of the wave vector k is, for a simple cubic

$$E(k) = E_o - 2I\{\cos k_x a + \cos k_y a + \cos k_z a\}, \tag{2}$$

where I is the transfer integral

$$I = \int \psi_n\, H\, \psi_{n+1}\, d^3x. \tag{3}$$

The band-width B is given by

$$B = 2zI \tag{4}$$

where z is the co-ordination number; for random positions we usually take z = 6.

If a random potential $V - \frac{1}{2} V_o < V < \frac{1}{2} V_o$ is applied to each well, the band is widened, the breadth being

$$\sqrt{(V_o^2 + B^2)}, \tag{5}$$

and a mean free path ℓ is introduced, given by (Mott and Davis[16])

$$\frac{1}{\ell} = 0.7\,\frac{1}{a}\left(\frac{V_o}{B}\right)^2 , \tag{6}$$

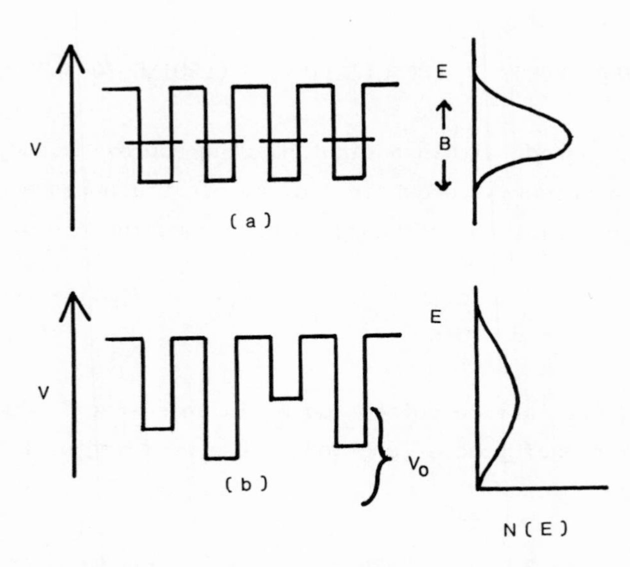

Fig. 4. The potential energy of an electron in the Anderson model (Anderson 1958) (a) without a random potential, and (b) with a random potential V_o. B is the band-width in case (a). The density of states N(E) is also shown.

When V_o reaches the value B, $\ell \sim a$ which is its minimum value according to Ioffe and Regel[17]. This is clear from Mott's hypothesis that the wave function is then of the form

$$\Psi_o = \sum_n c_n \exp(i\varphi_n)\psi_n , \tag{7}$$

where the c_n are real coefficients and the φ_n random phases. According to Anderson, states will become localized throughout the band for a critical value of V_o/B. Attempts to calculate it have obtained various results, we shall use a recent one, that of Elyutin et al[18], which gives

$$(V_o/B)_{crit} = 1.7 \ . \tag{8}$$

The wave functions of the localized states are thought to be of the form

$$e^{-r/\xi} \, Re[\psi_o] \ , \tag{9}$$

where Ψ_o is given by (7). The quantity ξ is called the localization length and will tend to infinity as V_o/B tends to the critical value.

If V_o/B lies below that critical value, it was first pointed out by the present author[19] that what we now call a "mobility edge" must exist. This is an energy E_c in the band such that, for $E < E_c$ states are localized, the wave functions being of the form (9). The localization length ξ is believed to behave like

$$\xi = \text{const.} \ a\{E_o/(E_c - E)\}^{\nu} \tag{10}$$

with ν probably unity and E_o a constant of the order of the band width.

If E lies below E_c, the mobility is zero; or if for a degenerate electron gas the Fermi energy lies below E_c, the zero-temperature conductivity is zero. This is not obvious; it was argued in the early days of the theory that an electron could tunnel to a state of equivalent energy, and that the conductivity could not vanish. When an ensemble average is taken over all wave functions, it is however found[20] that $\sigma(\omega) \sim \omega^2$, and tends to zero with the frequency ω.

3.

CONDUCTIVITY IN THE ANDERSON MODEL

A convenient quantity to consider is $\sigma(E)$, defined as the conductivity of a non-interacting electron gas at zero temperature if the Fermi energy is E. At a finite temperature the conductivity will be

$$\sigma = - \int \sigma(E)(\partial f/\partial E) \, dE \tag{10a}$$

where f is the Fermi-Dirac distribution function; the mobility μ of an electron is given by

$$\mu = \sigma(E)/eN(E) \ , \tag{11}$$

where N(E) is the density of states.

When the mean free path ℓ is large compared with a, we may write

$$\sigma = ne^2\tau/m \ , \tag{12}$$

where n is the number of electrons per unit volume, τ the time of relaxation and m the effective mass. Writing

$$\tau = \ell/v = m\ell/k_F\hbar$$
$$n = (8\pi/3)\ (k_F/2\pi)^3$$

and S_F, the Fermi surface area, given by $4\pi k_F{}^2$, (12) reduces to

$$\sigma = e^2 S_F \ell/12\pi^3\hbar \ . \tag{13}$$

In the Ioffe-Regel limit $\ell \simeq a$, and for a half-full band we find

$$k_F a = \pi \ .$$

Writing σ_{IR} for (13) at the Ioffe-Regel limit, we find

$$\sigma_{IR} = e^2/3\hbar a \ , \tag{14}$$

a value for a normal material of order 3000 $\Omega^{-1}cm^{-1}$, or about $\frac{1}{3}$ that of liquid mercury.

For greater values of V_o/B one has to take account of the change in the density of states. By the use of the Kubo-Greenwood formula[21, 22], the present author[23] showed that (14) must be multiplied by g_M^2, where g_M is the ratio

$$g_M = N(E_F)/N(E_F)_{cryst} \tag{15}$$

This has the value

$$1/g_M = 1.74 \{1+(V_o/B)^2\}^{\frac{1}{2}} \tag{16}$$

This result is obtained by calculating $\sigma(w)$, which depends on the optical transition probability for electrons in states just below the Fermi energy to states just above, and so is proportional to

$$\left|\int \psi_E \frac{\partial}{\partial x} \psi_{E'} \, d^3x \right|^2_{av} \{N(E_F)\}^2 \tag{17}$$

To obtain $\sigma(o)$, one allows the energy difference $E - E'$ to tend to zero. Thus one finds for the conductivity

$$\sigma = (\tfrac{1}{3} e^2/\hbar a) g_M^2 \tag{18}$$

The present author then deduced that for a metal a "minimum metallic conductivity" σ_{min} must exist, if (18) is valid when $E_F = E_c$, and that its value will be given by giving to g_M the value when $E_F = E_c$, which is about $\frac{1}{3}$. Thus

$$\sigma_{min} = 0.03 \, e^2/\hbar a, \tag{19}$$

If the band is not half full, one should write (Mott[24]) instead of a, a_E given by

$$(a/a_E)^3 = N(E_c)/\langle N(E) \rangle \tag{20}$$

However, the scaling theory of Abrahams et al showed in 1979 that σ for a metal must tend to zero as $E_F \to E_c$, so that in general no minimum metallic conductivity could exist. As will be shown in the lectures by Dr. Bergmann (see also ref. 25), my mistake in obtaining σ_{min} was in neglecting multiple scattering, which leads to the continuous decrease of σ to a zero value. In these lectures I shall make use of an equation obtained by Kawabata[26] (see also Mott and Kaveh[6]) and in the modified form used here by Mott and Kaveh[27], namely

$$\sigma = \sigma_B \, g_M^2 \left[1 - \frac{c}{g_M^2 (k_F \ell)^2} \left\{ 1 - \frac{\ell}{L} \right\} \right] \tag{21}$$

Here σ_B is the Boltzmann conductivity (13), c is a cut-off length not known exactly but about unity, and L is the size of the specimen or L_i, the inelastic diffusion length, that is the distance that an electron diffuses before it makes an *inelastic* collision with either another electron, or a phonon. We write

$$L_i = (D\tau_i)^{\frac{1}{2}} \tag{22}$$

τ_i is here the time between such collisions and D the diffusion coefficient. Equation (21) is deduced under conditions such that the last term is a perturbation, but we shall use it to extrapolate to the concentration at which the transition occurs, where it gives good results. Thus at the transition, since $c/g^2(k_F\ell)^2 = 1$ at that value of g, and putting $\ell = a$, we find

$$\sigma = \sigma_B \, g_c^2 \, a/L \tag{23a}$$

$$= \sigma_{min} \, a/\ell = 0.03 \, e^2/\hbar L. \tag{23b}$$

These equations can also be used if L is the size of the specimen, and also in a magnetic field H the cyclotron radius L_H given by

$$L_H = (c\hbar/eH)^{\frac{1}{2}}, \tag{24}$$

where c is the velocity of light and in an alternating current of frequency ω

$$L_\omega = (D\omega)^{\frac{1}{2}}$$

These equations show the rather surprising result that the conductivity *increases* with the inelastic scattering $1/\tau_i$. This is because the constructive interference between multiply scattered waves, which increases the scattering and so lowers σ, is less effective if the volume available is reduced (see also paper by Bergmann in this volume).

In the field of heavily doped semiconductors - the main subject of these lectures - there is much evidence for equations (21), (22), as there is also from the electrical properties of amorphous metals. These will be reviewed in §7; effects of long range electron-electron interaction (§6) are also important in these materials. Equations (11) and (23) have been used by the present author[24] to calculate the mobility of an electron with energy at the mobility edge in hydrogenated amorphous silicon; here L_i is the inelastic diffusion length resulting from collisions with phonons, and there is of course no term resulting from electron-electron interactions; fair agreement with experiment is claimed.

The work of G.A. Thomas' group (Ng et al[28] and this volume) show the effect of L_ω for a.c. fields; one might represent the result by

$$\sigma = \sigma_{Drude}\left[1 - \frac{C}{(k_F\ell)^2}\{1 - \frac{\ell}{L_\omega}\}\right]$$

To obtain agreement with experiment one needs however a large (~10) value of C. Also, since σ rises above σ_{Drude}, it may be necessary, as Ng et al suggest, to take into account fluctuations of Ψ of the kind illustrated in fig. 5, which would enhance $\sigma(\omega)$.

The use of equation (21), with g_M=1, leads to the function $\beta(G)$ of scaling theory having the form

$$\beta(G) = 1 - C_1/G.$$

G is here the dimensionless conductance and $\beta(G) = d\ell n\, G/d\ell n L$. C_1 should be a universal constant (Wölfle and Vollhardt[29] find $1/2\pi^2$). Our value is 0.03, and this depends on the constant $g_c k_F \ell$ and the cut-off constant c. g_c should depend on co-ordination number; the other constants must compensate for any variation.

Hikami[30] has shown that if one writes

$$\beta(G) = 1 - C_1/G + \Sigma\, C_n/G^n,$$

both C_2 and C_3 vanish. This strongly suggests that (21) may be a good approximation, even near the transition.

4.

THE ABSENCE OF A MINIMUM METALLIC CONDUCTIVITY IN THE KUBO-GREENWOOD THEORY; BEHAVIOUR OF LIQUIDS.

The scaling theory, equation (21) with $L = \infty$ and many experiments to be reviewed later indicate that $\sigma(T = 0)$ tends continuously to zero for a metal as $E_F \rightarrow E_c$, so that the quantity σ_{min} does not exist. Since equation (17), the Kubo-Greenwood formula, consists of a sum of squared terms, this is rather surprising. Mott[31] and Mott and Kaveh[27] have examined this problem. They conclude that, over a range ξ, a metallic wave function at a *small* distance ΔE above a mobility edge will not differ greatly from that at the same energy below it; so the functions will appear as in fig. 5. They will have long-range fluctuations, and the amplitude at the minima tends to zero as $\xi \rightarrow \infty$. ξ is here the localization length at an energy $|E_F - E_c|$ *below* E_c, and this statement, of course, only applies when this quantity is small, and certainly not when $\sigma \gtrsim \sigma_{min}$. Under these conditions, it was shown that

$$\sigma(T{=}o) \sim A_1 e^2/\hbar\xi, \qquad A_1 = 0.03 \tag{25}$$

Other values of the constant are given in the literature (e.g. Lesueur et al[32]), but it is not clear that ξ is the same. Ovadyahu[33] finds ξ below the mobility edge from an analysis of variable range hopping and then observes "metallic" conductivity for $L_i < \xi$, obtaining a value of ξ in this way. His value of the constant A_1 in (25) is somewhat greater.

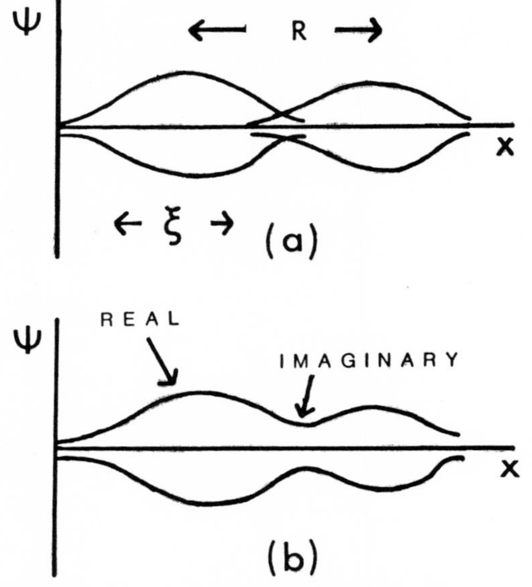

Fig. 5. A suggested form for the envelopes of the wave functions (a) below and (b) above E_c. R is the distance between centres, and ξ is the localization length.

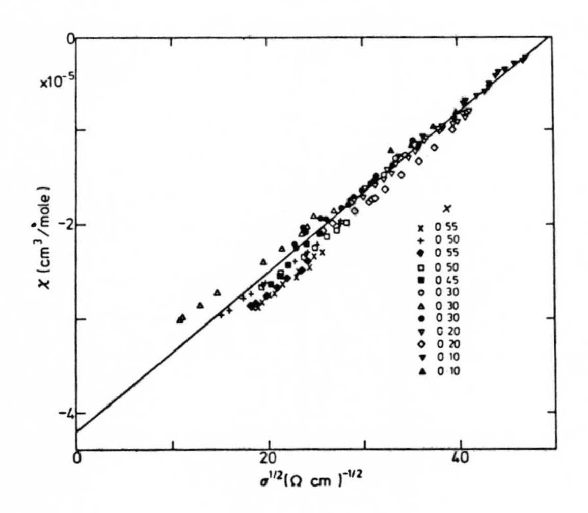

Fig. 6. Pauli susceptibility of $Te_{1-x}Tl_{2+x}$ liquid alloys plotted against $\sigma^{\frac{1}{2}}$ for various temperatures and composition (from Cutler[36]).

Another result of this analysis is that, in the Kubo-Greenwood formula the two functions Ψ_1, Ψ_2 are nearly identical. In obtaining the value of σ for $\ell \simeq a$ (at the Ioffe-Regel limit and below), it was originally supposed that because the wave functions lost phase memory in a distance $\underline{a}$, so will the product $\bar{\Psi}_E \Psi_{E'}$ in (17). This is so when $\sigma \gtrsim \sigma_{min}$, but near the mobility edge the integral will not increase with distance x. If to obtain the conductivity for $L = L_i$ we take the integral over a distance L_i, the Kubo-Greenwood integral will be decreased by $(a/L_i)^{\frac{1}{2}}$. We find that

$$\sigma = 0.03\ e^2/\hbar L_i \tag{26}$$

as in equation (23).

Mott[34] as stated above has examined the mobility of electrons in a-Si-P, and finds for electrons excited to the mobility edge, writing $\mu_c = \sigma/N(E_c)e$, that (26) is in fair agreement with the observations. τ_i is estimated from the known rate of scattering by phonons, and from observations of the lifetimes of photo-excited electrons.

For σ_{min} to exist as given by (18) the second term in equation (20) should vanish, as it will if all collisions are inelastic. It is argued by Mott[35] that this is so for liquids. This is shown by the system $Te_{1-x}Tl_{2+x}$ (Cutler[36]) in which χ, the Pauli susceptibility which should be proportional to g_M, is plotted against $\sigma^{\frac{1}{2}}$, giving a straight line, as shown in fig. 6. Ref. 6 gives further evidence that σ_{min} does exist in liquids.

When the inelastic mean free path is short, particularly in liquids, the mobility edge is broadened. The broadening ΔE is - perhaps surprisingly - rather small, and given (Mott[24], Mott and Kaveh[6]) by

$$\Delta E/E_o \simeq 0.03\ (a/L_i)^3$$

Here E_o is defined by

$$N(E_F) = 1/a^3 E_o$$

Thus even if $L_i \sim a$, as in liquids, $\Delta E/E_o$ is only a few per cent.

5.

THE INELASTIC DIFFUSION LENGTH IN METALS RESULTING FROM ELECTRON-ELECTRON COLLISIONS

The inelastic diffusion length is given by

$$L_i = (D\tau_i)^{\frac{1}{2}} . \tag{27}$$

τ_i, the rate of electron-electron collisions should be given by

$$(2\pi/\hbar)\ |M|^2\ a^9(k_BT)^2\ \{N(E_F)\}^2$$

where M is the matrix element of the electron-electron interaction. This should be given by

$$M = e^2/\kappa\lambda$$

where λ is the screening length and κ the dielectric constant. If we take Thomas-Fermi screening, this gives

$$M = 1/a^3N(E) ,$$

which yields for τ_i

$$1/\tau_i = (2\pi/\hbar)\ N(E_F)\ a^3\ (k_BT)^2 . \tag{28}$$

This is the Landau-Baber scattering rate, which has recently been reviewed by Kaveh and Wiser[37], and is valid if the change in k is larger than $1/\ell$. In this case

$$1/L_i = k_BT\{2\pi\ N(E)a^3/\hbar D\}^{\frac{1}{2}} .$$

A temperature-dependent correction to the conductivity is, according to equations (21) and (29), with $\sigma = Ne^2D$,

$$\Delta\sigma = \sigma_B\{c/k_B{}^2\ell)\}\, k_BT\, \sqrt{(2\pi e^2/\hbar\sigma)} \tag{30}$$

where c as before is of order unity.

For the case where the momentum change is less than $1/\ell$ - that is for short mean free paths, analysis by Schmidt[38] and by Kaveh and Wiser[37] we find a variation as $T^{3/2}$, observed in 3-dimensions by Ovadyahu[33] (see Mott and Kaveh[6] p. 350). However $1/\tau_i$ proportional to T^2 is observed more frequently than expected (Bergmann, this volume).

The linear increase in σ with T has been observed in many systems, as we shall see; at low temperatures, however, the correction resulting from electron-electron interaction varying as $T^{\frac{1}{2}}$ will be predominant. This is introduced in the next section.

We may remark here that when L_i results from interaction with phonons at temperatures above the Debye temperature, $1/\tau_i$ is proportional to T and $\Delta\sigma$ to $T^{\frac{1}{2}}$.

6.

CORRECTIONS TO THE DENSITY OF STATES AND TO THE CONDUCTIVITY RESULTING FROM LONG-RANGE COULOMB INTERACTION BETWEEN ELECTRONS

In discussions of electrons in the conduction band of a semiconductor, no consideration of electron-electron interaction is necessary. In amorphous metals and in impurity bands, on the other hand, the effect of such interactions is more important than in crystals; some authors (e.g. Finkelstein[4]) consider that a theory treating these interactions as a perturbation is not legitimate. In conduction bands we discuss next the effect of the Coulomb interaction $e^2/\kappa r_{12}$. In impurity bands - and in any case where a tight binding model can be used - we may distinguish between the intra-atomic interaction energy $e^2/\kappa r_{12}$ (the Hubbard U) and the long-range interaction. The former, we believe, only has a major effect for *uncompensated* semiconductors and for systems such as liquid Na_xNH_3, fluid Cs and so on, where each centre contains just one electron, as we shall see in subsequent sections.

We consider next long-range interactions.

Altshuler and Aronov[39] in 1979 were the first to show that in three dimensional metallic systems, with a finite elastic mean free path, long-range Coulomb interaction leads to a change, positive or negative, in the density of states near the Fermi level, of the form

$$\delta N(E) = B|E - E_F|^{\frac{1}{2}} \tag{31}$$

as illustrated in fig. 7. This introduces a term in the conductivity

$$\delta\sigma = AT^{\frac{1}{2}} ,$$

where A may have either sign. The equation for $\delta\sigma$ obtained by these authors (see also Kaveh and Mott[6] p. 360) is

$$\frac{\delta\sigma}{\sigma} = \left(\frac{\hbar}{mD}\right)^{\frac{3}{2}} \left(\frac{T}{T_F}\right)^{\frac{1}{2}} \left\{\frac{4}{3} - 2F\right\} ; \tag{32}$$

as shown by Fukuyama in this volume, the terms in braces may have other forms. Here D is the diffusion coefficient, and it will be seen that the term is large when D is small. K_BT_F is the Fermi energy; $\frac{4}{3}$ comes from the exchange and F from the Hartree interaction, which should be given by

$$F = x^{-1} \ln (1+x) , \qquad x = 2k_F\lambda \tag{33}$$

where λ is the screening length. Since F lies between 0 and 1, the correction can have either sign; a negative value will occur for $F > \frac{3}{4}$, and thus for small values of x, which means weak screening.

The first observation of the $T^{\frac{1}{2}}$ term was by Bronovoi and Sharvin[40] in bismuth; the correction to the resistivity of this material shows the Landau-Baber T^2 behaviour at low T when undeformed, but after cold work, which introduces dislocations giving strong elastic scattering, a $T^{\frac{1}{2}}$ behaviour of the resistivity ρ is observed, with

$$\rho = A + BT^{\frac{1}{2}}$$

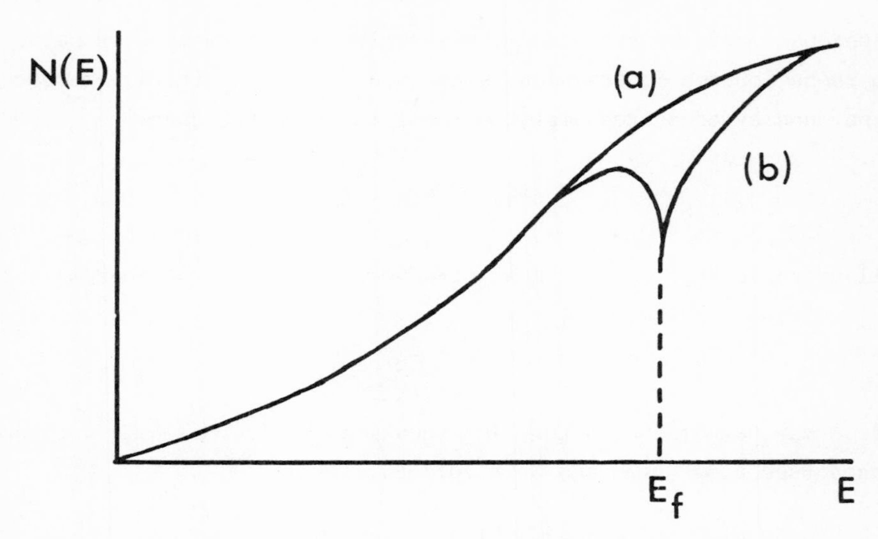

Fig. 7. Density of states of a degenerate electron gas (a) without, and (b) with long-range electron-electron interaction.

The electrical conductivity of some amorphous metals show this behaviour at low T, then as the temperature is raised the linear correction (30), and finally at high T another $T^{\frac{1}{2}}$ correction resulting from a value of L_i caused by interaction with phonons, when $1/\tau_i \propto k_B T$ (Howson[41], Howson and Greig[42]; see Mott and Kaveh[6] p. 390).

Another treatment of the effect of interaction is that of McMillan[43] who writes

$$N(E) = N(E_F)[1 + \{(E-E_F)/\Delta\}^{\frac{1}{2}}]$$

Here Δ is the width of the Coulomb gap described § 11. For the relationship to equation (31), see ref. 6, p. 370.

7.

THE METAL-INSULATOR TRANSITION IN DOPED AND COMPENSATED SEMICONDUCTORS

In this section, following § 1, we suppose that the transition takes place in an impurity band, and we can apply the Anderson model to the phenomenon, though there are here two kinds of disorder, diagonal and non-diagonal. Both have been discussed in the literature. By diagonal we mean the random potential V_o introduced by charged acceptors and donors, and by non-diagonal the variation of the transfer integral (eqn. 3) consequent on the random positions of the donors. Also we assume that the Hubbard U, namely the interaction between two electrons on the same donor, simply prevents double occupation but does not have any large effect on the conductivity. This means that we neglect short-range interaction and use first a theory of non-interacting electrons, adding the effects of long-range interaction described in the last section as a perturbation.

If we suppose that the compensation is of the order 50%, so that half the states in the band are singly occupied donors, an estimate of the concentration at which the transition occurs may be made as follows. In the absence of disorder the width B_o of the impurity band is 2zI, where z is the number of nearest neighbours to each centre and I the transfer integral, given by

$$I = (e^2/\alpha\kappa)\ (1 + \alpha R)\ \exp(-\ \alpha R) \qquad (34)$$

Here $\alpha=1/a_H$, where a_H is the Bohr radius and R the distance between centres. Localization will occur when $V_o = 1.7\ B_o$, where V_o is the spread in the energy of states resulting from the charges on the acceptors and donors and given by $c\ e^2/\kappa R$ where $c \sim 1$. So the condition for localization is

$$1.7 \times 12\ (e^2/\kappa R)(1 + \alpha R)e^{-\alpha R} = c\ e^2/\kappa R\ . \qquad (35)$$

If except in the exponential we put $\alpha R = 4$, this gives

$$\alpha R = \ell n(102/c)\ ; \qquad (36)$$

c is probably less than unity, but appearing in the logarithm it makes little difference; a value of αR of order 5 is indicated. Thus if N is the concentration of donors and a_H the Bohr radius, we find

$$N^{\frac{1}{3}}\ a_H \simeq 0.2 \qquad (37)$$

A more detailed calculation would estimate c and take into account the disorder resulting from random positions of the centres. Hirsch and Holcomb (priv. comm) find N_c *greater* than for uncompensated samples.

In this form of transition - a pure Anderson transition- the density of states should show no discontinuity, and the same appears to be true of the conductivity. At concentrations above the transition the conductivity is metallic, tending to a finite value as $T \to 0$; σ itself should behave like*

$$0.03\ e^2/\hbar\xi \tag{38}$$

where ξ, defined only near the transition, was introduced in §4.

In a magnetic field we have two effects to consider:-

1. The magnetic field B will shrink the orbits, ultimately to a radius of L_H where $L_H = (c\hbar/eB)^{\frac{1}{2}}$. If one starts with the metallic state, this will lead to a transition to the insulating state. The value of the metallic conductivity at which this occurs will be

$$\sigma_o = 0.03\ e^2/\hbar L_H\ . \tag{39}$$

However, if $L_H < a$, Mott and Kaveh[6] have proposed

$$\sigma_o = 0.03\ e^2/\hbar a\ , \tag{40}$$

though a zero value for strong fields is predicted by scaling theory. (39) has not been tested experimentally, but the work of Biskupski[44] and co-workers on doped and compensated InSb showed that (40) was correct for this system. Early results are shown in fig.8. By varying a the relationship (40) could be verified over a range of a, as shown in fig. 9. The observed value of the constant is 0.027, within the approximations of the theory equal to the author's theoretical value. The results quoted were obtained with temperatures down only to 0.1 K, but have been extended to 30 mK by Biskupski et al[45] and Long and Pepper[46]; no temperature-dependence of the metallic specimens is observed.

* *The constant 0.03 was obtained by Mott[31], while the considerations of Mott and Kaveh[6] give about 0.1.*

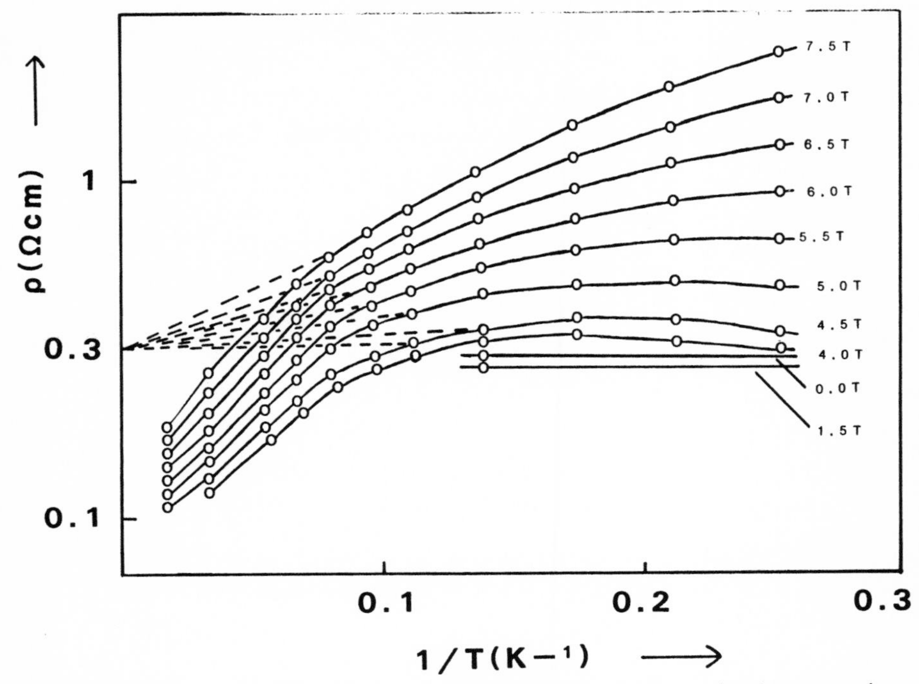

Fig. 8. Plot of resistivity ρ of n-type InP versus 1/T for varying magnetic fields (Biskupski 1982)[89].

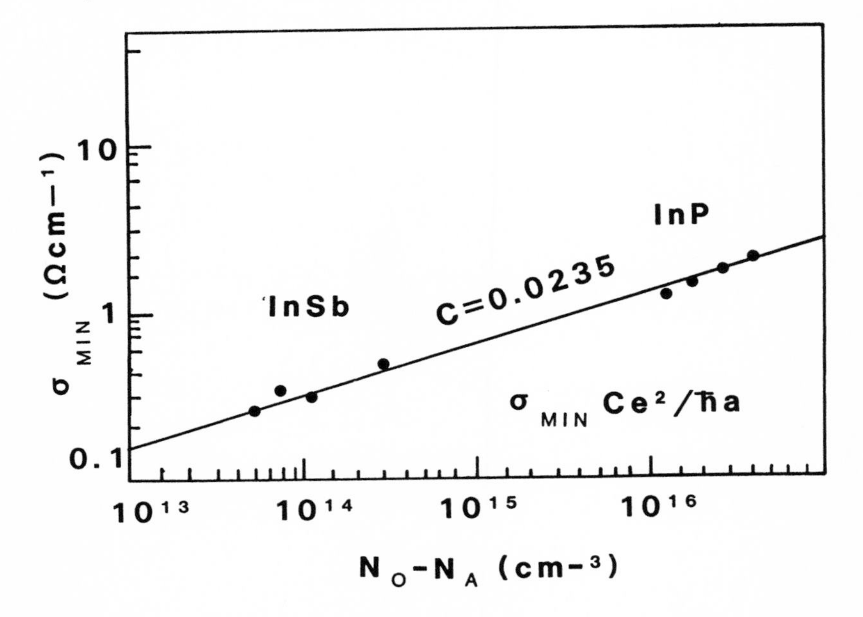

Fig. 9. Plot of σ_{min} versus distance between donors in InP (Biskupski 1982)[89].

Since the $T^{\frac{1}{2}}$ correction appears small, it is conjectured that for these materials the quantity $(\frac{4}{3} - 2F)$ in equation (32) is (fortuitously) small.

At low values of the magnetic field, on the other hand, the term $(1 - \ell/L_H)$ in equation (21) must *increase* the conductivity and Shapiro[47] first proposed that the metal-insulator transition must depend on field H and concentration n of donors (at T = 0) as illustrated in fig. 10.* Spriet et al[48] have recently shown that this is the case in an investigation of n type InP (compensated). The effect at small fields goes as $H^{\frac{1}{2}}$, and the effect for small H, being smaller, is not observed. The authors also show that an increase in field H for a just insulating specimen gives rise to *two* transitions, to metallic and back again. An investigation of hopping conduction for concentration where there is no metallic regime would be interesting; T_0 should decrease and then increase again with H.

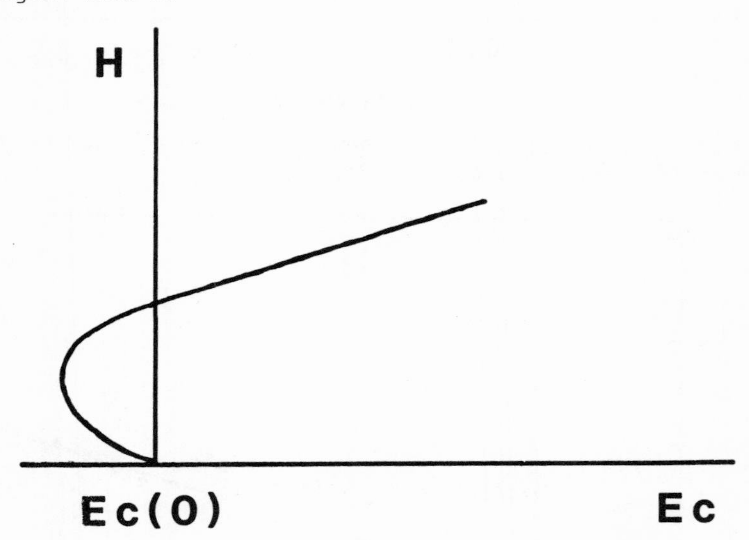

Fig. 10 Mobility edge in a magnetic field.

8.

THE METAL-INSULATOR TRANSITION IN UNCOMPENSATED SEMICONDUCTORS

The present author[49] in 1949 discussed the metal-insulator transition

* *Inclusion of interaction may lead to the result that, near the transition, σ does not increase with field, casting doubt on the result of fig. 10 (Fukuyama, priv. comm).*

in a crystalline array of one-electron centres, and gave an argument to show that, as the interatomic distance $\underline{a}$ was altered, the concentration n of free electrons would jump discontinuously from zero to $1/a^3$ and that this would occur at a concentration n_c of donors given by

$$n_c^{\frac{1}{3}} a_H = 0.25 \tag{41}$$

Here a_H is the hydrogen radius. The derivation of (41) given in 1949 has been replaced by another, and I was lucky to obtain the correct result (cf. ref. 6). However, equation (41), when applied to donors or acceptors in semiconductors, has been astonishingly successful over a wide range of materials, as pointed out by Edwards and Sienko[50], in spite of disordered arrangements of the centres. Edwards' and Sienko's results are reproduced in various publications - e.g. ref. 6.

The most satisfactory discussion of the discontinuity is as follows, following Brinkman and Rice[51], Rice[52]. Consider any intrinsic semiconductor with conduction and valence bands as in fig. 11. A plasma with a density of n electrons and n holes has, per unit volume, energy of the form

$$E = \frac{6\hbar^2 n^{2/3}}{5\, m_{eff}} - C\, e^2\, n^{1/3}/\kappa \tag{42}$$

The first term represents the kinetic energy of the condensed electron-hole gas, m_{eff} being some mean of their effective masses. The second represents the Coulomb interaction between the electrons and holes; to calculate the constant C account must be taken of correlation. κ is the background dielectric constant. The quantity E plotted against $n^{1/3}$ has a minimum as shown in fig. 12, at a value n_c of n estimated to be about

$$n_c^{1/3} = \text{const}\ e^2 m_{eff}/\hbar^2\kappa$$

and at which the energy E_o is given by

$$E_o \sim \text{const}\ m_{eff}\ e^4/2\hbar^2\kappa^2$$

If electrons and holes are excited optically, they will condense into "droplets" with this density, a phenomenon extensively investigated. The application to our problem is that if ΔE in fig. 11 can be reduced, either by alloying, change of volume or in any other way, then when $\Delta E = E_o$ there should be at zero temperature a discontinuous change of n

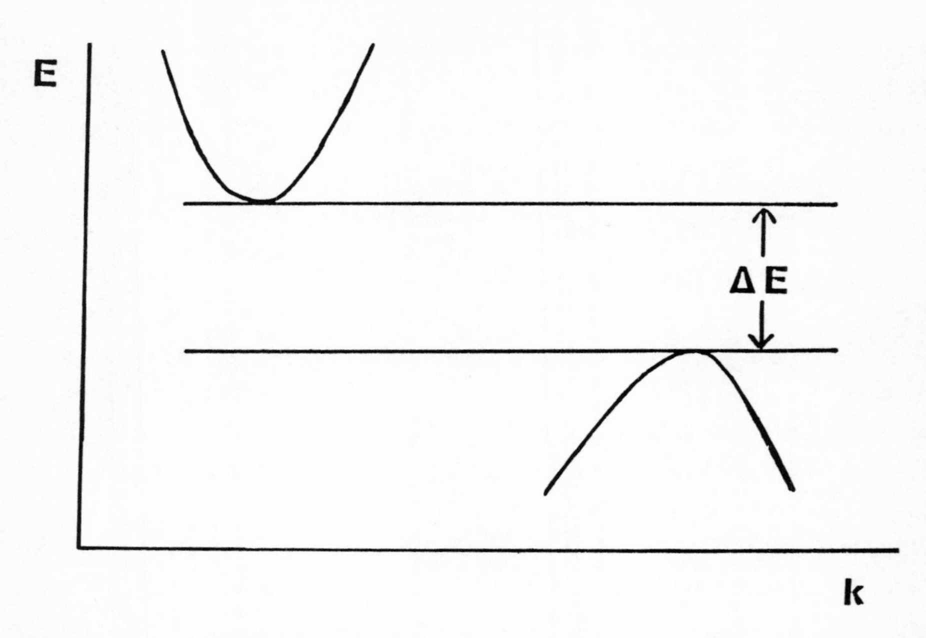

Fig. 11. Energy bands of an indirect semiconductor plotted against k.

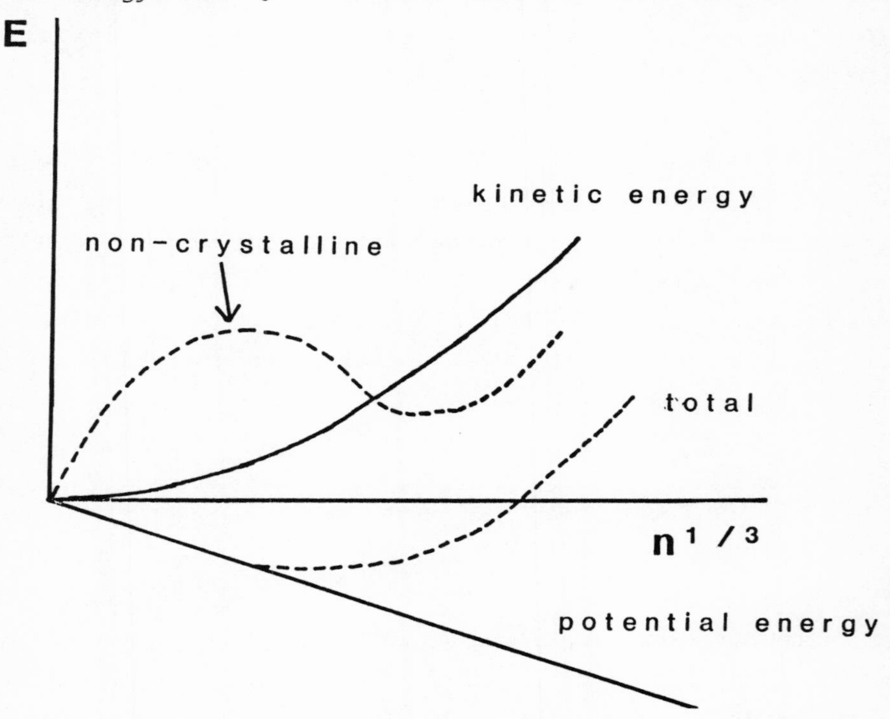

Fig. 12. Energy of an electron-hole gas plotted against $n^{1/3}$ for a crystal (solid line) and a non-crystalline system (broken line).

from zero to n_c. For the effect to be large, a small value of κ is needed.

For one-electron centres, the two bands will be Hubbard bands. The lower band is a "hole" band, the upper an electron band. These can be described as follows: If Ψ_n^e is the many-electron wave-function for all the electrons on the donors, including an extra one on atom n, then a wave function of the form

$$\sum_n \exp(ika_n)\bar{\Psi}_n^e \tag{42a}$$

will represent an electron in a state in the upper band. Similarly

$$\Sigma \exp(ika_n)\ \Psi_n^h \tag{43}$$

will represent the lower band, where Ψ_n^h describes the system with an electron missing from atom n. The gap between the bands will be

$$U - \tfrac{1}{2}(B_1 + B_2) \tag{44}$$

where B_1, B_2 are the widths of the two bands. For a crystalline lattice, the metal-insulator transition *neglecting the discontinuity* will occur when this vanishes.

Early calculations took B_1 and B_2 to be equal, so that the transition occurs when U = B. For U was taken

$$U = \frac{5}{8}\, e^2/\alpha\kappa\ , \tag{45}$$

where $\alpha = 1/a_H$, and for B we take

$$B = (2ze^2/\alpha\kappa)(1 + \alpha a)e^{-\alpha a} \tag{46}$$

This gives

$$\alpha a = \ell n\ 96 \simeq 4.6 \tag{47}$$

However, in doped semiconductors the upper band is much wider than the lower. Mott and Davies[53] took the half width of the lower band to be half the ionisation energy of the centres. This gives

$$\alpha a = \ln 120 \simeq 5 \tag{48}$$

Owing to the logarithm of a large number on the right, the assumptions here do not make much difference to the numerical result. The experimental value from Edwards and Sienko[50] is 1/0.27. It will be noted that the value differs little from that calculated for compensated materials. In equations (45), (46) κ was taken to be the background dielectric constant, not including that from the donors themselves. Since for two Hubbard bands a transfer of electrons between one and the other involves the movement from one site to another, κ caused by the centres themselves should be small.

Some authors[54, 55], following an early paper by Hertzfeld[90], use the Clausius-Mossotti formula,

$$\kappa = 1 + 4N\alpha/(1 - 4\pi N\alpha/3)$$

where α is the polarisibility of a donor, and suppose that the transition occurs when $\kappa = \infty$. In our view, however, as κ increases U will drop, and a transition as described here will always occur before κ diverges. None the less the Clausius-Mossotti formula may well give a fair numerical indication of the concentration at which the transition will occur.

For a crystalline material, and often for a non-crystalline one, we expect a discontinuity in n at the transition. Mott[56] has given a criterion for whether this exists in a non-crystalline situation. The evidence is that it does in fluids such as Cs and Na_xNH_3, to be discussed in §10, but not in doped semiconductors. If so, we may describe the transition in doped uncompensated semiconductors as follows. It will occur when the upper and lower Hubbard bands overlap *sufficiently for states at the Fermi energy to be delocalized.* The transition is thus of Anderson type, but the concentration at which it occurs will be given by (48), and is determined approximately by the Hubbard U.

That it is of Anderson type is shown by the existence of a variable range hopping near the transition in uncompensated Si:P (ref.16, P.123).

Near the transition on the metallic side it is clear that most sites are singly occupied, a few are unoccupied or doubly occupied. The system could therefore have some of the properties of the "highly correlated" electron gas described by Brinkman and Rice[57], cf. also Mott[58]. This possibility will be discussed in the next section; we think that the evidence is against it.

We should however expect such materials to have a large screening length (because for small overlap the number of carriers is small) and therefore a *negative* value of β in the expression (equation 28)

$$\delta\sigma = \beta T^{\frac{1}{2}} \tag{49}$$

This appears to be the case for Si:P, Si:B, Ge:As, Si As (cf. ref. 6, p. 361). On the other hand it is positive for a-Si-Nb (and for compensated materials generally, but appears to change sign near the transition.

The classic work of Thomas and co-workers[59, 60, 61, 62] on the approach to the transition in Si:P at very low temperatures shows that near the transition, using stress tuning

$$\sigma(T=0) \sim (n-n_c)^{\frac{1}{2}} \tag{50}$$

and this behaviour appears to be observed for all uncompensated materials investigated. On the other hand for a-Nb-Si and compensated materials[62], it is found that

$$\sigma(T=0) \sim (n-n_c) \tag{51}$$

Mott and Kaveh[6] suggested that the former behaviour results from electron-electron interaction, only when β in eqn,(49) is negative. We have already seen that compensated and uncompensated semiconductors are likely to differ in this respect. The argument was then as follows. There is, as well as (49), a term in the conductivity independent of T, and this is also proportional to $\frac{4}{3}$ - 2F (ref. 6, p. 360). The Kawabata formula (21) for zero temperature with interactions will then become

$$\sigma = \sigma_B \, g_M^2 \left[1 - \frac{C_1}{g_M^2 (k_F \ell)^2} + \frac{C_2}{g_M^2 (k_F \ell)^2} \right] \tag{52}$$

where C_2 is positive when the Hartree term predominates, that is (we assume) for uncompensated semiconductors. We suppose - as before - that owing to the first two terms in the square brackets σ tends to zero at the transition; For uncompensated semiconductors these authors assume that C_2 is negative and that, while the first term is of the form $Ae^2/\hbar\xi$ as before, the second can be written const/$\xi D(\xi)$, and as D is proportional to σ the second term predominates and

$$\sigma^2 \propto \xi$$

If $1/\xi$ even with interaction varies with energy according to (10), that is as $n-n_c$, equation (50) is explained. Shafarman and Castner[63] have deduced the index of ξ from an analysis of variable-range hopping in Si:As, using a method just applied by Pollitt[64]. They assert that in eqn. (50), $S=\frac{1}{4}$ fits the data better than $S=\frac{1}{2}$, which is perhaps to be expected near the transition, and this is what they assume in making their analysis; they find for the index of $1/\xi$, a value of about 0.8, near to unity.

This derivation, then, supposes that $1/\xi$ varies as $n-n_c$ and below the transition as n_c-n. Calculations of the dielectric constant without interaction show that the dielectric constant κ varies as ξ^2 (see Mott and Kaveh[6]). However experiment shows that κ is proportional to σ^{-2} near the transition. Lee (priv. comm.) maintains that κ should vary as ξ^2 *with* interaction. His argument is that $\kappa(q,\omega)$ is proportional to $D/(D^2 - i\omega)$ in the metallic range. But, as below the transition clusters of volume ξ^3 should behave like a metal, one could write there

$$\kappa \propto D/\{(D/\xi^2) - i\omega\} \, .$$

He argues that interaction will only affect D - but in any case, if we put $\omega = 0$, D goes out, so that

$$\kappa \propto \xi^2 .$$

If this is so, $\xi \propto 1/(n_c-n)^{\frac{1}{2}}$. We have no theory to explain this.

In favour of the first approach Mott[65] suggested an instability at $E = E_C$ if the index is less than $\frac{2}{3}$. If the index is ν, the wave functions should rise or fall exponentially as $e^{-\beta r}$ where $\beta r = (r/a)^{3(1-3\nu/2)}$, and if $\nu = \frac{1}{2}$ this gives $(r/a)^{3/4}$. The consequences of this are - in our view - that the properties of the system will not tend to a constant value with size.

An example when σ goes linearly to zero is a - $Nb_x\ Si_{1-x}$ investigated by Hertel et al[66]. Equation (21), without interaction, gives the right constant in the relation

$$\sigma = \sigma_o\ (n-n_c)$$

n_c being about 0.1 and σ_o about 1.5 σ_{min}.

We should expect

$$\frac{\sigma}{\sigma_B} = \frac{2}{9g_M^2}\ \frac{d\ell n g_M}{dn}\ (n-n_c)$$

If the NG atoms form a band, we could write $n=1/R^3$ where R is the distance between such atoms. Without disorder we take for the band width

$$2Z(e^2/\kappa R)\ (1 + \alpha R)e^{-\alpha R}$$

and with disorder due to random fields

$$\text{const } e^2/\kappa R$$

Thus $\quad g_M \sim \text{const } e^{-\alpha R}$

and $\quad \frac{d\ell ng}{dR} \sim -\ \alpha R \sim 4$

and $\quad \frac{d\ell ng}{dn} = \frac{4}{3}$

Thus

$$\sigma = \sigma_o (n-n_c)$$

with

$$\sigma_o = \frac{e^2}{3\hbar a} \frac{4}{81}$$

which is just above σ_{min}, as observed.

9.

EFFECTS OF INTRA-ATOMIC CORRELATION

It has been suggested in the last section that just above the transition an electron gas could have the properties of a correlated gas as described by Brinkman and Rice[57]. It is certainly true that a few (say 10%) of the donors will be doubly occupied, and an equal number unoccupied. But there is one important difference. In a crystal these doubly and unoccupied sites represent *mobile* electrons and holes. The current is due to them alone, the response to a field is thus reduced, which can be represented by an increased effective mass, leading to an enhanced specific heat and paramagnetism.

In the case of Si:P however in the impurity band the doubly and unoccupied states are not fully mobile, but fixed in space by the disorder. Therefore, we do not think that the mass enhancement should occur. There are however certain phenomena which have (tentatively) been ascribed to this enhancement and for which an alternative explanation must be sought.

The electronic specific heat of Si:P in the metallic state near the transition follows the free electron behaviour for electrons in a conduction band down to about 1.3 K; according to Kaveh and Liebert[15] this same behaviour is to be expected for an impurity band. However, below this temperature, γ (where $c_v = \gamma T$) increases by about 2. If our argument given above is correct, this cannot be caused by the Brinkman Rice enhancement; moreover this mechanism would not, as far as we can see, account for the drop in c_v above ~ 1 K. One possibility is that some donor sites must exist exceptionally far from other sites, and perhaps with low energy, which act as Kondo sites. At a Kondo site there should normally be a single electron, but an electron (or hole) jumps to it from states at the Fermi energy, and the frequency of such a process is

$$w_k = (\Delta/\hbar)\exp(-E_k/\Delta) \qquad (55)$$

where E_K is the energy necessary for the jump and Δ a self-energy. This in a "Kondo metal" can lead to a large enhancement of m_{eff} over a range of energies at the Fermi limit, of magnitude $\hbar w_k$. When $k_B T > \hbar w_k$, this enhancement will drop away. We make the hypothesis that this is the cause of the drop in c_v as the temperature rises.

Takemori and Kamimura[67] and Kamimura[68] have a different explanation. They use a model of a Fermi glass put forward by the present author in 1974 (see ref. 16, p. 111) in which below the Fermi level the states are singly occupied, so that a transition from two states which are singly occupied to a configuration where one is doubly occupied and one unoccupied, as in fig. 16, can occur. This gives rise to a Schottky bump in the specific heat, observed in the intermediate region (below the transition). Also they show that the random values, from zero upwards, of the energy for this transition, gives a *linear* term in T in the specific heat, which was observed.

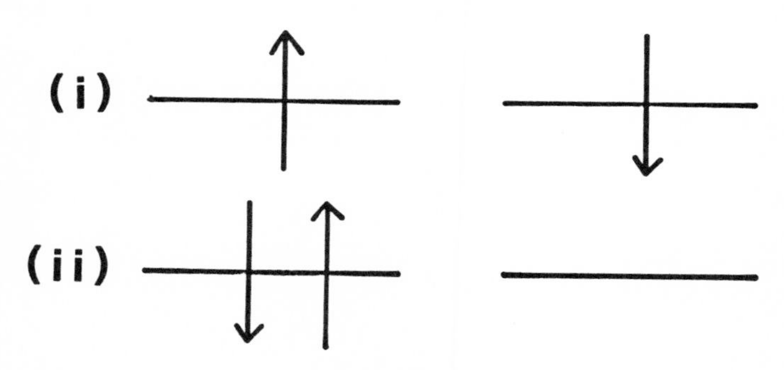

Fig. 16.

An estimate of how much of this linear term has this origin, and how much from the density of states at the Fermi surface would be of interest. We guess from the unpublished work of Wallis (ref. 16, p. 123) that in Si:P at $n = 2.45 \times 10^{18} cm^{-3}$ the slope of the $\ln\sigma - 1/T^{\frac{1}{4}}$ plot has dropped by 2 from $n = 3.05 \times 10^{18}$, which means that N(E) should have dropped by 8 - much more than the observed drop in the specific heat. So some way from the transition most of the linear term is from the spins.

Actually, at the transition, since below E_c states are still localized, it is to be expected that this effect will persist, so the enhancement of c_v below ~ 1°K may be explained by this model.

Another unexplained effect is that observed by Long and Pepper[69] and illustrated in fig. 13, in which the conductivity of Sb-doped Si (with a concentration of Sb of $2.7 \times 10^{18} cm^{-3}$) is plotted against magnetic field. We make the following hypothesis to account for this behaviour. The random field which determines the charged sites will include their Coulomb interaction, which will tend to order their position in space. As the temperature rises, their positions will become more nearly random, so the disorder increases. This will lower σ, as observed. The magnetic field, by orienting the spins so that the lower Hubbard band is full, could decrease the dielectric constant, and consequently the tendency of the charged centres to order is increased.

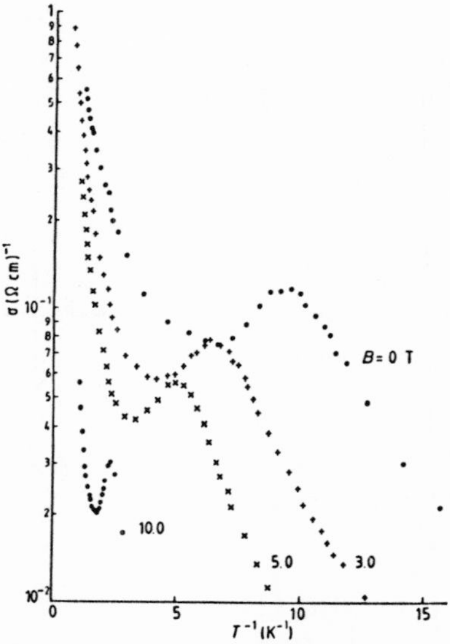

Fig. 13.

Conductivity σ plotted against 1/T for Si:Sb with dopant concentration $2.7 \times 10^{18} cm^{-3}$ (Long and Pepper[69]) for varying magnetic field B.

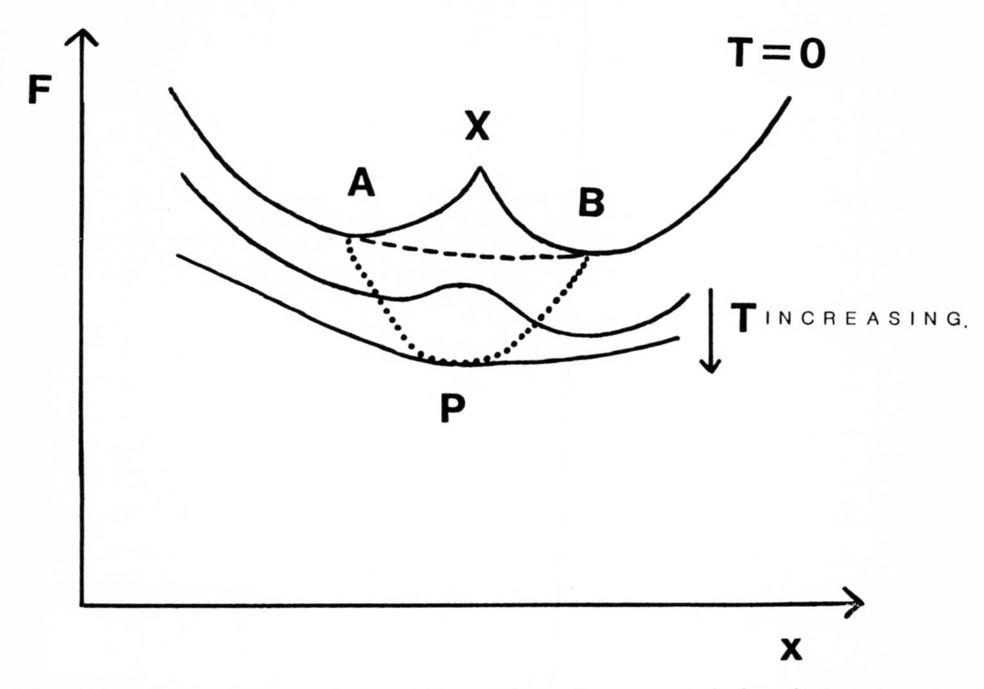

Fig. 14. Free energy of a system undergoing a metal-insulator transition A-B against a parameter x. P represents the consulate point and X the value of x at the transition.

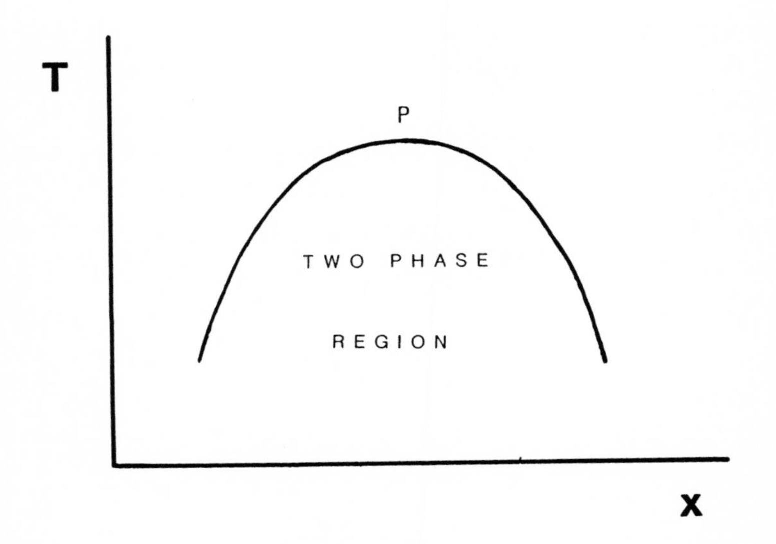

Fig. 15.

10.

METAL-INSULATOR TRANSITIONS IN LIQUIDS

We have seen that in any array of one-electron centres, all occupied, the metal-insulator transition with change of volume is discontinuous. The free energy plotted against volume x will therefore appear as in fig. 14. The volume between A and B will be unstable. Under pressure the volume will jump discontinuously from A to B. In solutions, such as that of Na in NH_3, we expect a two-phase region, such as that illustrated in fig. 15, with a critical point P. Thus in fluids of one-electron centres, we expect a critical point at the concentration of the metal-insulator transtion*. We have already seen that for fluids the quantum corrections to the conductivity do not occur, so the conductivity at P should be of the order 200 Ω^{-1} cm^{-1}, falling off rapidly as c becomes smaller and becoming temperature-dependent. Extensive experimental work from the school of Professor Hensel[71] in Marburg on fluid caesium shows that this is so, as is also the case for metal-ammonia solutions. Interestingly, at the critical point of mercury the conductivity is much smaller, c. 10^{-3} Ω^{-1} cm^{-1}. I have suggested[58] that ideally another critical point must exist at the metal-insulator transition, but because of the large background dielectric constant this may well lie at temperature below the fluid region. The idea of two critical points, one associated with a metal-insulator transition and one with inter-atomic forces as in rare gases, goes back to 1943 (Landau and Zeldovitch[72]).

If χ is the compressibility of a fluid, an equation near the critical point T_c of the form

$$\chi = |(T_c - T)/T_c|^{-\gamma} \qquad (56)$$

with critical exponent γ of order unity is observed for Cs, and also for mercury. This is what we expect from mean field theory, and is *not*

* *See also ref 76 and a discussion by Holzhey and Schirmacher[70] of the phase separation in molten salt mixtures.*

observed for rare gases. This suggests that the interaction is very long-range. Its observation is perhaps surprising for Hg, which is not metallic at the critical point (Hensel et al[71]). Perhaps $N(E_F)$ is finite (as in a metal) but states are localized; if the localization is weak, forces might still be long-range.

For metal-ammonia solutions there is evidence, from the indices at the critical points, of the long-range nature of the forces (see ref. 58 p. 250), and Damay et al [73, 74].

11.

HOPPING CONDUCTION

This concept has its origin in the work of Miller and Abrahams[75], and was brought forward to explain the phenomenon of low temperature "impurity conduction" in doped semiconductors, investigated in detail by Fritzsche and co-workers[76, 77]. This is a process of charge transport, occurring for any concentrations of donors or acceptors below the metal-insulator transition, essentially for compensated semiconductors. We now consider it possible also for uncompensated material for any concentration for which $N(E_F)$ is finite, and thus for a range of concentrations below n_c; this has been observed (Mott and Davis[16], p. 123). Miller and Abrahams considered thermally activated hopping to nearest neighbour sites only, leading to a conductivity of the form

$$\sigma = \sigma_3 \exp(-\varepsilon_3/kT) \tag{57}$$

with $\varepsilon_3 \sim e^2/\kappa a$. Mott[78] first showed that, if hops to more distant sites are included (variable-range hopping), a conductivity of the form

$$\sigma = A \exp\{(T_o/T)^s\} \tag{58}$$

is to be expected, with

$$k_B T_o = \text{const}/N(E_F)a^3$$

and $s = \frac{1}{4}$. Hopping of this kind has been extensively observed, though experimental difficulties of determining the correct value of s are considerable.

Efros and Shklovskii[79] showed that, if Coulomb interaction between the sites is taken into account, the one-electron density of states will have the form of Fig. 17, vanishing at E_F. This does *not* mean that the relaxed density of states and thus c_v vanish. This form of the density of states leads to the index $s = \frac{1}{2}$ in equation (58), for the conductivity.

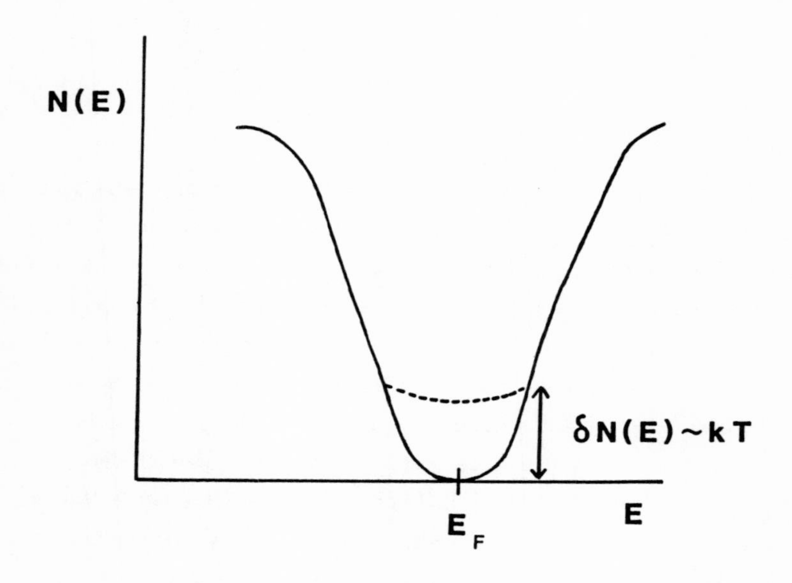

Fig. 17. The Coulomb gap in the density of states, according to Efros and Shklovskii (1975). The effect of temperature is shown by the dashed line.

Experimental evidence for this form of the density of states was obtained by Davies and Franz[80] in a study of photoemission in tungsten bronzes.

Of great interest is the question of when $s = \frac{1}{4}$ is to be expected and when $s = \frac{1}{2}$. Two parameters enter the theory; the disorder parameter V_0 and the Coulomb interaction (Δ). In doped and compensated semiconductors they are comparable; thus in our view the Coulomb gap will disappear only when the electron gas becomes non-degenerate. According to calculations by Davies et al[81], for Si:P at a phosphorus concentration of 0.01 n_c, this should occur at 10K. Mott and Kaveh[6] give an argument to show that, as the gap begins to fill up, $s = \frac{1}{2}$ remains the correct index. This is illustrated in fig. 17. If temperature leads to a density of states proportional to T and $s = \frac{1}{4}$, then the variation of $N(E_F)$ will lead to an observed index equal to $\frac{1}{2}$.

Near the transition, the minimum in the density of states should be narrow, and we should expect the index $\frac{1}{4}$ except at very low temperatures.

Timp et al [82] show that $s = \frac{1}{3}$, the value for two-dimensional problems without interaction, is observed for hopping conduction in the inversion layer of a sodium-doped MOSFET. They deduce that there is no evidence for the Coulomb gap in this system.

Entin-Wohlman and Ovadyahu[83] in experiments on the conductivity of indium oxide films find $s = \frac{1}{2}$, but that the conductivity can be significantly reduced by putting the surface in contact with a metal. It is surmised that the image force reduces the Coulomb gap.

It should be remembered that all theoretical work lies in the regime where there is little overlap between the wave functions of the localized states. It is not clear that a Coulomb gap should exist when there is strong overlap, that is near the M-I transition, though Davies (priv. comm.) believes that $N(E_F)$ (without relaxation) will always vanish, but over a very narrow region. It must be emphasized, however, that *with* relaxation $N(E_F)$ will not normally vanish near the transtion.

Benzaquem and Walsh[84] in a careful investigation of n-GaAs find that s lies in the range 0.2 to 0.27.

Voegele et al[85] investigated hopping conductivity betwen 15 and 300 K in amorphous silicon modified by the implantation of Au or Si ions. They found values of s in the range $\frac{1}{4}$ to 1 and sought to explain these results through variable-range hopping with a correlation term. Finlayson and Mason[86] find $s = \frac{1}{2}$ in InP.

Vignale, G., Shinozuka, Y. and Hanke, W.[87] in a recent preprint, study the effect of quantum hopping on the one-particle density of states. They find, contrary to the previous classical treatment, that in two dimensions there is no Coulomb gap, in agreement with results of Timp et al[82]. They find in three dimensions that the Coulomb gap disappears when the hopping rate exceeds a certain critical value.

Note added in proof.

Professor Kaveh's explanation of the relationship (50) observed for uncompensated semiconductors is discussed in detail in ref. 88. If this explanation is correct, the screening length for these materials must be small, while the model given here in §8 suggests the opposite. Professor Kaveh (priv. comm.) points out that the enhancement of the specific heat observed at low temperatures may lead to this result, as may also the many-valley nature of the semiconductors in which this effect has been observed, if this affects the upper Hubbard band, as suggested in §1.

REFERENCES

1. Abrahams, E., Anderson P.W., Licciardello, D.G. and Ramakrishnan, T.V. Phys. Rev. Lett. 42, 673.
2. Efetov, K.B., Sov. Phys. JETP 61 (3) 606 (1985).
3. Altshuler, B.L. and Aronov, A.G., Solid State Commun. 30, 115 (1979).
4. Finkelstein, A.M., Soviet Phys. JETP 57, 47 (1983).
5. Efros, A.L. and Shklovskii, B.I., J. Phys. C Solid State Phys. 8, L49 (1975).
6. Mott, N.F. and Kaveh, M., Adv. in Phys. 34, 329 (1985).
7. Anderson, P.W., Phys. Rev. 109, 1492 (1958).
8. Davies, J.H. and Franz, J., Phys. Rev. Lett 57, 475 (1986).
9. Jain, K., Lai, S. and Klein, V., Phys. Rev. B 13, 5448 (1978).
10. Tunstall, D.P. and Sohal, G.S., J. Phys. C: 16, L251 (1983).
11. Davis, E.A. and Compton, A.H., Phys. Rev. A 140, 2183 (1965).
12. Jerome, D., Ryter, C., Schulz, H.J. and Friedel, J. Phil. Mag. B 52, 403 (1985).
13. Meyer, J.R., Bartoli, F.J. and Mott, N.F., Phil. Mag. B 52, L57 (1985).
14. Meyer, J.R. (in press).
15. Kaveh, M. and Liebert A. (to be published).
16. Mott, N.F. and Davis, E.A., Electron Processes in non-crystalline Materials, Oxford 1979.
17. Ioffe, A.F. and Regel, A.R., Prog. Semicond. 4, 327 (1960).
18. Elyutin, T.V., Hickey, B., Morgan, G.J. and Weir, G.F., Phys. Stat. Solidi (b) 124, 279 (1984).
19. Mott, N.F., Adv. Phys. 16, 49 (1967).
20. Mott, N.F. Phil. Mag. 20, 7 (1970).
21. Kubo, R. Can. J. Phys. 34, 1274 (1956).
22. Greenwood, D.A., Proc. phys. Soc. 71, 585.
23. Mott, N.F., Phil. Mag. 26, 1015 (1972), see also ref. 16.
24. Mott, N.F., Phil. Mag. B 43, (1985), 941.
25. Bergmann, G., Phys. Rev. B 28, 2941 (1983).
26. Kawabata, A., Solid St. Commun. 38, 823 (1981).
27. Mott, N.F. and Kaveh, M., Phil. Mag. B 52, 177 (1985).
28. Ng, H.G., Capizzi, M., Thomas, G.A., Bhatt, R.N. and Gossard, A.C., Phys. Rev. B33, 7329 (1986).
29. Wölfle, P. and Vollhardt, D., in Anderson Localization, ed. by Y. Nagaoka and H. Fukuyama (Berlin: Springer-Verlag p. 26).

30. Hikami, S., Phys. Rev. B 24, 2671 (1981).
31. Mott, N.F., Phil. Mag. B 49, L75 (1984).
32. Lesueur, J., Dumoulin, L. and Nedellec, P., Phys. Rev. Lett. 55 2355 (1985).
33. Ovadyahu, A. Phys. Rev. Lett. 52, 569 (1984).
34. Mott, N.F., Phil. Mag. B 51, 19 (1985).
35. Mott, N.F., Phil. Mag. B 52, 169 (1985).
36. Cutler, M., Liquid Semiconductors, 1977 Academic Press, New York.
37. Kaveh, M. and Wiser, N., Adv. Phys. 33, 257 (1984).
38. Schmidt, A., Z. Physik 271, 251 (1974).
39. Altshuler, B.L. and Aronov, A.G., Solid St. Commun. 30, 115 (1979).
40. Bronovoi, I.L. and Sharvin, Y.V., JETP Lett. 28, 117 (1978).
41. Howson, M.A., J. Phys. F, 14, L25 (1984).
42. Howson, M.A. and Greig, D., Phys. Rev. B (in press).
43. McMillan, W.L., Phys. Rev. B24, 2739 (1981).
44. Ferré, D., Dubois, H. and Biskupski, G., Phys. Stat. Solidi, B 70, 81 (1975).
45. Biskupski, G., Dubois, H., Wojkievicz, J.L., Briggs, A., and Remenyi, G. J. Phys. C. 17, L411 (1984).
46. Long, A.P. and Pepper, M., J. Phys. C 17, 3391(1984).
47. Shapiro, B., Phil. Mag. B50, 241 (1984).
48. Spriet, J.P., Biskupski, G., Dubois, H., Briggs, A., Phil. Mag. B (in press) (1986).
49. Mott, N.F., Proc. Phys. Soc. A 62, 416 (1949).
50. Edwards, P.P. and Sienko, M.J., 1978, Phys. Rev. B17, 2575 (1975).
51. Brinkman, W.F. and Rice, T.M. Phys. Rev. B 4, 1656 (1971).
52. Rice, T.M., Solid St. Phys. 32 1 (1977).
53. Mott, N.F. and Davies, J.H., Phil. Mag. B 42, 845 (1980).
54. Castner, T.G., Phil. Mag. B 42, 643 (1979).
55. Edwards P.P. and Sienko. M.J., Phys. Rev. B7, 2575 (1978).
56. Mott, N.F., Phil. Mag. B 44, 265 (1981).
57. Brinkman, W.F. and Rice, T.M., Phys. Rev. B7, 1508 (1973).
58. Mott, N.F., Metal-Insulator Transitions, Taylor & Francis, London, 1974.
59. Rosenbaum, T.F., Andres, J., Thomas, G.A. and Bhatt, R.N., Phys. Rev. Lett 45, 1723 (1980).

60. Rosenbaum, T.F., Milligan, R.F., Paalanen, M.A., Thomas, G.A. and Bhatt, R.N., Phys. Rev. B 27, 7509 (1983).
61. Thomas, G.A., Physica B 117, 81, (1983).
62. Milligan, R.F., Rosenbaum, T.F., Bhatt, R.N. and Thomas, G.A., "A review of Electron-electron interactions in disordered systems, eds. A.L. Efros and M. Pollak, North Holland 1985.
63. Shafarman, W.N. and Castner, T.G. Phys. Rev. B 33, 3570 (1986).
64. Pollitt, S., Commun. Phys. 1, 207 (1977).
65. Mott, N.F., Phil. Mag. B 44, 265 (1981).
66. Hertel, G., Bishop, D.J., Spencer E.C., Rowell, J.M. and Dynes R.C., Phys. Rev. Lett 50, 743 (1983).
67. Takemori, T. and Kamimura, H., Adv. Phys. 32, 715 (1983).
68. Kamimura, H., in Electron-electron Interaction in disordered Solids 1985 p.555, eds A.L. Efros and M. Pollak, North Holland 1985.
69. Long, A.P. and Pepper, M., J. Phys. C 17, (1984) L425.
70. Holzhey, C. and Schirmacher, W., J. de Phys. (Colloque) 1986.
71. Hensel, F., Jungst, S., Noll, F. and Winter, R. "Mott Festschrift", Inst. of Amorphous Studies, Vol. 2, p. 109 (1985).
72. Landau, L.D. and Zeldvich, G., Acta Phys. Chim. USSR 18, 194 (1943).
73. Damay, P., Le Clercq, F. and Devoider, P., J. Phys. Chem. 88, 3760 (1984).
74. Damay, P., Le Clercq, F. and Chieux, P., J. Phys. Chem. 88, 3734 (1984).
75. Miller, A. and Abrahams, S., Phys. Rev. 120, 745 (1960).
76. Fritzsche, H., J. Phys. Chem. Solids 6, 69 (1958).
77. Fritzsche, H., Proc. 13th Scottish Universities Summer School in Physics, p.55 (1973).
78. Mott, N.F., J. non-cryst. Solids, 1, 1 (1968).
79. Efros, A.L. and Shklovskii, B.I., J. Phys. C: Solid State Phys. 8, L49 (1975).
80. Davies, J.H. and Franz. J., Phys. Rev. Lett 57, 475 (1986).
81. Davies, J.H., Lee, P. and Rice, T.M., Phys. Rev. B 29, 4260 (1984).
82. Timp, G., Fowler, A.B., Hartstein, A. and Butcher, P.N., Phys. Rev. B 33, 1499 (1985).
83. Entin-Wohlman, G. and Ovadyahu, Z., Phys. Rev. Lett 56, 643 (1986).
84. Benzaquem, M., and Walsh, D., Phys. Rev. B 30, 7284 (1984).

85. Voegele, V., Kalbitzer, S. and Bohringer, K., Phil. Mag. 52, 153 (1985).

86. Finlayson, D.M. and Mason, P.J. J. Phys. C: Solid State Phys. 19 L299 (1986).

87. Vignale, G., Shinozuka, Y. and Hanke, W., Phys. Rev. B 34, (August in press) (1986).

88. Kaveh, M., Phil. Mag. 52, L7 (1985).

89. Biskupski G., Dubois, H. and Laborde O., "Application of High Magnetic Fields in Semiconductor Physics". Lecture notes in Physics (Berlin Springer 1983 p. 411).

90. Hertzfeld, K.F., Phys. Rev. 29, 701 (1927).

ELECTRONIC PROPERTIES IN TWO DIMENSIONAL SYSTEMS

P.J. Stiles

Physics Department, Brown University, Providence RI, USA

1.
INTRODUCTION

This treatise is meant as a primer for those people who wish to begin to work in the area of the electronic properties of two-dimensional (2D) systems and is to be used in conjunction with the excellent review article by Ando, Fowler and Stern (AFS)[1]. It is not meant to be as complete as is AFS as it is much shorter, but aims to offer some insights as to why certain things are done and to bring to the attention of the reader some aspects of the field at the time of the writing. This short work will attempt to give a phenomenological description using orders of magnitude, dimensional analysis and clues from our understanding of the real three-dimensional (3D) world. Since the mid sixties there has been an international conference series of the work described here. This series labelled as "Electronic Properties of Two-Dimensional Systems"[2] should be consulted as well as standard condensed matter physics textbooks such as Ashcroft and Mermin[3].

One might well ask why the interest in two-dimensional systems. It is not just because they are different. One of the most remarkable attributes of these two-dimensional systems is the fact that in a one carrier system one is able to obtain very high mobilities with reasonably high densities. Thus one can study electron-electron interactions with reasonably sharp energy levels.

In a semiconductor with a 3D system, there are three choices as to how to obtain high densities of carriers. One is to raise the temperature and thermally excite carriers. This results in two kinds of carriers and lots of scattering due to the presence of phonons. The second is to shine light to generate the carriers and to do so at low temperatures. There are still two kinds of carriers, and the density is not more constant than the light intensity. Further, carriers are continually recombining. The third technique is to dope the system. However, although we are able to obtain only one kind of mobile charge, the scattering from the coulomb center left behind ruins the mobility. In essence, the compensating charge for our carriers is in the same region of space as the carriers and hence the scattering is large. The two-dimensional systems have the best of both worlds. There is a single kind of carrier, and the compensating charge is removed from the carriers. This is true for both systems, the MOSFET and heterostructure, to be described later.

An attempt is made to give first a philosophical discussion of what constitutes a "quasi-two-dimensional system", how it is realised in practice, what the basic electronic properties are, a simple discussion of conductivity, then of course magnetoconductivity and then some specific subjects. The latter will centre around the density of states (DOS) and how it relates to localization. Within this part, a biased view of what constitutes localization in these systems will be presented.

2.

THE PHYSICAL SYSTEMS: CONFIGURATIONS

As described above, the feature that leads to a very low level of scattering is the separation of the "free carriers" and their compensating charge. Because of a focus on two dimensional systems, the interest is in structures that produce a uniform electric field perpendicular to the flat two-dimensional sheet of carriers. Therefore the paradigm is the flat capacitor. The best known such system is the Metal-Oxide-Semiconductor Field-Effect-Transistor (MOSFET) and similar structures. The basic structure is illustrated in Figure 1.

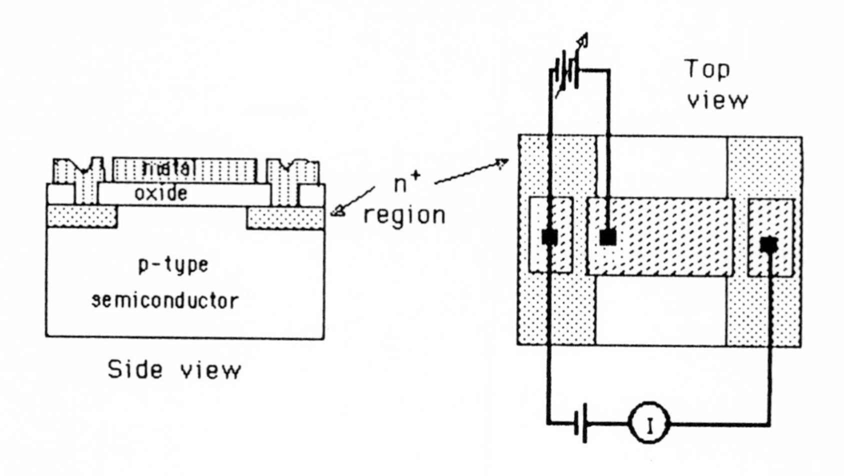

Figure 1

In describing the behaviour of these systems, it is convenient to consider that the sample is at T=0 unless otherwise mentioned. The n^+ regions are conducting and the bulk is not. The positive voltage applied to the upper electrode (called the gate) relative to the n^+ regions results in a net positive charge on the gate electrode and an equivalent areal density of electrons at the interface of the semiconductor. Approximate configurations and values are given in Figure 2.

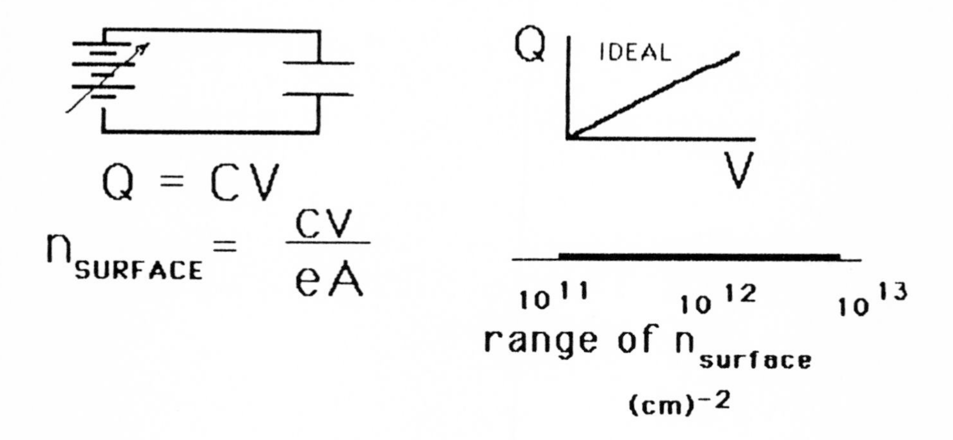

Figure 2

The three types of devices used are illustrated in Figure 3. The <u>Corbino disk</u> structure is an easy one to use, as the conducting 2D part

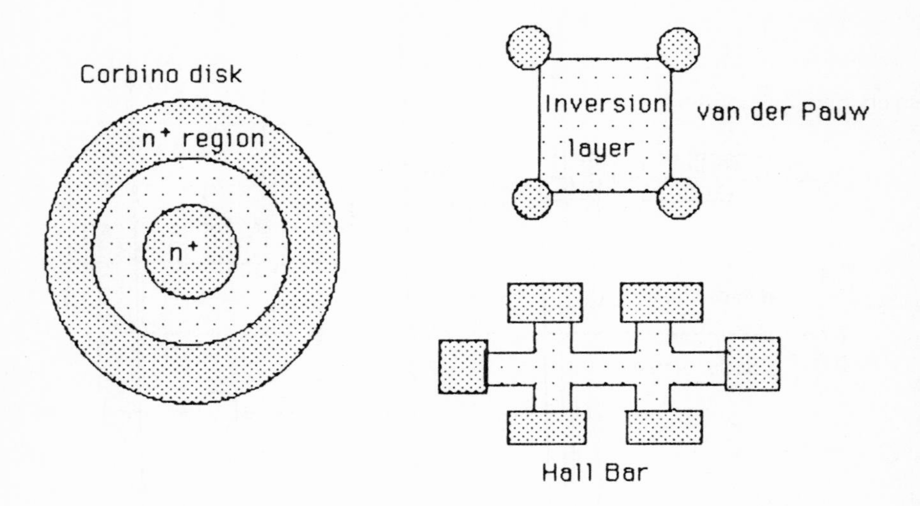

Figure 3

used is completely enclosed on the surface, so that there is no possible leakage paths in parallel with the two n^+ regions. Its disadvantage is that it is a two terminal measurement having a possible contact resistance in series with the channel and hence confusing the measurements. In addition when measuring a two terminal resistance with the current and the voltage using the same contacts, one cannot separate the two components of the resistivity tensor. The standard Hall bar configuration is illustrated in Figure 3. Except for the possible leakage paths in parallel with the channel, and the fact that the resistance of the device is higher because of the larger length to width ratio, it is the ideal structure. There are complications in reducing the voltage measurements to components of the resistivity tensor due to the geometry. The problems can be minimized by having the current contacts far away from the voltage probes. The van der Pauw configuration is a very convenient structure to use because of its simplicity. Leakage is possible, and it has geometric considerations for the analysis of the data. Further it has restrictions on the conductivity itself. However it allows the measurement of the full conductivity tensor with the fewest contacts.

An important point to remember from the experimental point of view, if one has separate voltage and current contacts, high resistance contacts are not a problem as long as the resistance is not high compared to the input of the voltmeter.

Measurement techniques: There are two standard measurement circuits (which can be used with either a.c. or d.c. voltage); one employs a constant voltage across the sample which results in a current determined by the sample, and the other circuit uses constant current which develops a voltage across the sample that depends on the sample resistance. There are advantages to both circuits. There are many applications where the phenomena depends on the electric field along the sample. In this case, one should use constant voltage. On the other hand, the Hall bar configuration usually uses a constant current. One then measures the parallel and transverse voltages. In either case, the advantage is to be able to measure the minimum number of parameters at each point. Of course, for best signal to noise, one would use neither of the circuits but something about half way in between.

In addition to the above techniques, there are two others that have been employed. Rather than have resistive coupling to the sample, the other two employ capacitive or inductive coupling. In the case of heterostructures, capacitive coupling is quite easy. The technique is illustrated in Figure 4 and the equivalent circuit is given in the figure also, both for the case of a Si MOSFET and a heterostructure. The case for the heterostructure is obviously easier as there is no parallel conductivity from a gate structure. However if one wishes to change the density in the heterostructure one needs a gate.

Inductive coupling has been used successfully(4). The arrangement is the same in principle as a transformer. If the sample is put within the field of the primary of a transformer, the coupling to the secondary will be altered due to the conductivity of the sample. By looking at the voltage induced in the secondary as a function of some parameter, the conductivity variations can be measured. The primary and the secondary of the "transformer" were thin film loops over the inner and the outer edges of the 2D system in a Corbino structure.

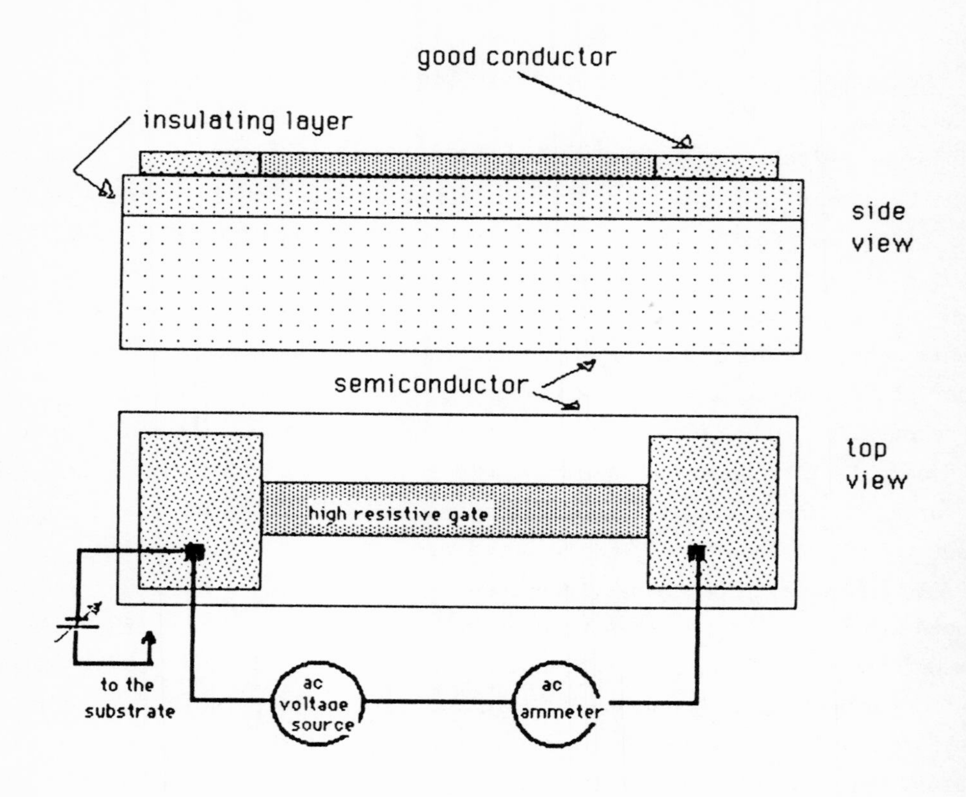

Figure 4

3.

THE ELECTRONIC SYSTEM: ELECTRIC QUANTIZATION

In discussing the electronic system, it is recognized that one is dealing with a one-dimensional potential and hence utilize a one dimensional Poisson equation

$$d^2\varphi(z)/dz^2 = 4\pi p(z)/\kappa_{sc} \qquad (3.1)$$

where $\varphi(z)$ vanishes in the bulk (complete screening of the gate electric field) (see AFS for a treatment where temperature is important). The potential vs distance into a p-type semiconductor is given in Figure 5 for the cases of flat band (obvious), depletion (fewer majority carriers at the surface, they are depleted in number), accumulation (where more majority carriers have accumulated) and inversion (where we have inverted the type from majority to minority). For n-type just turn the figure upside down.

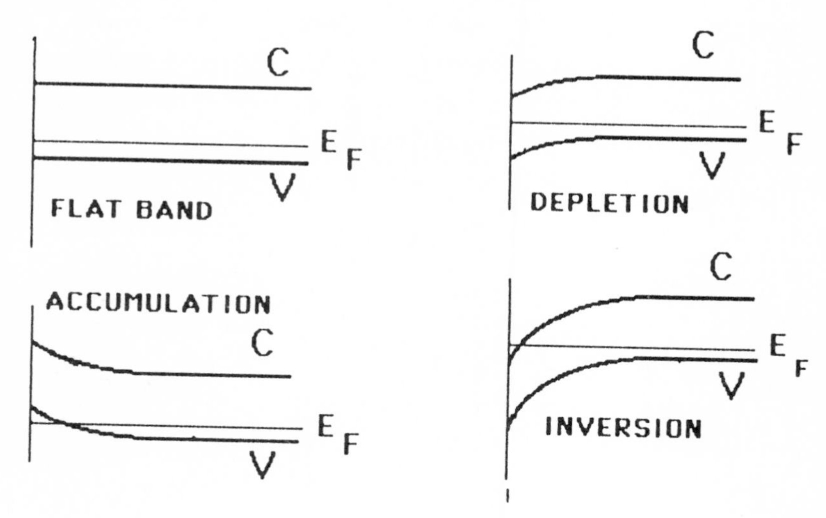

Figure 5

Because one may have sheet charge in the oxide, care must be taken as to how surface charge is defined. First, the definition of depletion and inversion charge densities:

$$n_d = N_a^- \times \ell_d \qquad (3.2)$$

where n_d is the depletion charge density (cm^{-2}), N_a^- is the 3D density of negatively charged acceptors, while ℓ_d is the length of the depletion region (from the surface to the point in the semiconductor where the bands are flat).

$$n_s = C(V_g - V_t)/eA \qquad (3.3)$$

where C/A is the capacitance per unit area, V_g is the gate voltage relative to the n^+ regions and V_t depends on workfunction differences, oxide charge and thickness and other things as well.

Figure 6 illustrates the MOS structure with a gate voltage, V_g, applied. This diagram is utilized to illustrate another consideration of the potential and that refers to the substrate voltage, V_{sub}. As a voltage is applied between the gate and the n^+ regions causing a difference in the electrostatic potential, one can apply an additional voltage between the n^+ regions and conducting plane at the parallel back surface of the bulk. When one generates carriers in the bulk either thermally or by shining light in the bulk, the end result for the potential in the bulk after a sufficiently long time will be as illustrated in Figure 6.

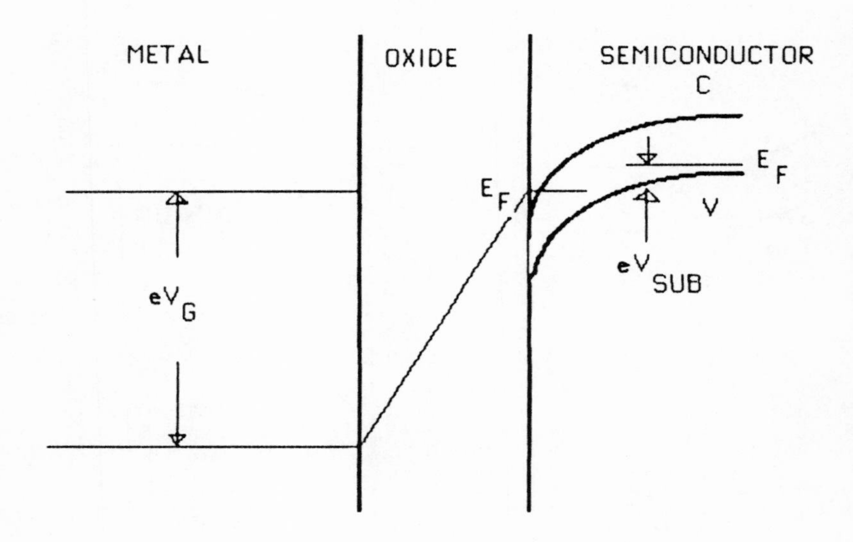

Figure 6

Quantum mechanical effects play a significant role even at room temperature and a dominant one at temperatures near 0 K. At what level must the problem be treated quantum mechanically? Rather than a first principles approach, the simplest effective mass approach is assumed, even simpler than that used by AFS. The kinetic energy operator as given in AFS is:

$$T = -\frac{\hbar^2}{2} \sum_{i,j} W_{ij} \frac{\partial^2}{\partial x_i \partial x_j} \tag{3.4}$$

where the W_{ij} are the elements of the reciprocal effective-mass tensor. For our case they are all equal to $(m^*)^{-1}$. The envelope of the wave function in the Schrödinger equation satisfies the following equation,

$$\frac{\hbar^2}{2m_z}\frac{d^2\varphi_i}{dz^2} + [E_i - V(z)]\varphi_i(z) = 0 \tag{3.5}$$

and the energy levels are given by,

$$E(k_1,k_2) = E_i + \frac{\hbar^2}{2} W_{11}(k_1^2 + k_2^2) \tag{3.6}$$

where the E_i are the energies corresponding to the degree of freedom for motion into the semiconductor, z. Ignoring spin and degeneracies due to the band structure of the crystal in question, the beginnings of the two dimensional system are seen.

The potential in eq. (3.5) is separable as the potential is only a function of z. A product wavefunction is obtained. Therefore one can separate the real three-dimensional world into 1D and 2D worlds. Energy levels are shown in Figure 7. If one has a many electron system at T=0, electrons will occupy all the available states below the Fermi energy. The Fermi energy E_F is referred to as the difference between the electrochemical potential and the 1st electrically quantized level, E_0. To have the system behave as if it were a two-dimensional system, all the processes should exclude changes of the quantum number corresponding to the degree of freedom in the z direction. This requires that all thermal and scattering "energies" kT and h/τ be small compared to E_1-E_F.

Referring to Figure 7 one can make estimates of how these levels depend on different parameters. In the absence of any carriers to cause screening via Poisson's equation, the bottom of the conduction band rises in energy as the electric field, E_e, times the distance z. The kinetic energy depends on momentum but one does not have any "momentum". An estimate of that can be made by using the uncertainty principle, $\hbar=\Delta z \cdot \Delta p_z$, what one has to pay in energy to localize the state. One obtains,

$$E_0 \sim F^{2/3}/m^{1/3} \quad , \qquad z \sim 1/F^{1/3}m^{1/3} \tag{3.7a,b}$$

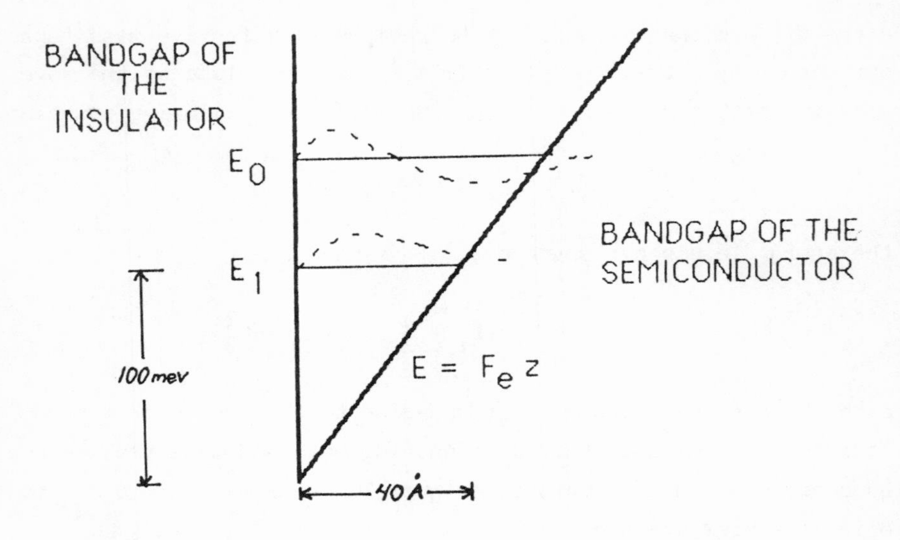

Figure 7

It is noticed that the energy rises as the electric field increases and is lower for heavier masses perpendicular to the surface. The spacial extent of the wave function also is lower for heavier masses perpendicular to the surface but decreases as the electric field is increased. To increase n_s one increases the normal electric field; this results in an increased E_0 and in forcing the electrons closer to the surface. Typical values are given in Figure 7.

The density of states (DOS) of the system in question is of great interest. The number of states from the electric sub-band to energy E, is given by,

$$n = \frac{1}{2\pi^2} \Sigma k_x k_y \tag{3.8}$$

assuming that the number of states is large and that one can replace the usual sum with the integral. Using the energy dispersion relationship we obtain,

$$n = g_s g_v m^* E_F / 2\pi\hbar^2 \qquad D(E) = g_s g_v m^* / 2\pi\hbar^2 \tag{3.9a,b}$$

where D(E) is the DOS, g_s is the spin degeneracy, g_v is the degeneracy due to band structure effects, and m* is the effective mass of the electrons in the plane. As expected it is independent of energy. Cautionary notes are; 1, that if the effective mass is dependent of energy then so is the density of states; 2, that the effective mass is the geometric mean of the two principal masses in the plane; 3, and that if more than one electrically quantized levels are occupied one must sum up the contributions (as is given above) from each band.

The states derived from the simple potential above leads to the simple description of the density of states. What of the case of a contribution from a random potential? Here short range potentials are considered. They can have an effect on both the real part and the imaginary part of the energy of the state. Therefore each such state can have the centre of its level shifted as well as having it broadened. If these random potentials are both randomly positioned and of random strength, in order to calculate the density of states, one would have to have the distribution function of these potentials. Is it "white", "Lorentzian" or what?

An important aspect of real systems is the fact that the distribution of these potentials may be bimodal or multimodal. Rather than consider all cases, focus on the fact that in macroscopic samples there may be macroscopic variations which dominate certain experiments. Does one consider states in a region of the sample at the Fermi energy which are separated by an impenetrable and insurmountable barrier when considering electrical conduction? No, but how about thermal conduction, where phonon exchange may take place? Yes. Thermodynamic behaviour? Yes. The barrier described above can never be such a hard rock and in fact at sufficiently low temperatures conduction takes place by hopping over barriers and tunneling through them as well.

4.

ELECTRICAL TRANSPORT: B=0

In Figure 8 electrical transport is discussed as the conductivity and what it depends on and for what it can be used. One point should be perfectly clear. Although one discusses things as if the sample were at T=0, the effects of localization set in at some low temperature and make this 3D like description inappropriate. This will be discussed

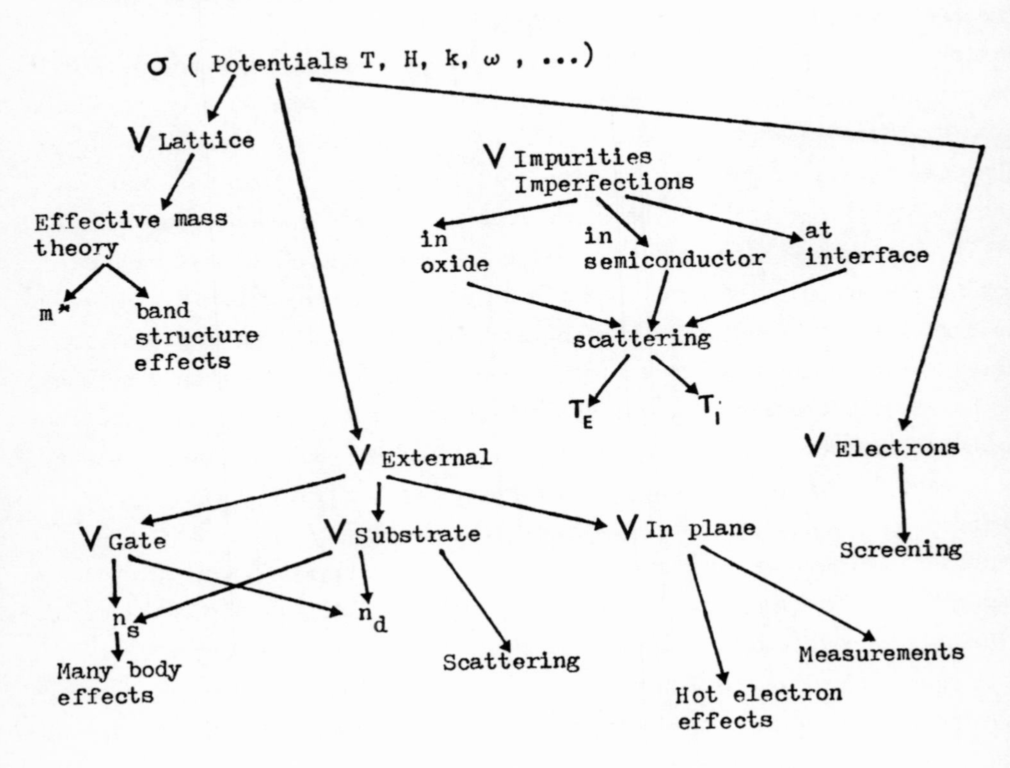

Figure 8

The above is not exhaustive, but is an attempt to illustrate how the different parameters determine just what we measure. Therefore, transport measurements can be used to determine the potentials that are present as well as the properties of the material itself.

later and in accompanying articles.

In keeping with the 3D-like description typical conductivities are plotted as a function of the number of carriers at different temperature in Figure 9. Room temperature: One sees that the conductivity is relatively independent of the density for almost all of the accessible range of densities. The major contribution to the scattering is due to phonons. The tendency to saturation at high densities is due to the fact, mentioned earlier that as the density is increased by increasing the gate voltage, the carriers are being pushed closer to the surface. As the real surface is not perfectly smooth, the surface roughness limits the conductivity. Liquid nitrogen temperature: Over most of the range, the conductivity is larger than that for room temperature. This is because the scattering due to phonons is decreased at lower temperatures. The range in which surface roughness limits the conductivity is even larger than at room temperature. Liquid helium temperature: The striking difference at this lower temperature is noted. Phonon scattering is gone, surface roughness is quite limiting, and the conductivity is lower than at liquid nitrogen temperature for low densities. It is

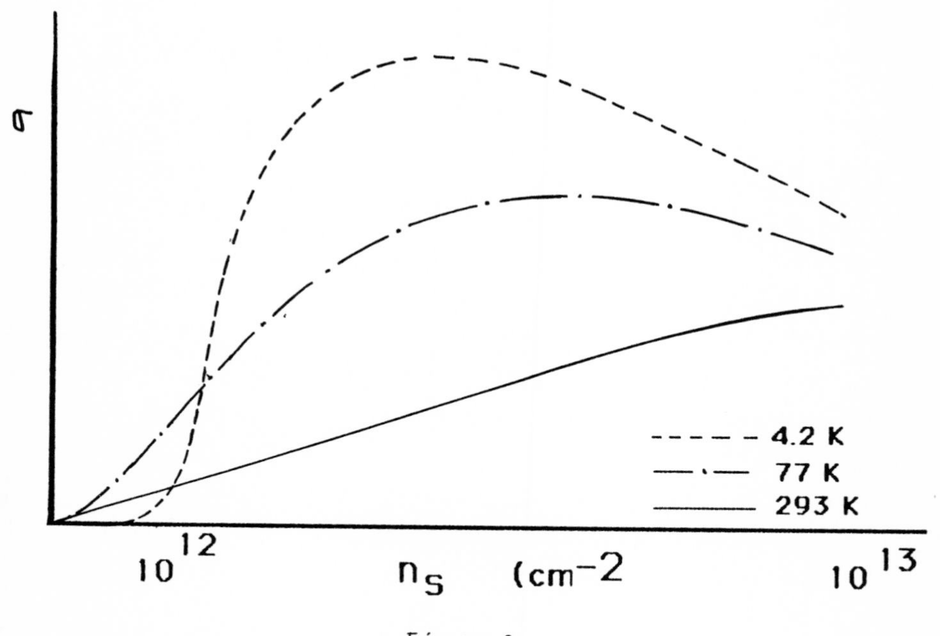

Figure 9

in this range, that the carriers are in bound states at low temperatures, whereas they had enough thermal energy to be free at the higher temperatures. In addition, the fluctuating potential at the interface, most often due to charge in the oxide, is limiting the mobility at the lower densities. As the density increases, the Fermi momentum of the carriers increases as does the screening, and as a result the mobility rises.

Figure 10 focuses on the effects at low temperatures. The density is divided into regions to discuss the dominant effects.

Region 1: The threshold voltage, V_t, is measured by the beginning of conductance at room temperature or 77 K but not at 4.2 K where strong localization dominates the conductivity.

$$V_t = V_{t,measured} - V_{t,theory} \sim (N_+ - N_-) \tag{4.1}$$

with Q=CV and assuming that the charge is in the oxide near the interface Variations in threshold between samples with other parameters the same, are due to differences in the net charge of the fluctuations in the surface potential and therefore are affected by the total number of charges. This is the region of strong localization. It happens at low temperatures when one has from a varying potential profile, $k_F \ell$ about 1. Then with

$$k_F \sim n_s^{1/2} \text{ and } \ell = p_F \tau/m = \hbar\, k_F \tau/m \text{ one obtains}$$
$$k_F \ell = k_F^2/m = n_s e^2 \tau/m = \sigma \tag{4.2a,b,c}$$

which introduces the concept of minimum metallic conductivity.

Region III: Here the conductivity is limited by scattering from the charges in the oxide modified by screening. In Figure 11 the calculated mobility as a function of electron density for different amounts of charge in the oxide and different surface roughness[5] is illustrated. The variations at low densities are due to oxide charge while the differences at high densities are due to surface roughness (region V).

Region IV: This and similar structure that is often observed is due to band structure effects. Earlier one considered only a simple parabolic band as contributing to the electrically quantized ground state. However, most surfaces of the semiconductor have a much more complicated band structure.

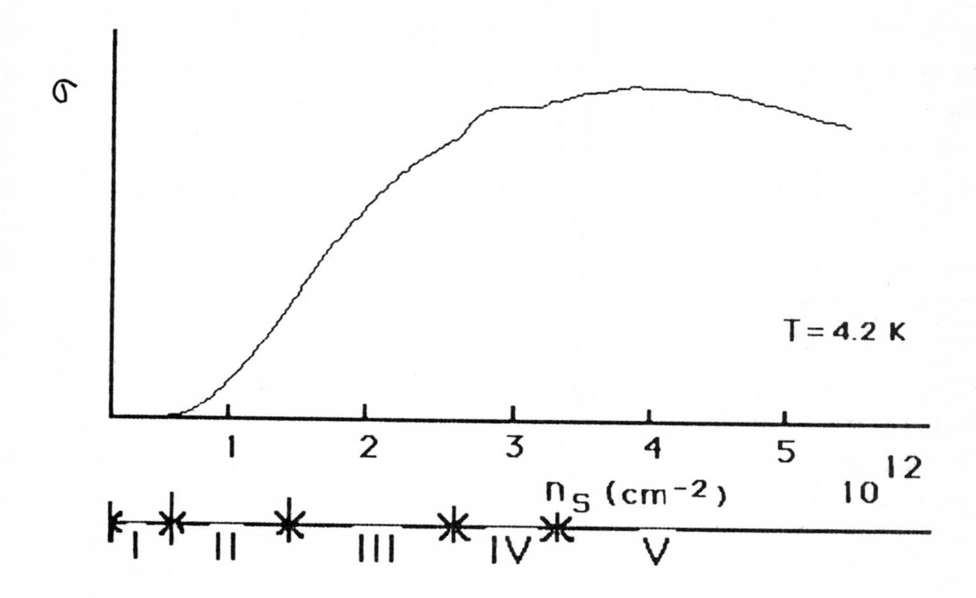

Figure 10

Region V: This is the region discussed in region III where the surface roughness limits the behaviour. The parameters relating to surface roughness in Figure 11 are explained in Figure 12. For this region, the scattering time decreases more rapidly than n_s^{-1} so that the conductivity decreases as the number of electrons increases!

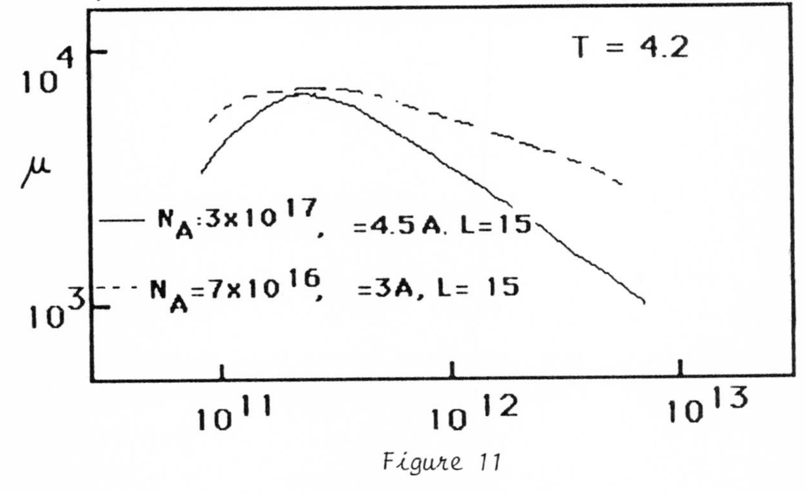

Figure 11

As the Si MOSFET structure has oxide charge and a rough interface between the semiconductor and the oxide that is grown at about 1000 C why not use an epitaxially grown insulator on the semiconductor? Such is being tried in many laboratories at this very time. The early MEISFETs (EI → Epitaxial Insulator) have not shown superior qualities as of yet[6]. We illustrate in Figure 13 a structure that does eliminate the behaviour that limits the mobility in Si MOSFETs. It is the generic heterostructure. It achieves a near perfect interface by growing each atomic layer at low temperatures where the energetics favour completing a layer before starting another layer. In this structure, the compensating charges for the 2D electrons are separated from the electrons by a buffer layer, are immobile and are positively charged donors. The buffer layer is chosen to be thick enough to minimize scattering. In the Si MOSFETs the oxide charge appears to be right at the interface rather than hundreds of Angstroms away as in the case of heterostructures. Electron density can be varied in heterostructures by utilizing the equivalent of a gate or substrate electrode.

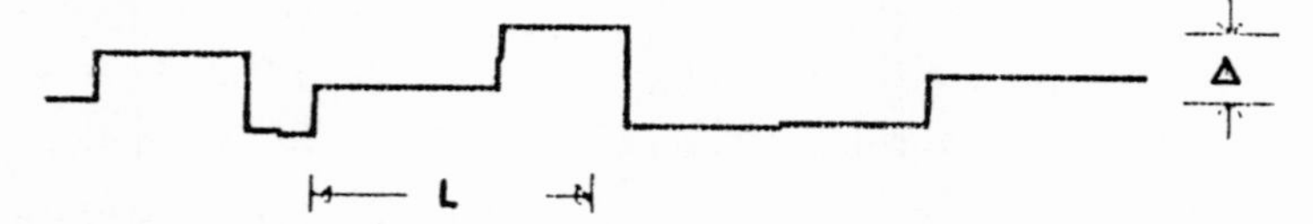

Δ is the mean height of the fluctuations and (3 to 5 A)
L is the correlation length (15 to 40 A)

Figure 12

Illustrated at the bottom of Figure 13 is the band structure as the two semiconductors as brought together. The un-ionised donors in the larger bandgap material can exist at lower energies in the conduction band of the narrower gap semiconductor. Therefore there is a transfer of charge to equate the Fermi levels. Compare this diagram to Figure 6. The negative charge in the narrow gap material is our 2D system in analogy to the Si MOSFET case, while the positive charge sheet of donors is equivalent to the positive charge metal gate in the Si MOSFET system. The main difference between the two systems is that the charged donors are fixed and localized while the positive charge in the MOSFET is

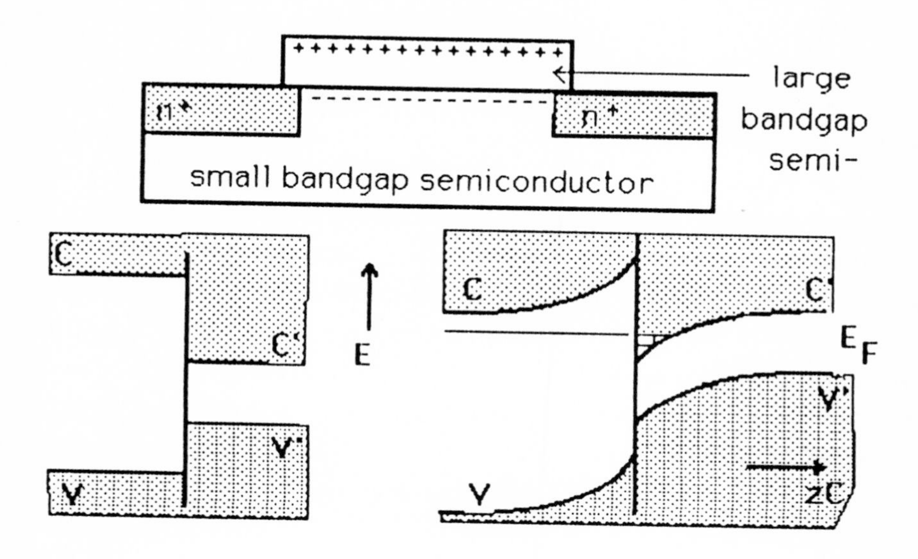

Figure 13

delocalized. There is scattering from the former, but not from the latter.

5.

THE ELECTRONIC SYSTEM: MAGNETIC FIELD EFFECTS

The introduction of a magnetic field perpendicular to the 2D system introduces vector potential terms into the quasi-momentum in equation (3.5). The situation is that the "electric force" due to quantization is much larger than the "magnetic force" which is only an important consideration when the magnetic field is not perpendicular as in one case considered later. Effectively one finds that one is quantizing the flux through the orbit in real space.

The problem is still separable into a 1D and 2D problem and the energy is given by,

$$E = E_0 + (n + 1/2)\hbar\omega_C + sg^*\beta H + \Delta E_V \qquad (5.1)$$

where E_0 is due to the electric quantization in the z direction perpendicular to the surface, $(n+1/2)\hbar\omega_C$ is the energy due to the magnetic field quantization of the two remaining dimensions, n is the Landau manifold quantum number, ω_C is the cyclotron frequency, $+sg^*\beta H$ is the energy due to the spin of the carrier, s is $\pm$ 1/2, β is the Bohr magneton, ΔE_v represents the energy splitting of states which are degenerate in the 3D band structure. For semiconductors like Si, where spin-orbit coupling is very small, the effect of a magnetic field applied at an angle to the normal 2D system is easy to understand. The orbit, restricted to be parallel to the surface, quantizes the normal component of the magnetic field. The spin splitting, not communicating with the orbit, experiences the total field.

The DOS is no longer independent of energy, but in this approximation is a series of functions at energies given by eq. (5.1). The degeneracy of each state is most easily calculated as,

$$N_L = \hbar\omega_c \times g_s g_v m^*/2\pi\hbar^2 = eB/hc \tag{5.2}$$

In Figure 14, the density of states (ignoring spin and valley degeneracies) is illustrated for different strengths of imperfections in the sample. As the density of states is broadened in the absence of a magnetic field (14a to 14b) similar things are expected to happen with

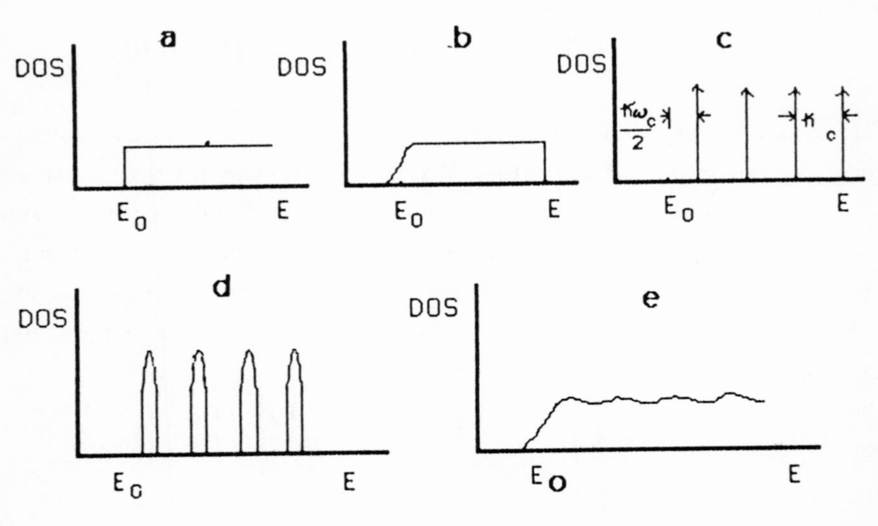

Figure 14

the magnetic field present. However there is now an energy relationship to consider. If the width of the Landau levels is given approximately as $h/(\tau)$, then this energy must be compared with $\hbar\omega_c$. For the former zero, much smaller and comparable to the latter, one should expect situations like (14c), (14d), and (14e), respectively. One way to visualize this difference is to look at E_F for the non-interacting picture as a function of magnetic field for the 2D system as illustrated in Figure 15.For the idealized system with no broadening of the levels, E_F is discontinuous. For the 3D system, E_F is continuous. One of the questions that needs to be addressed, is what is the real DOS; are there energies where there are no states expected?

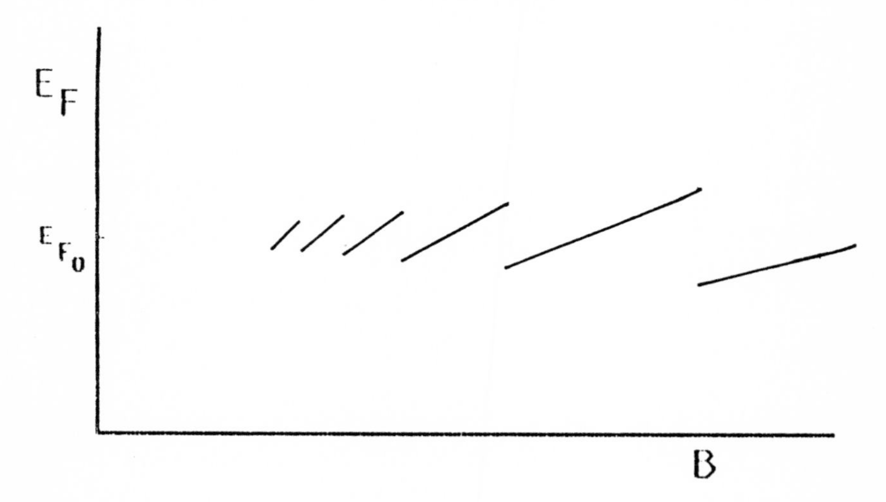

Figure 15

The general properties of the system in the presence of a magnetic field change from 3D-like in the low field region ($\omega_c\tau \ll 1$) to 2D-like in the high field region ($\omega_c\tau > 2$). The intermediate field region, where the oscillations in the density of states lead to oscillatory behaviour in some of its properties is nearly 3D-like, but the oscillations are much bigger than would be the case in 3D. This results from the fact that if one calculates the density of states for a 3D system that is similar to a 2D system, major changes in the density of states due to the quantization of the motion parallel to the magnetic field do not exist as there is a continuous variation in energy of states due to motion along the magnetic field.

6.
ELECTRICAL TRANSPORT: MAGNETIC FIELD EFFECTS

In two dimensions the conductivity tensor is given as,

$$\bar{\bar{\sigma}} = \begin{bmatrix} \sigma_{xx} & \sigma_{xy} \\ \sigma_{yx} & \sigma_{yy} \end{bmatrix} \tag{6.1}$$

In 3D, and in 2D for weak fields, $\omega_c\tau << 1$ the following is true for the simple case,

$$\begin{aligned} &\sigma_{xx} = \sigma_{yy}, \text{ and } \sigma_{xy} = -\sigma_{yx} \text{ and } \sigma_0 = ne^2\tau/m^* \\ &\text{we have } \sigma_{xx} = \sigma_0/(1 + (\omega_c\tau)^2) \text{ and } \sigma_{xy} = \omega_{xx}\tau_c \end{aligned} \tag{6.2}$$

It is easier to describe the simple experimental results in the contex of simple ideas and theoretical predictions. Some illustrative results[7 are shown in Figure 16 and 17[8]. Curve A in Figure 16 is for B=0 and each succeeding curve is for a higher magnetic field. One sees the near classical behaviour as the field is increased, the conductivity at a given density decreases on average.

The system begins to behave in a manner that is strictly 2D instead of 3D when the oscillations become strong as in curves D, E and F. However a simple understanding of why each oscillation of the same number (choose #5 in Figure 16 for example) has about the same magnitude of conductivity independent of magnetic field and density, is based on a simple diffusion picture[9,10]. In a strong magnetic field, which is where the system acts 2D, the conduction takes place as an electron jumps from one cyclotron orbit centre to another along the direction of the electric field. The jumps would be dominated by nearest neighbour jumps with the usual short range scatterers. The diffusion constant would be the distance squared divided by the lifetime of the orbit as,

$$D^* \sim (2n+1)\ell^2/\tau \tag{6.3}$$

where ℓ is the magnetic length. Using the concept of the broadening of the orbit with lifetime as before and with the degeneracy as before one obtains

$$D(E_n) \sim \tau/\ell^2 \tag{6.4}$$

and so the Einstein relation is,

$$(\sigma_{xx})_{peak} \sim e^2 D(E_n) D^* \sim e^2(n+1/2)/h \tag{6.5}$$

Therefore the peak value of the conductivity depends only on n and not on magnetic field or density. A discussion of the curves D, E, and F will be given later

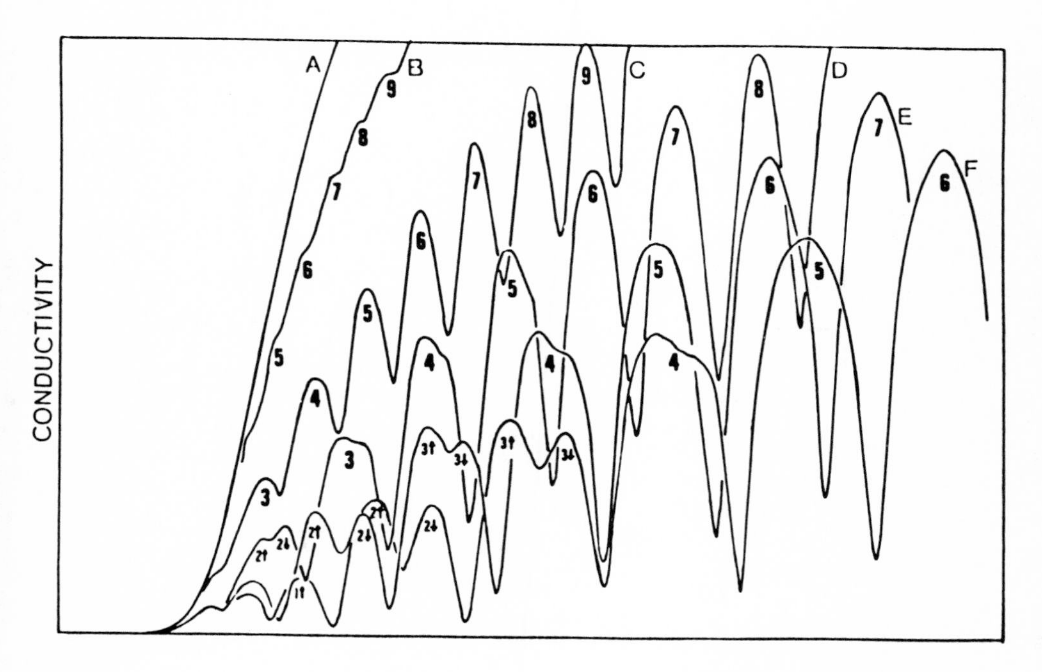

Figure 16

Ando and Uemura go on to point out that the diffusion of the centre of mass in the same situation is proportional to the fluctuation of the gradient of the local potential energy. This leads to the following expression for the transverse conductivity,

$$\sigma_{xy} = -n_s ec/H + \sigma_{xx}/\omega_c\tau \ . \tag{6.6}$$

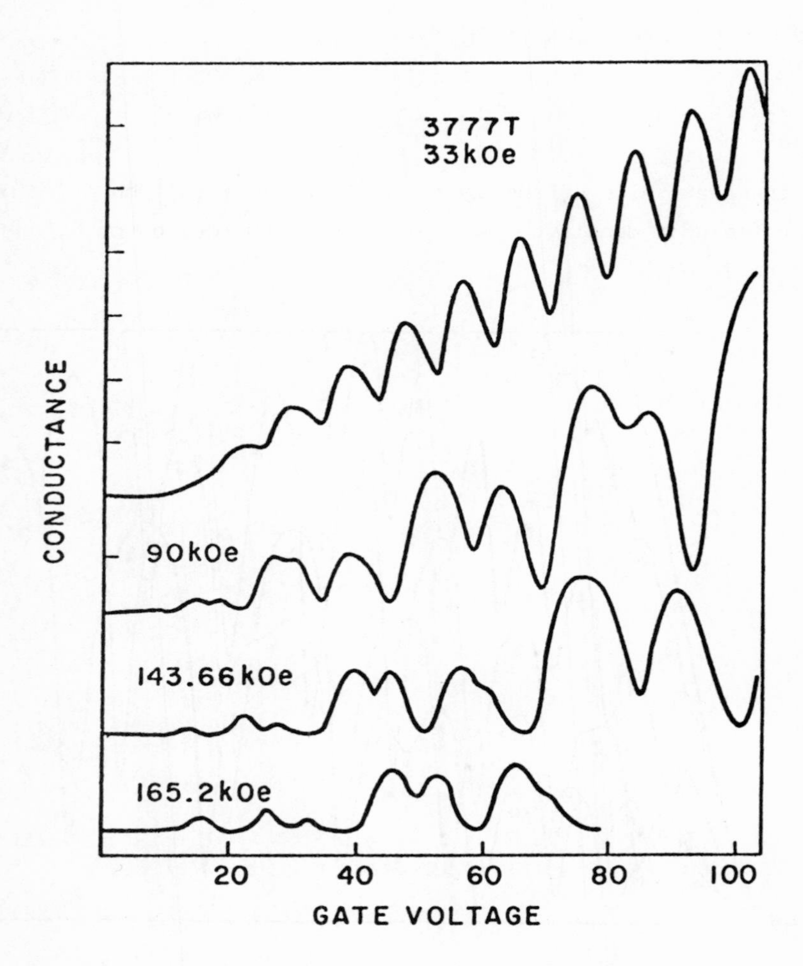

Figure 17

Figure 17 is from the first such work(8) and illustrates well the behaviour at <u>intermediate</u> and <u>high fields</u>. It was also the first evidence that the system was 2D. This is because it is periodic for fixed magnetic field. That is what is expected as the degeneracy of each level depends only on the magnetic field. Further the degeneracy for the 33 kOe field curve was as expected, four times that for a single level, $g_s g_v eB/h$. By comparing the basic periodic structure in the 90 kOe curve to that in the 33kOe curve it seems that the period in gate voltage or density increases by the ratio of the magnetic field strengths. However in the higher field curve the basic period is halved due to the separation of the spin levels. In addition, at lower densities, it is again halved due to the double degeneracy due to band structure effects. Structure is observed when the separation between any two energy levels is comparable or larger than the broadening of those levels.

The high field region is where the quantum Hall effect can be observed. It will be discussed in detail in accompanying articles. Of particular interest is the intermediate field region. Ando[11] gives for the longitudinal component of the conductivity,

$$(\sigma_{xx}) = \frac{n_s e^2 \tau}{m} \frac{1}{1+(\omega_c\tau)^2} \left[1 - 2 \underbrace{\frac{(\omega_c\tau)^2}{1+(\omega_c\tau)^2}}_{\text{I}} \times \underbrace{\frac{2\pi^2 k_B T}{\hbar\omega_c} \operatorname{csch} \frac{2\pi^2 kT}{\hbar\omega_c}}_{\text{II}} \underbrace{\cos \frac{2\pi E_F}{\hbar\omega_c}}_{\text{III}} \underbrace{\exp\left(-\frac{\pi}{\omega_c\tau}\right)}_{\text{IV}} + \ldots \right] \tag{6.7}$$

The coefficients of the bracket are the usual effects in classical magnetoconductivity, and are the same in 3D and in 2D (except in 3D σ is in units of $(\text{ohm-cm})^{-1}$ and $(\text{ohm})^{-1}$ in 2D). The oscillatory part is composed of the product of the terms outside of the bracket and four others as indicated in eq. (19). Term I contains just the product $\omega_c\tau$ so that it represents just a magnetic field dependence. Term II is basically a temperature damping of the oscillations term. What is compared is two energies, the thermal energy, kT, with the energy separation of Landau levels. Therefore one is able to do spectroscopy by utilizing the well defined energy, kT (which will be illustrated later). Term III is the basic oscillatory term, here written as the ratio of E_F to the separation of the Landau levels. Although that is true in this simple case, the basic ratio is the density, n_s, to the magnetic field as in the degeneracy of the levels. Term IV is the exponential damping due to the broadening. It should be noted that the τ in the exponential is not the conductivity τ, but the oscillatory one. The distinction is that any momentum scattering event will destroy the coherence of the orbit, while those that are basically forward scattering will not affect the conductivity τ to any degree. We can determine τ_{osc} by knowing the effective mass and the temperature and examining the magnetic field dependence of the oscillations. With short range scatterers, the DOS[1] is given by

$$DOS = (m^*/2\pi\hbar^2)[1-2\cos(2\pi E_F/\hbar\omega_c)\exp(-\pi/\omega_c\tau_f)] \quad (6.8)$$

7.

TILTED MAGNETIC FIELD

As mentioned above, studying the oscillatory behaviour of Si MOSFETs with the magnetic field applied at an angle, θ, relative to the normal 2D system should exhibit intriguing effects. In the following[12] discussion, we are ignoring the effects of valley splitting, as it is not resolved under the conditions discussed. Figure 18, illustrates

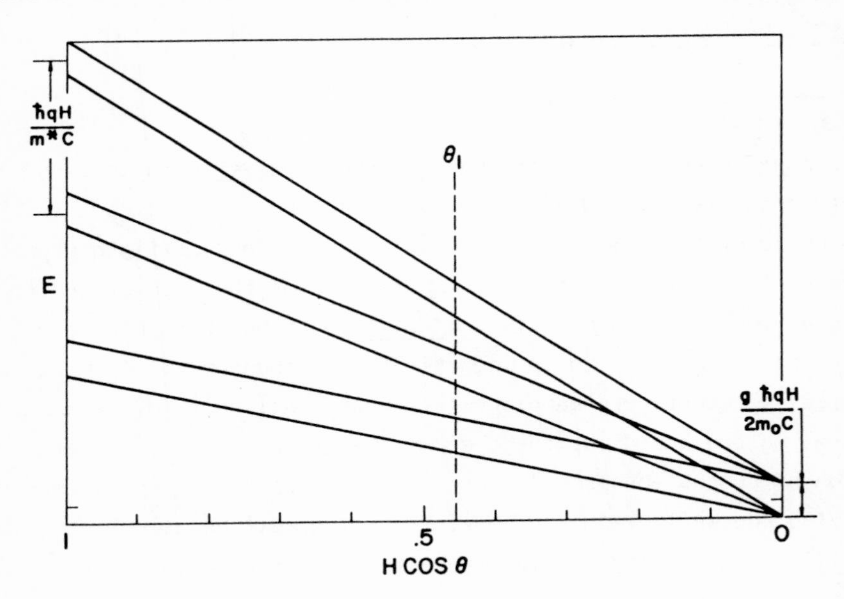

Figure 18

how one expects the energy levels to depend on the angle between the applied magnetic field and the normal. One can see that the centre of the broadened Landau levels would be on the Landau levels for small angles, but between levels for most angles greater than θ_1, the angle where the levels are evenly spaced. If for a given field the broadening is such that the spin levels are resolved, then one expects behaviour like that illustrated in Figure 19. The derivative of the conductivity with

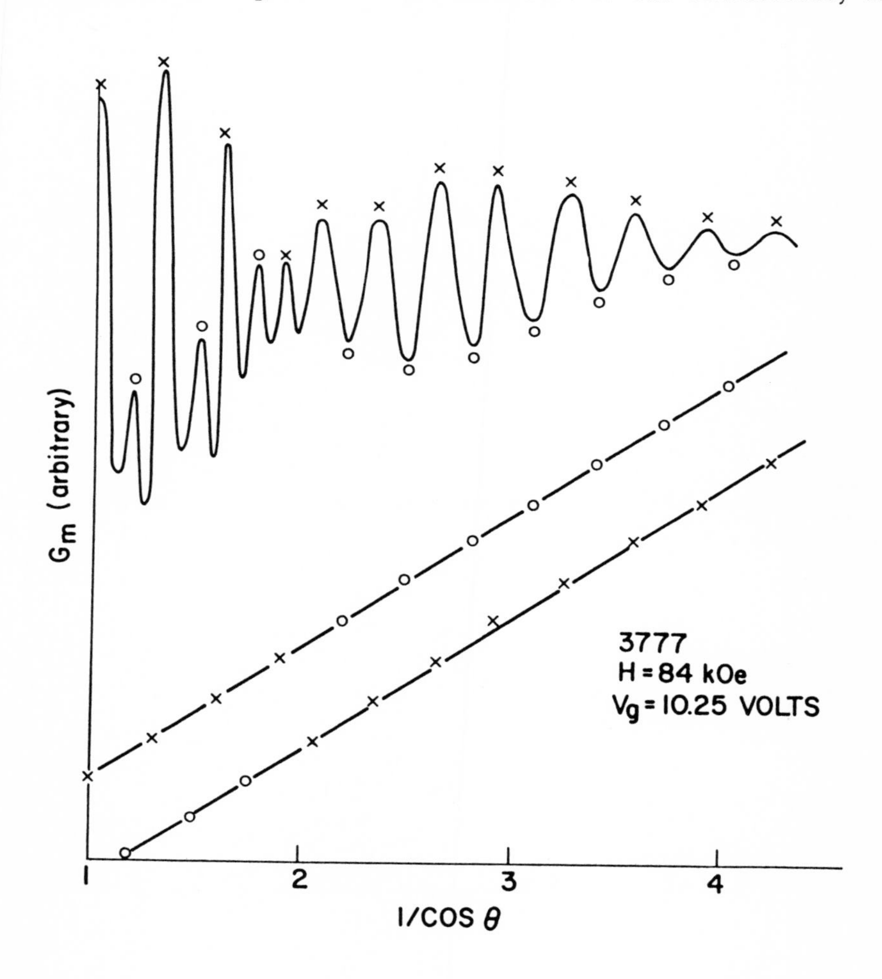

Figure 19

respect to the density is plotted versus $(\cos\theta)^{-1}$ in the upper part of the figure while the index of the oscillations are plotted versus $(\cos\theta)^{-1}$ in the lower half. One notes that the oscillations are indeed periodic and confirm that it is the normal flux that is being quantized. The switching of the maxima "o" and "x" are due to the reversal as one crosses the angle θ_1 as in Figure 18.

Having determined θ_1 for a given V_g, we know that the spin splitting is 1/2 of the Landau splitting at that angle. Therefore from

$$eH \cos\theta_1/m^*c = g^*\beta H \tag{7.1}$$

one determines g^* for the V_gs chosen. The results for g^* as a function of gate voltage is shown in Figure 20. When this was first observed, it was unexpected. After all, g in silicon is almost exactly 2, and here it is not. Further, it depends on electron density. The first essentially correct description of the behaviour was that many-body effects dominated the behaviour[13]. These effects depend on a ratio of energies, the Coulomb interaction to the kinetic energy.

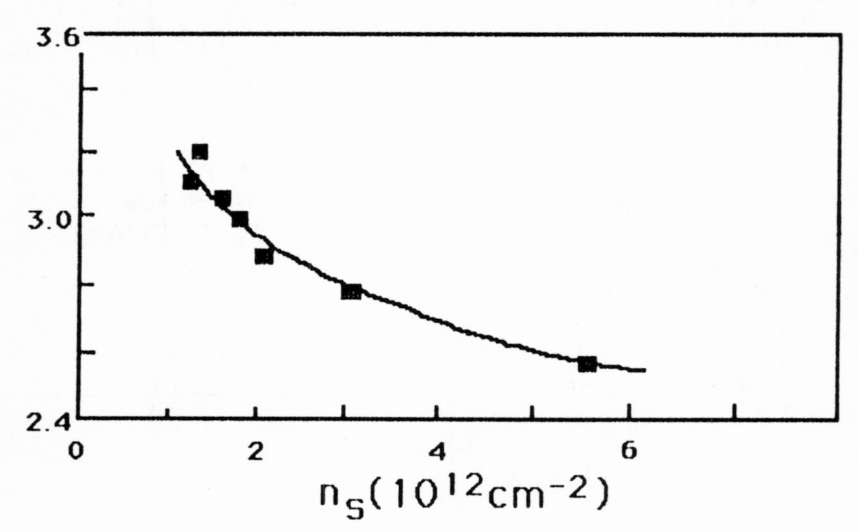

Figure 20

With

$$E_{coulomb} \propto r^{-1}, \quad \text{and} \quad E_{kinetic} \propto r^{-2} \qquad (7.2a,b)$$

one would expect the larger deviation from single particle behaviour at smaller densities as is seen in Figure 20.

8.

TEMPERATURE SPECTROSCOPY: EFFECTIVE MASSES AND OTHER ENERGY SPACINGS

By measuring the temperature dependence of oscillations in the region where eq. 6.7 is applicable, one may interpret the results in a manner that allows a determination of the effective mass of the carriers involved. Such a procedure has been applied by many researchers[1,2]. The technique is illustrated with some results[14] in Figure 21. This illustrates the dependence on temperature of the amplitude of the damped sinusoidal oscillations. With the assumption that the formula applies, then determinations of the energy splitting to 1% is not difficult.

The circles in Figure 21 are data and the three theoretical lines are for different assumed energy splitting. The solid line, which obviously fits the data the best of the curves illustrated is for an energy splitting either for an effect mass of $0.185m_o$ or a spin splitting corresponding to a g value of 2.8. It is obvious that a change of mass of about 2% is a decidedly worse fit and the 4% difference in the g value is even worse. The g value determined from the fit is the same as that determined by the change in phase in the oscillations as in Figure 19.

The fit for the effective mass is different from that obtained when the magnetic field is normal to the surface. In those circumstances, the effective mass is about $0.21m_o$. A weak dependence on density was found[14]. However the literature contains numerous accounts of the success of this technique as well as disagreements[15,16] on fine points as to whether there is a universal dependence of the effective mass on density. The major problem is that the system is really an interacting one, not the single particle one that is assumed.

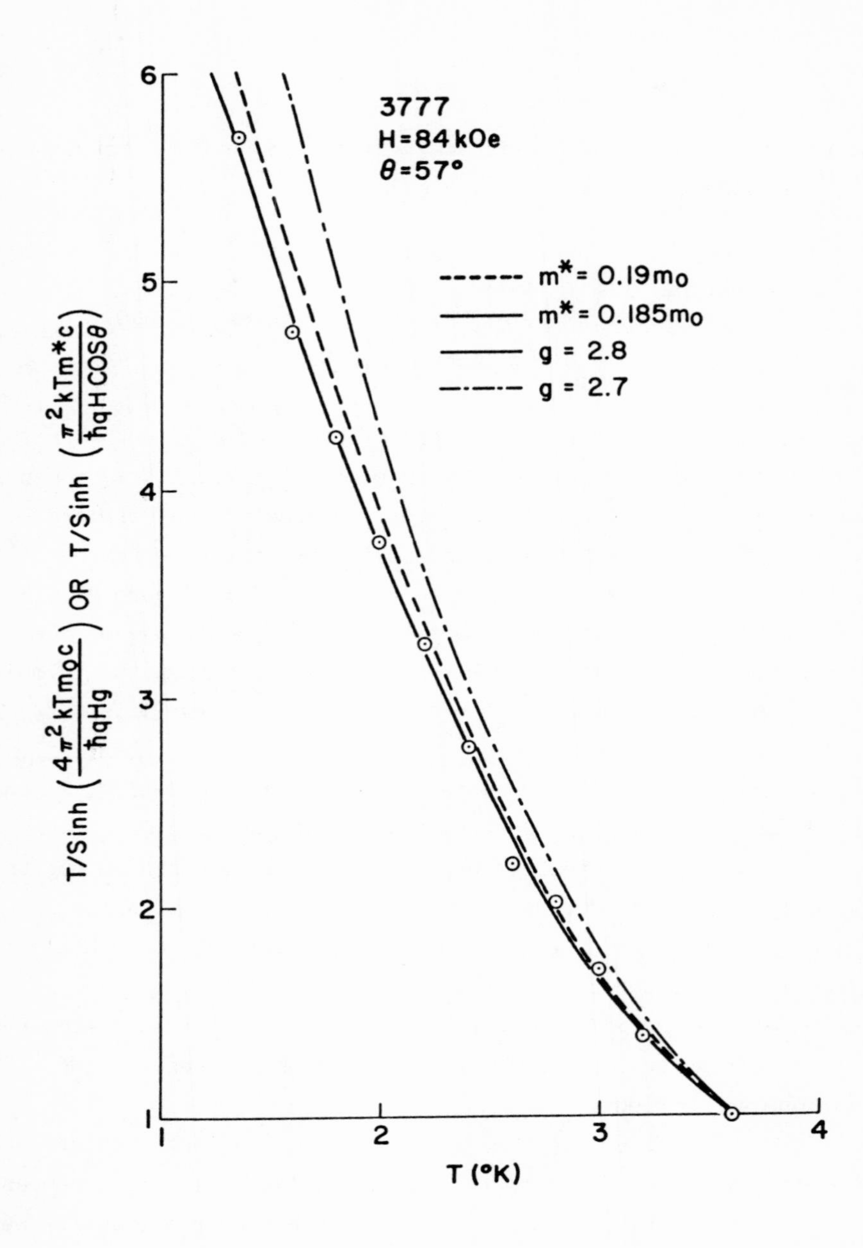
3777
H=84 kOe
θ=57°
m* = 0.19m₀
m* = 0.185m₀
g = 2.8
g = 2.7
T/Sinh (4π²kTm₀c / ħqHg) OR T/Sinh (π²kTm*c / ħqH COSθ)
T (°K)
6
5
4
3
2
1
1
2
3
4

Figure 21

9.

CAPACITANCE: WHAT IS LEARNED FROM IT?

Of what value is it to study the capacitance of a MOSFET or a heterostructure? Is it not just the usual relationship between charge and voltage such as $C = Q/V$? Of course, but important information is hidden in that relationship. Think of the following example. Suppose that the conductors are separated by an insulator; they have a voltage across it and a charge Q on it. This leads to a value of the capacitance as above. If the voltage is increased and there are no new states for electrons to occupy within this energy range, eV, on the negative electrode, then the charge will stay the same while the voltage has increased. This results in a different value of the capacitance at this new voltage.

If the differential capacitance is defined as,

$$C \equiv dQ/dV \tag{9.1}$$

then it is easily seen of what value measurements of the capacitance are. The case described in the paragraph above is used again. In the region where the voltage is increased and there is no additional charge on the electrodes, dQ is 0 and hence so is the differential capacitance. When there are states, an increase in voltage results in additional charge so that the differential capacitance is not zero. Such considerations imply that there is a correspondence between the two. A discussion follows. An important fact to remember is that C is now used for the differential capacitance.

The total capacitance may be written as,

$$C_t^{-1} = C_b^{-1} + C_i^{-1} \tag{9.2}$$

where $C_b = G_{ox}A/t_{ox}$. The channel capacitance, C_i, is given by

$$\frac{A}{C_i} = \frac{z_o\gamma}{\epsilon_{sc}} + e^2\frac{dn}{d\mu} \tag{9.3}$$

assuming a simple variational approximation, ignoring image and many-body effects and penetration into the barrier. Here z_0 is the average position of the electrons in the channel, γ is a constant that is a little over 0.5 and ϵ_{sc} is the relative dielectric constant of the semiconductor, and $dn/dE|\mu$ is what is sought, the thermodynamic DOS at the Fermi energy.

The question arises as to how measurements of the capacitance are carried out. The structures involved are analogous to a lossy stripline where one electrode is resistive. The question of the effect of frequenc must be addressed.

We will use the concept of distributed capacitance to look at the problem in a "finite element" approach as we treat the case of a Si MOSFET. This approach is very similar to that first taken by Stern[17]. The sample is divided into p strips, with each strip element given as,

$$c_o = C_{ox}/p, \quad c_d = C_d/p \quad \text{and } r = R/p \qquad (9.4a,b,c)$$

where C_{ox} is the oxide capacitance, C_d is the depletion capacitance and R is the resistance of the strip. Here a simpler problem is addressed at low temperature in a region where the depletion region is fully formed. As the depletion charge will not change, the depletion capacitance can be ignored. It is assumed that the resistance of the gate electrode is negligible.

Figure 22 illustrates the distributed capacitance model. It also illustrates the current at each node. At each node the sum of the currents must be zero. Therefore,

$$(V_{n+1}-2V_n+V_{n-1})r + ic^*(V_n-V_o) = 0 \qquad (9.5)$$

where $c^*=c_oc_i/(c_o+c_i)$. The limit as p approaches infinity leads to the following differential equation

$$d^2V/dx^2 + i\omega RC^*V/L^2 = i\omega RC^*V_o/L^2 \qquad (9.6)$$

Equation 9.6 is the one dimensional diffusion equation as is expected. It is easy to visualize this as the charge must diffuse from one end along the resistive channel in order to charge the capacitor. The voltage that drives the diffusion is the difference between what the region

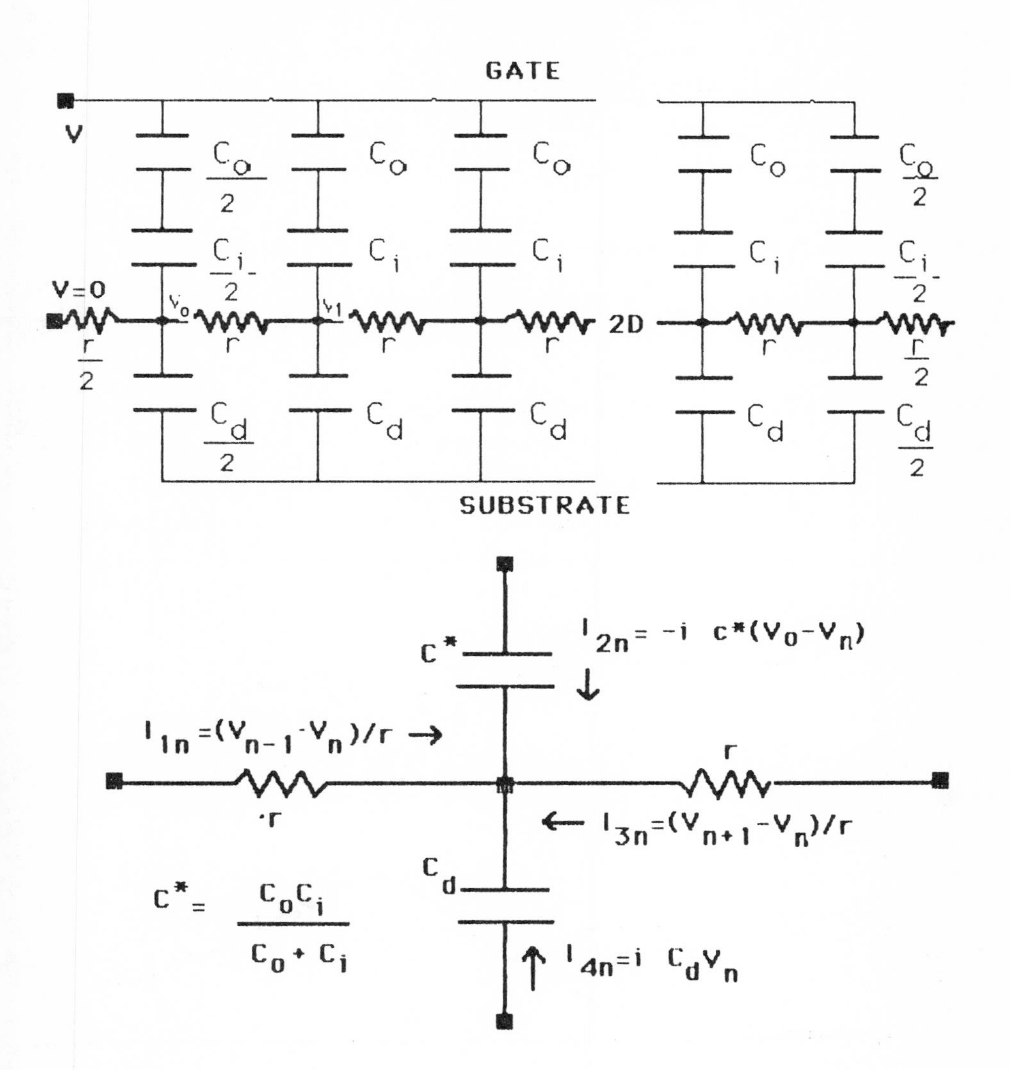
GATE
V
V=0
SUBSTRATE
2D
C_o
C_i
C_d
r
C^*
$I_{2n} = -i\ c^*(V_0 - V_n)$
$I_{1n} = (V_{n-1} - V_n)/r$
$I_{3n} = (V_{n+1} - V_n)/r$
$I_{4n} = i\ C_d V_n$
$C^* = \frac{C_o C_i}{C_o + C_i}$

Figure 22

should be charged and what it is at a given time. With the voltage zero at each end of the sample the solution for the voltage is given by

$$V_{(x)} = [1-\cos(ax/L) - \tan(a/2)\sin(ax/L)]V_o \qquad (9.7)$$

and the following is obtained,

$$C_{eff}=Q(ac)/V(ac) = C^*C_{ox}/(C^*)\cdot[1+(2C^*\tan(a/2))/a] \qquad (9.8)$$

where $a = i\omega R(C^*)$. The behaviour as a function of either frequency or resistance (as the function changes as the product of the two variables) is shown in Figure (23). Here is illustrated the ideal curve as well as the behaviour for different frequencies for two different long samples. The "theoretical" curve represents well the experimental situation.

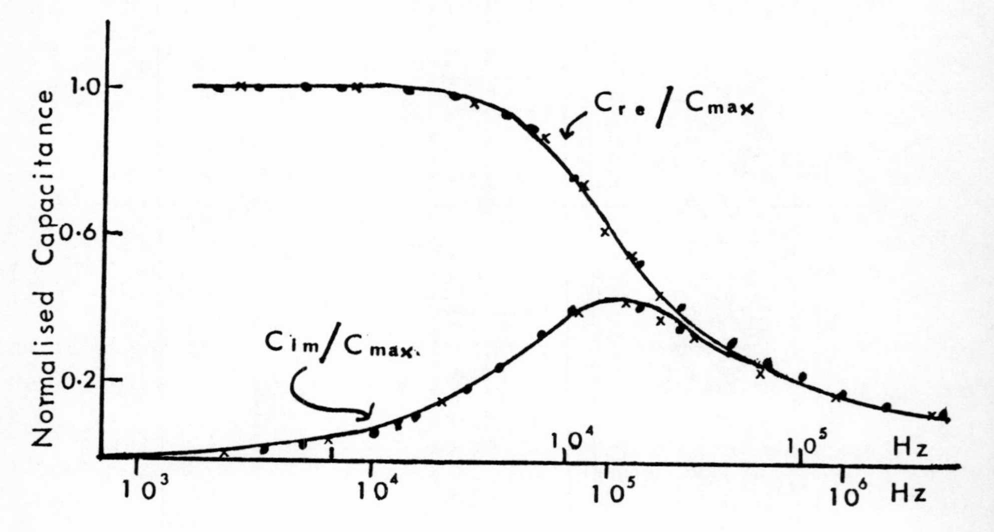

Figure 23

One of the major drawbacks of the techniques as applied above is that when the density of states is low, not only is the capacitance low, but the resistance is high. The question is whether the capacitance drops faster than the resistance increases. It is a question of time constant. One way to visualize the problem is to look at Figure 24.

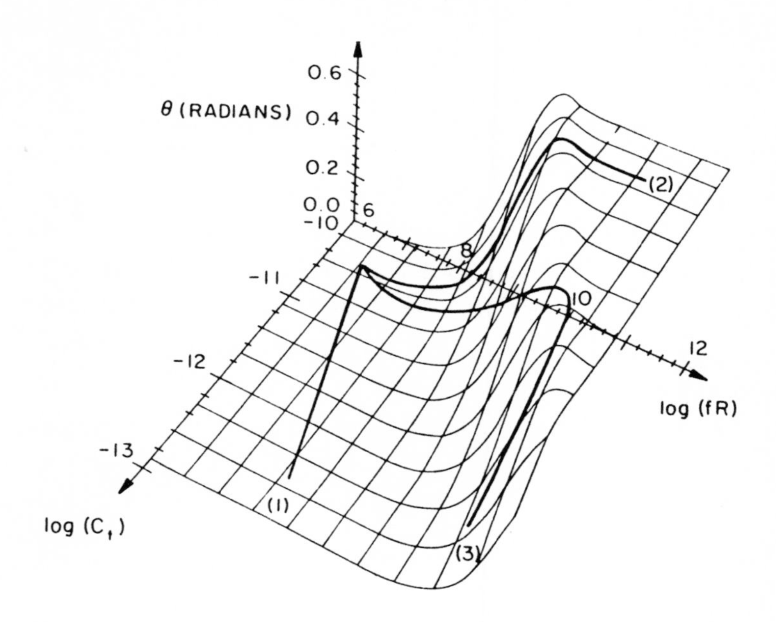

Figure 24

This is a three dimensional plot of the phase angle of the measured capacitance versus both the product of the frequency times the resistance and the value of the capacitance[18]. Starting at the common point of paths 1, 2 and 3, and the capacitance and resistance changing as some external parameter is varied the paths represent the different kinds of behaviour that may happen. Curve 2 represents the change of resistance only, and one observes the rise of the phase angle to the plateau value. If both the resistance and the capacitance change simultaneously, path 3 may be followed with a rise and then a decrease in the phase angle. The ideal curve, where there is no observable change in the phase angle is obviously curve 1. In this case, the measured capacitance is the real capacitance.

When the question of frequency effects clouds the interpretation, a new configuration can and has been used[19]. The basic concept, is that if trying to inject carriers into the 2D electron gas perpendicular to the magnetic field results in overwhelming magnetoresistance effects, inject them along the field. In the particular instance referred to above, an electrode was grown within a tunneling distance from the 2D electron gas. The parameters were chosen such that the tunneling resistance was the order of a hundred ohms. This resistance is such that there are no magnetoresistance effects; one can operate at frequencies of tens of kilohertz instead of the tens of hertz and low magnetic fields that the earlier experiments were forced to do to eliminate resistance effects. Further, this new technique allows studies in the highest fields even where the fractional quantum Hall effect is observed. Perhaps the results will be discussed in accompanying articles.

10.

DIRECT MEASURE OF CHANGES IN THE ELECTROCHEMICAL POTENTIAL

A general technique for measuring changes in electrochemical potential (ECP) was developed[20] and applied to 2D systems of interest. This technique, one of detecting changes in ECP between two electrodes subject to a change in some parameter, is a variation of measuring contact potentials.

It is applied to the 2D systems in the following manner. One electrode is the 2D electron gas while the other is a counter-electrode, such as the gate in a Si MOSFET. For heterostructures it is a similar conductor on the outer non-conducting (Ga,Al)As layer (this layer is analogous to the oxide in Si MOSFET). The systems have some fixed charge on them (it can be Q=0 as in the case of a heterostructure). Then when some parameter is varied and the 2D system changes its ECP and the "inactive" counter-electrode does not change (or its behaviour has been calibrated) the voltage across this fixed charge system is measured.

Details of the measurement are straightforward except for considerations of this "perfect" voltmeter. In theory one can achieve (by potentiometric methods) as near to a perfect voltmeter as is required. In practice, the input of the volmeter should have an input capacitance that is very much less than the capacitance of the sample and an input resistance which allows no decay of the voltage during the duration of the experiment. This requires a capacitance of about 1pf and $R > 10^{15}$ ohms.

This is, in general, not too difficult to achieve with a low temperature amplifier.

Results have been reported by two groups[20,21] although the comments here are only on the ones involving both Si MOSFETs and (Ga,Al)As heterostructures. Recall what the Fermi energy for fixed charge as a function of magnetic field would look like as in Figure 15. The discussion continues here in the same vein of a non-interacting single particle system. Figure 25 illustrates, over a limited magnetic field range, how the system would look for different deviations from the perfect system. The typical experimental results are not shown, but an average of some results are shown in Figure 26 for a Si MOSFET with mobilities of about 15,000cm^2/v-s. The major structure is what one would expect with the deviation on the rising part of the curve due to emerging of the spin splitting.

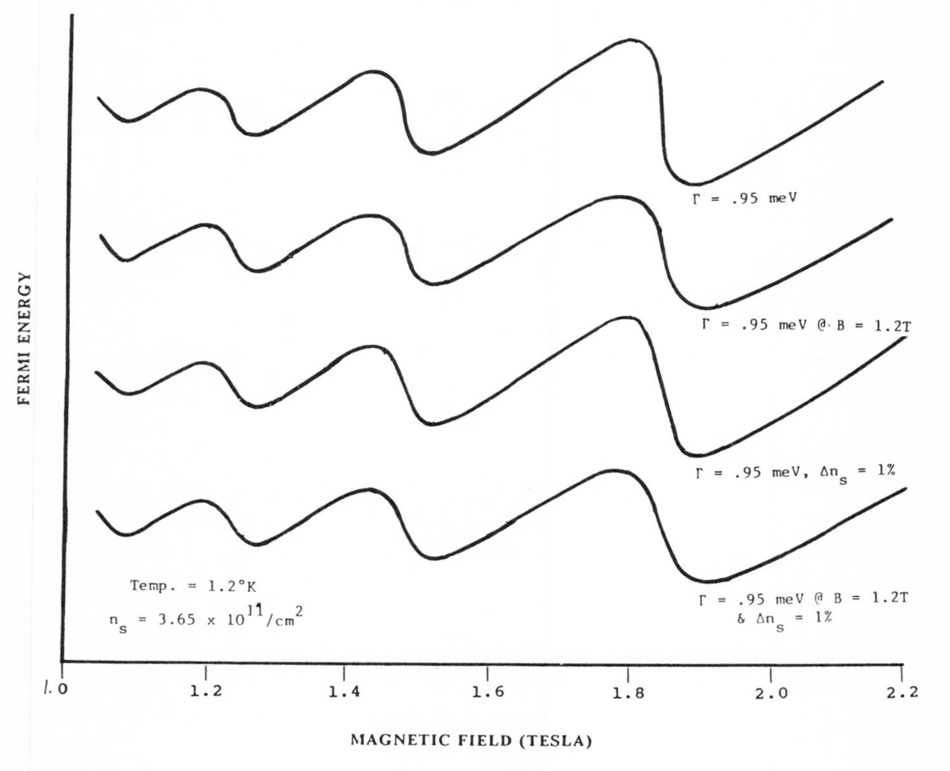

Figure 25

The overall variation is of the order of an mev, much less than the expected change of about 3 mev at 6T. In addition the drop between Landau levels is not vertical as expected for Gaussian broadening as expected from other measurements. There are strong indications that the idealized system does not exist. A more realistic system would be one with interactions and one which included more than short range scatterers.

Figure 27 represents fairly typical results for heterostructures studied so far. At magnetic fields just above 2T the type of structure seen is the same as observed in Si MOSFETs. However at higher fields, the structure is decidedly different. Although it is not illustrated in this figure, this odd structure is hysteretic, that is the sign depends on which direction the field is changing. It has to do with currents that are set up in the sample to shield the change in flux. These currents are long-lived, lasting at least 10^3 seconds. They do not relate to the density of states directly but rather to the quantum Hall effect, which will be discussed in accompanying articles.

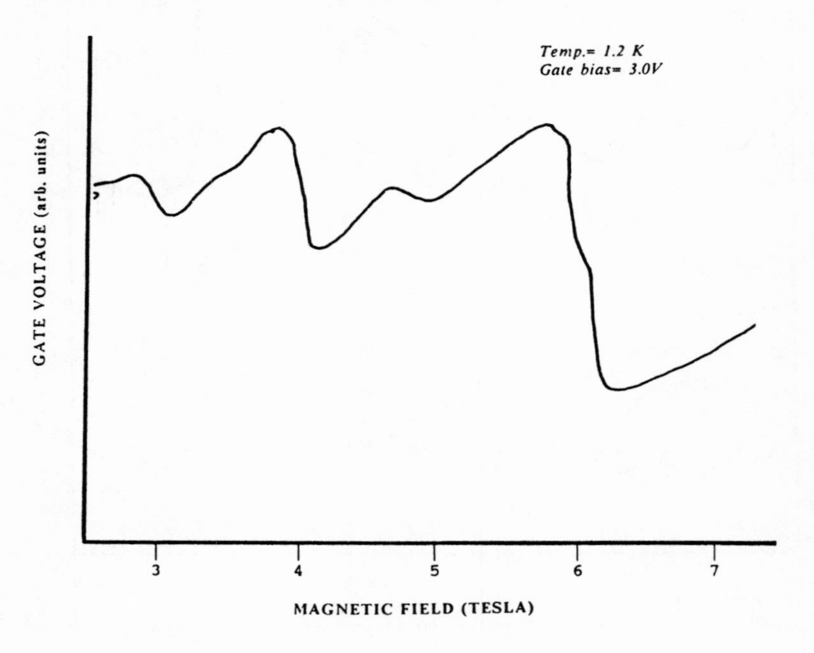

Figure 26

11.

DENSITY OF STATES IN HIGH MAGNETIC FIELDS

What conclusions are drawn from these experimental studies described above? Capacitance measurements give $d\mu/dn_s$ while ECP measures μ directly. How do these techniques and results compare to others? Two other types of experiments performed to determine the density of states are the de Haas-van Alphen effect and the specific heat.

The de Haas-van Alphen (dHvA) effect is the existence of oscillations in the magnetization of a material as a function of magnetic field. It has a long history for metals and semiconductors in 3D materials. The essential difference between 3D and 2D materials comes from the density of states. In 2D, with Landau levels well separated from each other, the density of states has oscillations of the order of 100%. In 3D, even with well separated levels (other parameters being held constant) there is a continuous variation in the momentum along the magnetic field. The only discontinuity is when the sum over states for the density of states for a particular Landau level stops. Therefore the density of states has no zeros, the variation in the DOS is orders of magnitude less and abrupt changes only occur in the derivatives of the free energy.

Although spin effects obviously play a role, the simple description that follows ignores spin effects. The definition of the magnetization is

$$M = d(F)/dH \qquad (11.1)$$

and the free energy is given by

$$F = n_s\mu - kT\int D(E)\ell n(1+\exp((\mu-E)/kT)dE \; . \qquad (11.2)$$

F and M at T=0 as a function of H are illustrated in Figures 28 and 29. The effect of temperature is to wash out the oscillations. Inhomogenieties tend to wash out the oscillations as well.

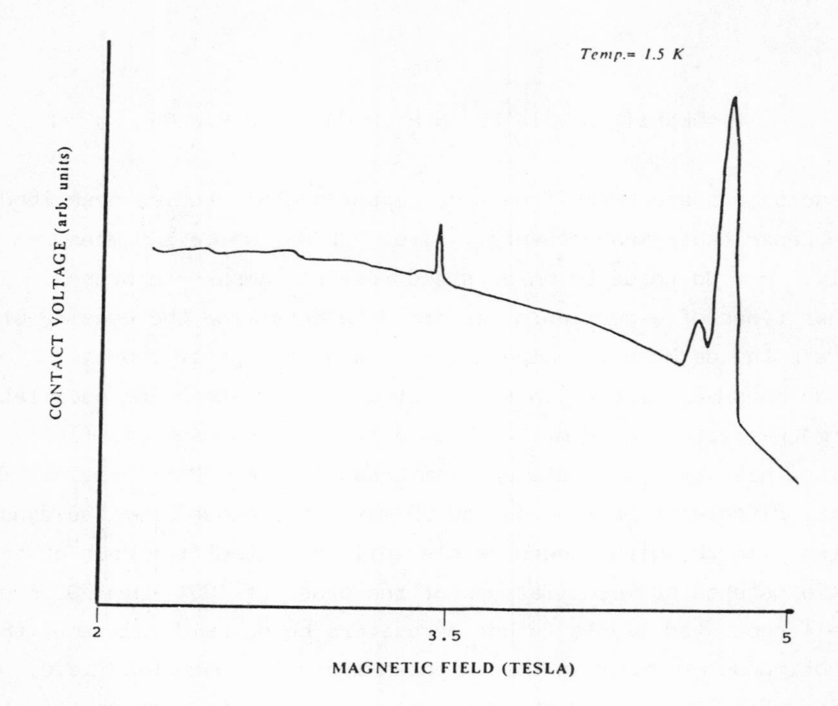

Figure 27

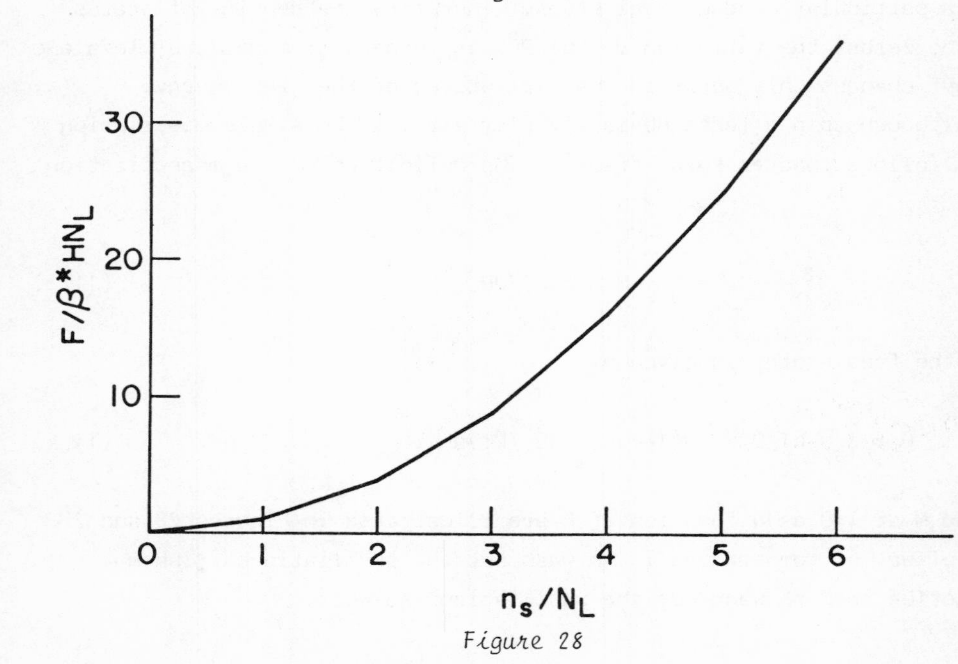

Figure 28

The specific heat is defined as

$$C_v = \int_0^\infty \frac{df}{dT} (E-\mu)D(E)dE \qquad (11.3)$$

Figure 30 by Zawadski and Lassnig[22] illustrates the type of behaviour expected. Notice that there are two types of oscillations corresponding to two kinds of states to be occupied with an increase in temperature, inter-level and intra-level. In theory, there is a wealth of information in the specific heat.

Both the dHvA and the specific heat measurements are bulk effects and the signal for 2D systems is extremely weak. The number of electrons in 1 cm^2 is only of the order of 10^{12} while in a 1 cm^3 3D sample there are 10^{23}!!! The discontinuity in the DOS helps; however the experimentalists must be clever. The first experiments were done on "hundreds"[23,24] of samples at once, that is on multiple quantum wells. A sample is prepared with many heterolayers near enough to each other for the effect in question to add but far enough apart so that wave functions do not overlap.

Conclusions from all these experiments are in agreement that there seems to be too many states between Landau levels at high magnetic fields. The disagreement is only whether this excess is a few or twenty percent, on whether sample inhomogenieties play a significant role, and what is the cause of the excess states. Experiments measuring activated conductivity between Landau levels reach similar conclusions[25].

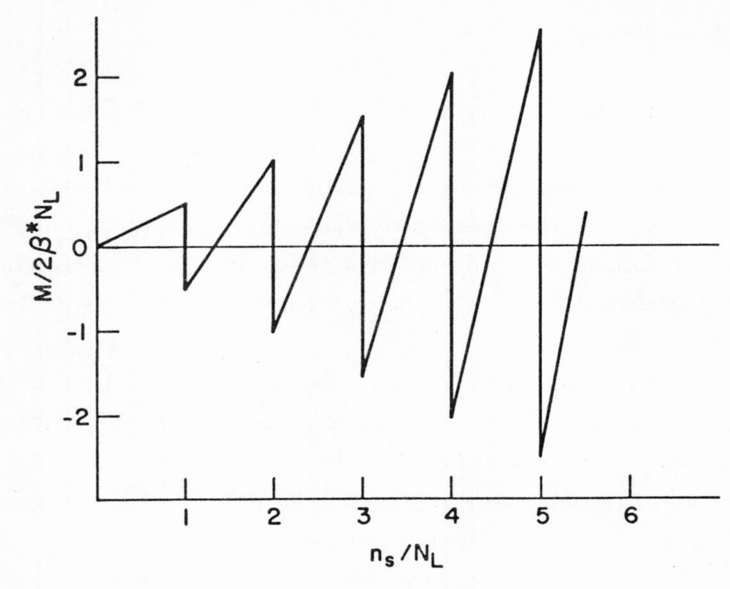

Figure 29

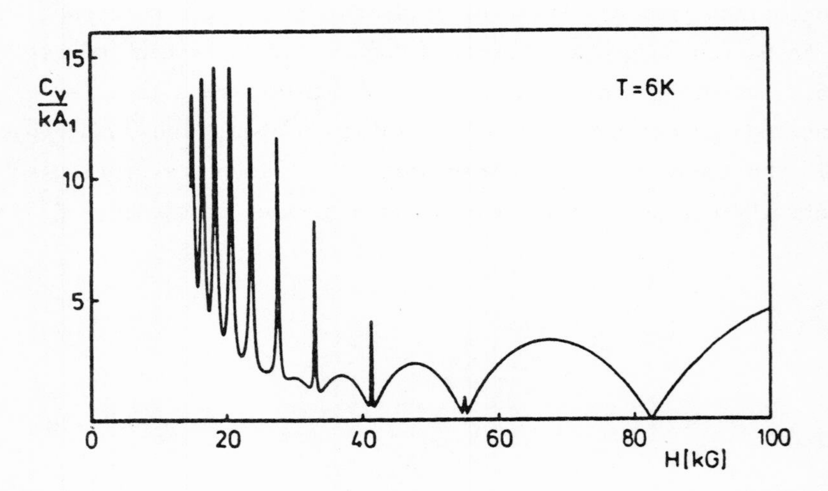

Figure 30

APPENDIX A

Although there has been very little mentioned here about the quantum Hall effect, in particular as to whether the system is behaving 2D-like or 3D-like, it seems appropriate to examine the essential dimensionality of the quantum Hall effect. It will be seen that it comes from systems which are essentially 2D. For an introduction to this phenomena see accompanying articles.

Consider the case of a single quantum well in the presence of a high quantizing magnetic field. However the magnetic field splitting is much less than energy splittings in the isolated quantum well. One would expect behaviour and properties as described in this article previously. Now consider two identical quantum wells separated by a distance such that there is no effective overlap of the wave functions. In this new system all the levels that existed for the single well are now doubly degenerate. If the wells are moved close enough together so there is a weak overlap of the wave functions in the z direction, sufficient to split the levels by an energy ΔE, a system similar to the Si 100 MOSFET would result; one where there is a small splitting of a "band structure" effect. Similar to the Si MOSFET case, if the splitting is less than the broadening, one only observes one "peak" in the density of states that is doubly degenerate. If the splitting is larger than the broadening, then there would be two peaks.

Now if there were j systems in parallel and not interacting, a single j-fold degenerate state would result. If a non-periodic array is moved close enough to interact with neighbouring layers and give an interaction energy "w", a density of states as illustrated in Figure A.1a would result. If the broadening was larger than w or larger than the average spacing per level, a single broadened j-fold level would result as shown in Figure A.1b.

The most intriguing case to consider is if the j layers are periodic and interact. In that case if j is sufficiently large, a quasi-continuous wave vector in the z direction along the magnetic field would result in a band (say of width "w") which would be like a cosine band. The density of states as a function of E and k_z is shown in A.1c. If the broadening was small or comparable to w it would smooth out the structure as in Figure A.1d. Those "levels" in A.1d are j-fold degenerate as are the ones in A.1b. What is the difference between A.1b and A.1d? Both will have conductivity in the z direction when and only when the Fermi energy is in the midst of the non-zero density of states. However fundamentally, they are two dimensional like when the Fermi energy is in the gap (say between n=1 and n=2) and they behave the same. Further, in this situation the only difference between their behaviour and a single well is the j-fold degeneracy. It is these zeros in the density of states that are characteristic of 2D-like behaviour.

If j is small enough and the broadening is less than the interactions there may be other zeros in the density of states. They are still two dimensional. The quantum Hall effect may exist in all such 2D-like materials (subject to the usual conditions).

APPENDIX B

One other aspect of conductivity which is worth examining is the behaviour of the conductivity in the range of $0.2K < T < 20K$ for Si MOSFETS with high mobilities. The effect due to phonon scattering essentially disappears upon lowering the temperature down to 20K. However a marked temperature dependence exists at lower densities.

Without trying in any way to be thorough in coverage of this phenomena an attempt is made to delineate some of the parameters. In Figure B.1 representative behaviour of a Si 100 MOSFET mobility is shown as a function of density for different temperatures. It is seen that the mobility at high densities is not changed very much by a further decrease of temperature while at the lower densities the change with temperature is quite dramatic. The mobility is most enhanced where the predominate scattering mechanism is that associated with charge in the oxide. Not surprisingly it is the temperature dependence of screening which accounts for this behaviour[B.1,2,3]. A simple understanding of the behaviour is as follows. For a pure system at T=0, the normalized polarizability is 1 for q/k_F less than 2. Then for higher values of q/k_F it drops rapidly. As the temperature is raised the polarizability drops in all regions reducing the screening.[B.4,5]

The presence of disorder lowers the polarizability without temperature[B.6] and results in less screening. Therefore the temperature dependence will be smaller in the presence of disorder as the screening is smaller to start with. This qualitative description covers most aspects but is not meant to be quantitative.

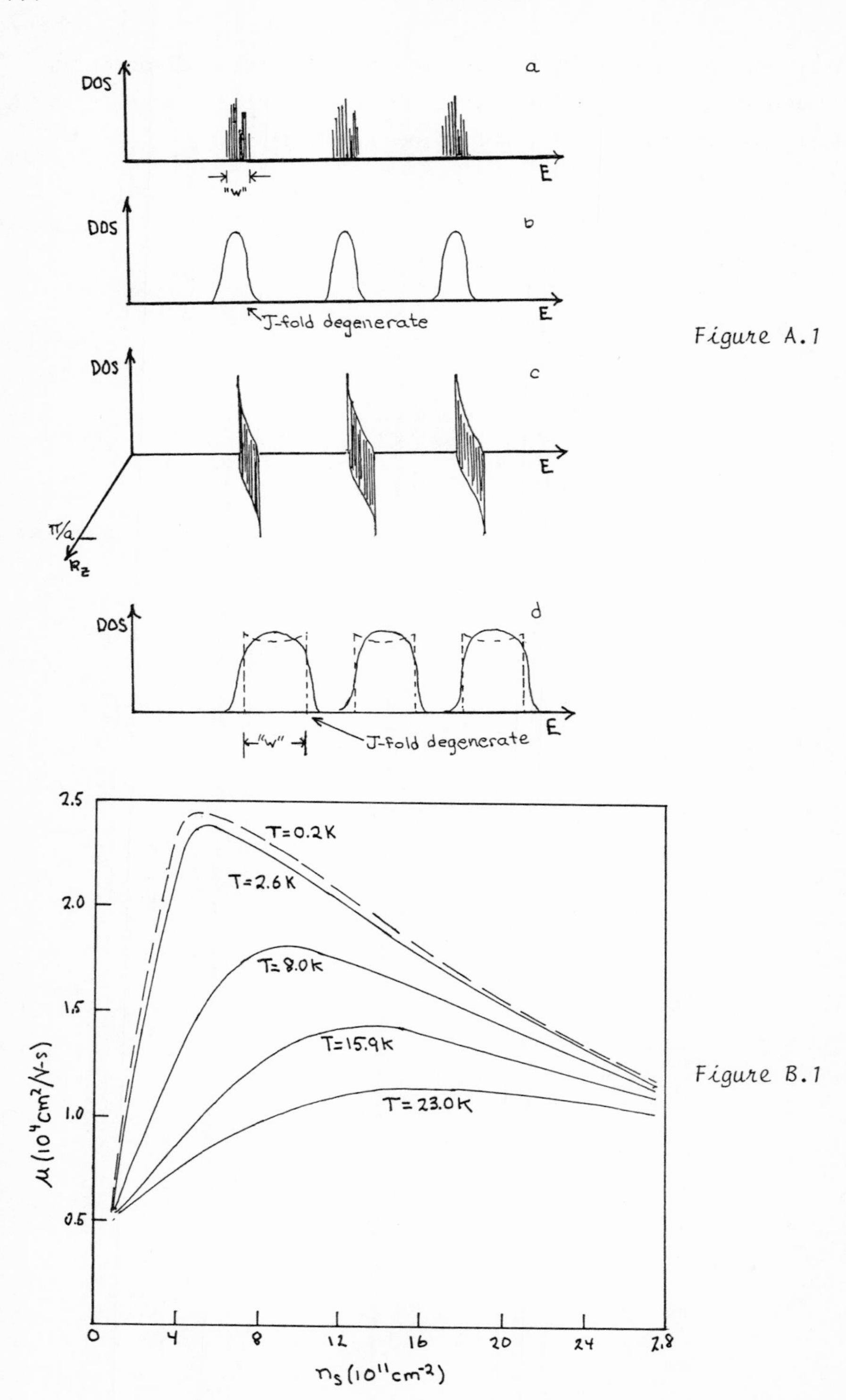

Figure A.1

Figure B.1

REFERENCES

1. T. Ando, A.B. Fowler and F. Stern, Rev. Mod. Phys. 54, 437 (1982).
2. The proceedings of the conference series "Electronic Properties of Two Dimensional System" are published as well as special editions of "Surface Science" volumes 58, 73, 98, 113, 142 and to be published.
3. Solid State Physics, N.W. Ashcroft and N.D. Mermin, (Hott, Rinehart and Winston, New York).
4. F.F. Fang, E. Pakulis, R.P. Smith and P.J. Stiles (private communication).
5. Y. Matsumoto and Y. Uemura, Jpn. J. Appl. Phys. Suppl. 2, 367 (1974).
6. T.P. Smith, III, B.B. Goldberg, M. Heiblum and P.J. Stiles, Surface Science, 170, 304 (1986).
7. A.A. Lakhani and P.J. Stiles, Sol. State Comm. 16, 993 (1975) and Phys. Rev. B13, 5386 (1976).
8. A.B. Fowler, F.F. Fang, W.E. Howard, Jr., and P.J. Stiles, Phys. Rev. Lett 16, 901 (1966).
9. T. Ando and Y. Uemura, J. Phys. Soc., Japan, 36, 959 (1974).
10. Y. Uemura, Jpn. J. Appl. Phys. Suppl. 2 Pt 2, 17 (1974).
11. See pg. 540 of ref. (1).
12. F.F. Fang and P.J. Stiles, Phys. Rev. 174 823 (1968).
13. J.F. Janak, Phys. Rev. 178, 1416 (1969).
14. J.L. Smith and P.J. Stiles, Phys. Rev. Lett 29, 102 (1972).
15. J.L. Smith and P.J, Stiles, Low Temp. Phys. LT13 (edited by D. Timmerhaus, W.J. O'Sullivan and E.F. Hammel (Plenum. N.Y.)) 4, 32 (1974).
16. F.F. Fang, A.B. Fowler and A. Hartstein, Phys. Rev. B16, 4446 (1977).
17. F. Stern, unpublished IBM internal report (1972).
18. T.P. Smith, III and P.J. Stiles, Sol. State Comm., 52, 941 (1984) and T.P. Smith, III, B.B. Goldberg, M. Heiblum and P.J. Stiles, Proc. of VI Int. Conf. EP2DS to be published in Surf. Sci.
19. T.P. Smith, III, W. Wang and P.J. Stiles, to be published in Phys. Rev.

20. R.T. Zeller, F.F. Fang, B.B. Goldberg, S.L. Wright and P.J. Stiles Phys. Rev. B33, 1529 (1986).

21. V.M. Pudalov, S.G. Semenchinsky and V.S. Edelman, Sol. State Comm. 51, 713 (1984).

22. W. Zawadzki and R. Lassnig, Sol. State. Comm. 50, 537 (1984).

23. E. Gornick, R. Lassnig, G. Strasser, H.L. Störmer and A.C. Gossard in Proc. of VI Int. Conf. EP2PS to be published in Surface Science and references therein.

24. J.M. Eisenstein, H.C. Störmer and V. Narayanamarti in Proc. of Int. Conf. EP2DS to be published in Surface Science and references therein.

25. D.E. Weiss, E.Stahl, B. Weimann, K. Ploog and K.v. Klitzing, Proc. VI Int. Conf. EP2DS to be published in Surface Science and references therein.

INTERACTION EFFECTS IN IMPURE METALS

Hidetoshi Fukuyama

Institute for Solid State Physics, University of Tokyo
7-22-1, Roppongi, Minato-ku, Tokyo 106, Japan

1. INTRODUCTION

The effect of the mutual interactions among electrons in metals has been a subject of extensive investigations for a long time. The interactions obviously play vital roles when these result in such phase transitions as superconductivity and itinerant magnetism. Even in cases where these phase transitions are absent and then the Fermi liquid is stable, the low lying excitations are affected by the mutual interactions. Conventionally these interaction effects are treated in the framework of Landau Fermi liquid theory,(1) where electrons or quasi-particles are treated essentially as if free but with various parameters being affected by the mutual interactions. The possible onset of phase transitions at temperatures far below the Fermi energy is due to the residual interactions between these quasi-particles. Basic understanding behind this treatment is that in the temperature range of interest far below the degeneracy temperature, or the Fermi energy, ε_F, the temperature dependences of various one-particle properties are very weak, i.e. usually proportional to $(T/\varepsilon_F)^{(2)}$. In this view the temperature dependences of various physical quantities should be same as those of free electrons; the spin susceptibility, χ, is given by Pauli paramagnetism independent of temperatures and the electronic contributions to specific heat, C_e is linear in temperature and the deviations from these dependences should be very weak.

On the other hand the effects of randomness, which is typically represented as an impurity scattering and is the origin of the residual resistance, have been treated as elastic scattering processes for these quasi-particles. Due to the elasticity of the scattering the effects of randomness have been believed to be also weakly dependent on temperature, which is actually the case within the conventional treatment of Boltzmann transport equation.

Consequently within the foregoing conventional framework χ, C_e/T and the resistivity due to impurity scattering, ρ, are essentially independent of temperature below ε_F, as schematically in Fig.1.

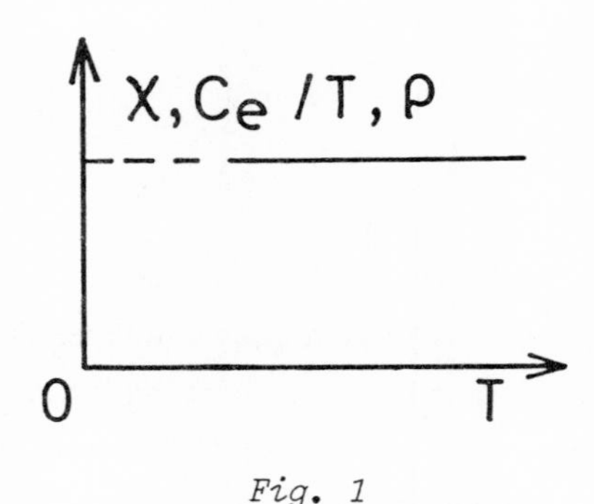

Fig. 1

Such a wide-spread belief in weak temperature dependences of physical quantities in metals has been challenged when Abrahams, Anderson, Licciardello and Ramakrishnan[2] proposed a scaling theory for the Anderson localization. This theory, which is originally designed to examine the cases of non-interacting electrons, has resulted in remarkable conclusions. At the same time it has opened a new systematic way to investigate the electronic properties of solids in the presence of randomness, which have followed.[3-6] In the scaling theory it is indicated that the conductance in two dimensional (d=2) systems should have a logarithmic temperature dependence instead of temperature-independent residual conductance. This is due to the inelastic scattering hindering the growth of the interference between Bloch waves.[7] Soon after this scaling theory it was found[8-11] that interactions have more direct effects on conductivity with the same logarithmic temperature dependence in d=2. It has later been clarified that the coexistence of randomness and mutual interactions generally lead to the singular temperature dependences in various physical quantities; $\sqrt{T}$ in d=3 and ℓnT in d=2. These findings will be compared to the Kondo effect[12] which has disclosed the logarithmic temperature dependences (independent of d) associated with dynamical scattering by paramagnetic spins. Actually the origins of these singular temperature dependences are common: the coexistence of sharp Fermi energy and the dynamical scattering processes.

As can be easily expected these recent developments are modifying our views on interaction processes in impure metals at low temperatures and it is the purpose of this lecture to give brief reviews on this subject. We take unit of $\hbar=k_B=1$.

2.
RANDOMNESS PARAMETER, DIFFUSON AND COOPERON

In the original paper by Abrahams et al.[2] the conductance, instead of conductivity, has been examined as a function of the system size, L, and the scaling theory has been derived. By this theory one is natually led to the clear identification of the expansion parameter to treat randomness from the metallic limit. This parameter, λ, is defined by

$$\lambda \equiv \hbar / 2\pi\varepsilon_F\tau ,$$

where ε_F and τ are the Fermi energy, the characteristic value of the kinetic energy in the condensed system and the life time of Bloch wave due to impurity potential. Here $\hbar$ is recovered to stress the quantal nature of this parameter. In the classical Boltzmann transport equation for the conductivity λ is considered to be vanishing since electrons are treated purely as classical particles there. The finiteness of λ on the other hand implies that electrons should not be viewed as classical any longer but the quantum mechanical nature as Bloch waves are to be properly taken into account in the course of scattering from impurity potentials. The regime where λ is small and then perturbational treatment with respect to λ is valid is called weakly localized regime. This regime is schematically shown in Fig.2 for each dimension, d, where $\sigma_0=ne^2\tau/m$, n being the electron density, is the classical conductivity.

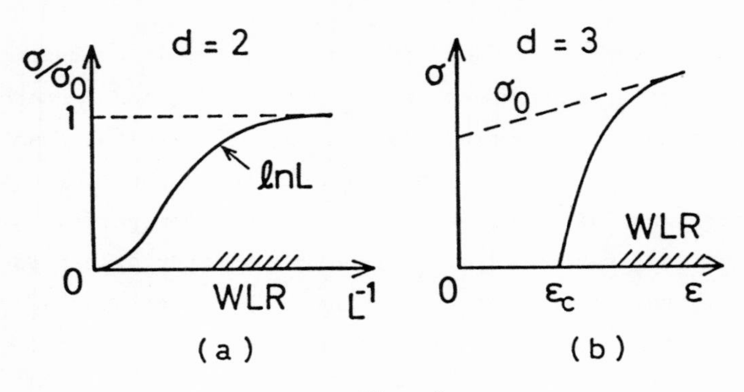

Fig. 2

The clear identification of the expansion parameter, λ, is especially important in the investigation of the effect of randomness in the presence of the mutual interaction, V, which through its dynamical nature scatters electrons between different eigenstates. Actually it is this particular aspect of the mutual interaction which makes it to play essential roles in the localization process. There exist two different features of effects of randomness chatacterized by finite λ; the interference effect between Bloch waves and the change of matrix elements of mutual interactions due to the absence of the translational symmetry in random systems, respectively. The former is diagrammatically represented as in Fig.3, where solid lines are, $\mathcal{G}(k,\varepsilon_n)$, ($\varepsilon_n=(2n+1)\pi T$; T being the temperature) given by

$$\mathcal{G}(\boldsymbol{k},\varepsilon_n)=\left[i\varepsilon_n-\xi_{\boldsymbol{k}}+\frac{i}{2\tau}\mathrm{sgn}\varepsilon_n\right]^{-1}, \qquad \frac{1}{\tau}=2\pi N(0)n_i v_0^2. \tag{2.1}$$

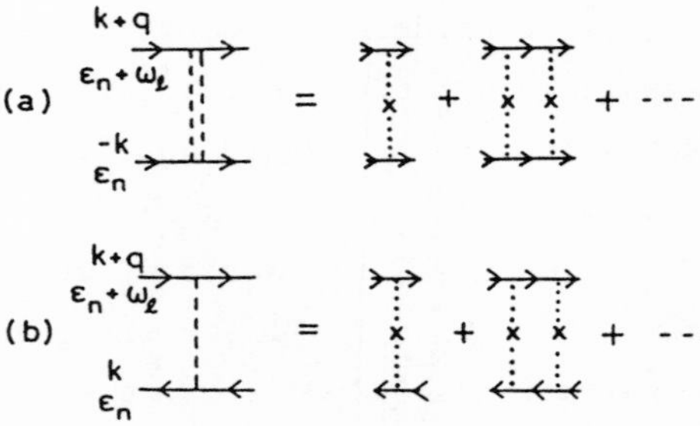

Fig. 3

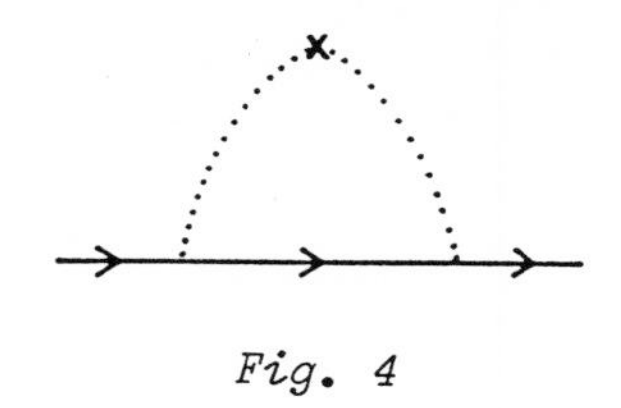

Fig. 4

Here $\xi_k = k^2/2m - \varepsilon_F$ and N(0) is the density of states per spin at the Fermi energy, i.e., $N(0)=m/2\pi \equiv N_2(0)$ for d=2 and $N(0)=mk_F/2\pi^2 \equiv N_3(0)$ for d=3 where k_F is the Fermi momentum. In Figs.3 the dotted lines (indicating the impurity potential, v_0) with crosses represent the average procedure over the configuration of impurities, whose average concentration is n_i. The life time given by eq.(2.1) is due to the self-energy shown in Fig.4. The series of diagrams in Fig.3(a) is summed as

$$n_i v_0^2 + (n_i v_0^2)^2 X + \cdots\cdots = \frac{n_i v_0^2}{1 - n_i v_0^2 X}, \tag{2.2}$$

where X is defined by

$$X = \sum_{\boldsymbol{k}} \mathcal{G}(\boldsymbol{k}+\boldsymbol{q}, \varepsilon_n + \omega_l)\, \mathcal{G}(-\boldsymbol{k}, \varepsilon_n)$$

$$= N(0)\int_{-\infty}^{\infty} d\xi \int d\Omega \left[i\varepsilon_n + i\omega_l - \xi - \boldsymbol{v}\cdot\boldsymbol{q} + \frac{i}{2\tau}\mathrm{sgn}(\varepsilon_n + \omega_l)\right]^{-1} \left[i\varepsilon_n - \xi + \frac{i}{2\tau}\mathrm{sgn}\varepsilon_n \right]^{-1}. \tag{2.3}$$

Here $\xi_{-k}=\xi_k$ is noted and $v_\alpha = k_\alpha/m$. The Fermi velocity v_F is defined by $v_F=|v|$. The integration over ξ in eq.(2.3) has a remarkable feature that X is vanishing for small $|\varepsilon_n|$ and $|\varepsilon_n+\omega_\ell|$ unless $\varepsilon_n(\varepsilon_n+\omega_\ell)<0$. If $\varepsilon_n(\varepsilon_n+\omega_\ell)<0$, X is evaluated as follows for small q and ω_ℓ:

$$X = 2\pi N(0)\,\tau[1 - Dq^2\tau - |\omega_l|\tau], \tag{2.4}$$

which is a function of q and ω_ℓ. Here $D=v_F^2\tau/d$ is the diffusion constant, and eq.(2.4) is valid as far as Dq^2, $|\omega_\ell|<\tau^{-1}$. Hence the r.h.s. of eq.(2.2), defined as $C(q,\omega_\ell)$, is given as

$$C(\boldsymbol{q}, \omega_l) = \frac{1}{2\pi N(0)\,\tau^2}\,\frac{1}{[Dq^2 + |\omega_l|]}, \quad \varepsilon_n(\varepsilon_n + \omega_l) < 0\,. \tag{2.5}$$

This function is often called a Cooperon, since this represents the particle-particle correlation function as in the discussion of the Cooper instability in the theory of superconductivity. Equation(2.5) is the expression of Cooperons in the absence of any dynamical interactions which scatter electrons between states with different energies. However, in the presence of such dynamical interactions the interference effect will be suppressed and described by a kind of life time τ_ε; $|\omega_\ell| \to \omega_\ell + \tau_\varepsilon^{-1}$. This τ_ε has been introduced originally by Anderson et al.[2] on the phenomenological ground and has been playing a key role in the discussion of magnetoresistance, where the limiting procedure, $i\omega_\ell = \omega \to 0$, is taken, ω being the external frequency of the electric field.

On the other hand the change of matrix elements is diagrammatically represented by Fig.3(b), whose series is given as

$$n_i v_0^2 + (n_i v_0^2)^2 Y + \cdots\cdots = \frac{n_i v_0^2}{1 - n_i v_0^2 Y}, \tag{2.6}$$

where

$$Y = \sum_{\boldsymbol{k}} \mathcal{G}(\boldsymbol{k}+\boldsymbol{q}, \varepsilon_n + \omega_l)\, \mathcal{G}(\boldsymbol{k}, \varepsilon_n) = X, \tag{2.7}$$

by $\mathcal{G}(-k, \varepsilon_n) = \mathcal{G}(k, \varepsilon_n)$. Hence for $\varepsilon_n(\varepsilon_n + \omega_\ell) < 0$ the r.h.s. of eq.(2.6) defined as $D(q, \omega_\ell)$ is given by

$$D(\boldsymbol{q}, \omega_l) = \frac{1}{2\pi N(0)\tau^2} \frac{1}{[Dq^2 + |\omega_l|]}, \qquad \varepsilon_n(\varepsilon_n + \omega_l) < 0. \tag{2.8}$$

This function representing the particle-hole correlation in contrast to the Cooperon is called a diffuson.

The Cooperon and diffuson, which will be written as $R(q, \omega_\ell)$, are convenient theoretical tools to examine the effects of randomness systematically. It is to be noted that both are divergent as $q \to 0$ and $|\omega_\ell| \to 0$ and that randomness yield particular contributions from the region of small q and ω_ℓ. Based on this fact a perturbation theory with respect to λ is constructed. This is due to the fact that the product of any number (say n) of $R(q, \omega_\ell)$ carrying the same q and ω_ℓ has contributions of the order of

$$\sum_q R(q,\omega_l)^n = \frac{1}{[2\pi N_2(0)\tau^2]^n}\frac{1}{4\pi D}|\omega_l|^{-n+1} \tag{2.9}$$

for d=2 and

$$\sum_q R(q,\omega_l)^n = \frac{1}{[2\pi N_3(0)\tau^2]^n}\frac{1}{2\pi^2 D^{3/2}}|\omega_l|^{-n+3/2} \tag{2.10}$$

for d=3. As seen both are vanishing if $\lambda=o$, i.e., $D=\infty$, and for d=2 and d=3 eqs.(2.9) and (2.10) lead to the corrections described by $\lambda=[(2\pi)^2N_2(0)D]^{-1}$ and $\lambda^2=[6\sqrt{3}\pi^4N_3(0)D^{3/2}]^{-1}$, respectively. If a process contains two $R(q,\omega_\ell)$ with different and essentially independent wave-vectors the corrections are of the order of λ^2(d=2) or λ^4(d=3), and so on. This is how one can perform perturbation with respect to λ systematically. From the discussion it will be evident that such a perturbation theory with respect to λ does not literally work in d=1 since the q-integration will result in $(N(0)\sqrt{D})^{-1}$ which is of the order one (note $N(0)\propto k_F^{-1}$ and $D\propto k_F^2$) and does not vanish even $\lambda\to0$. This is the reason why one should employ more elaborate treatment for d=1.

In this context it should be noted that the case of metallic films and wires shown in Fig.5(a) and (b) whose thickness, b, and the width, w, are larger than the mean free path, ℓ, (and then much larger than k_F^{-1} unless the system is strongly random as in amorphous metals) is neither purely two-dimensional nor one-dimensional. This is because the electronic properties can be viewed as essentially three-dimensional and then the diffusion constant is isotropic together with three-dimensional density of states although the contributions of $R(q,\omega_\ell)$ can be considered as either two- or one-dimensional depending on the temperature. In the case of films if the thickness is so thin that the quantization energy with

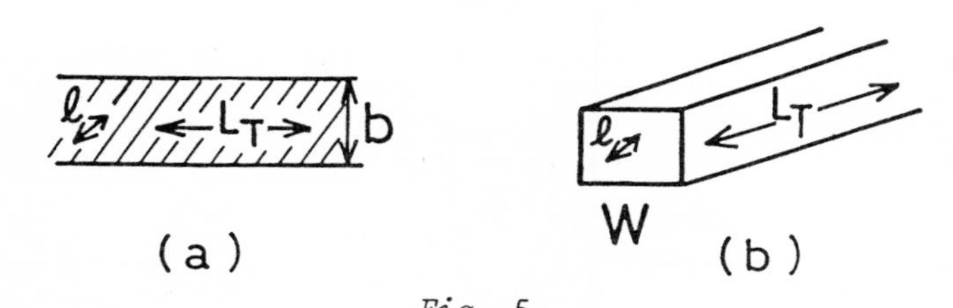

Fig. 5

respect to the momentum in the perpendicular direction, i.e. D/b^2, D being the three-dimensional diffusion constant, exceeds the characteristic energy, ω_ℓ, i.e. T, (namely if $b<L_T=\sqrt{D/T}$ is satisfied L_T being the thermal diffusion length) the summation over q in eq.(2.8) results in $(N_3(0)Db)^{-1}$ $\propto \lambda(3\pi/2k_F b)\equiv\lambda^*$ which is the expansion parameter in this case. Similarly if b, $w<L_T$ in the case of wires, the corrections due to randomness are of the order of $[N_3(0)bw\sqrt{D}]^{-1}\propto[(k_F b)(k_F w)]^{-1}$ which can be a small expansion parameter. (In contrast, $k_F b \sim k_F w \sim 1$ in strictly one-dimensional systems.)

Equation (2.5) for a Cooperon implies that there exist strong coupling between the incoming wave with wave-vector k and the outgoing wave with wave-vector $k'\simeq -k$ i.e. strong backward scattering (Fig.6).

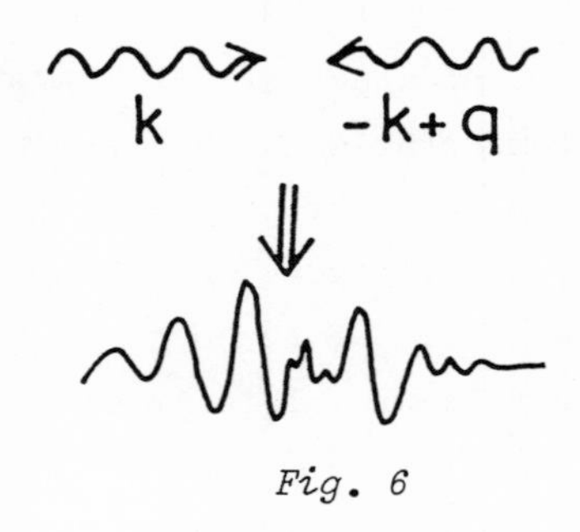

Fig. 6

On the other hand the diffuson is directly related to the density-density correlation function, $N(q,\omega_\ell)$, as follows (the spin degeneracy factor 2 is included) by use of Y, eq.(2.7),

$$N(\boldsymbol{q},\omega_l)=-2T\sum_{\epsilon_n}\left[Y+\frac{n_i v_0^2}{1-n_i v_0^2 Y}Y^2\right]$$
$$=-2T\sum_{\epsilon_n}\left[\frac{Y}{1-n_i v_0^2 Y}\right]. \tag{2.11}$$

In the summation over ε_n, a special care must be paid to the region of $\varepsilon_n(\varepsilon_n+\omega_\ell)>0$ where the existence of finite τ^{-1} is irrelevant. The contribution from such region, then, should be the same as in clean systems and by noting eq.(2.8) one obtains for small q and ω_ℓ as

$$N(\boldsymbol{q},\omega_l)=-2\Big(T\sum_{\epsilon_n(\epsilon_n+\omega_l)>0}+T\sum_{\epsilon_n(\epsilon_n+\omega_l)<0}\Big)Y/[1-n_i v_0^2 Y],$$
$$\simeq 2N(0)-2T\sum_{-|\omega_l|<\epsilon_n<0}Y/[1-n_i v_0^2 Y],$$
$$=2N(0)\left[1-\frac{|\omega_l|}{Dq^2+|\omega_l|}\right]=2N(0)\frac{Dq^2}{Dq^2+|\omega_l|}. \tag{2.12}$$

Hence the pole of the diffuson corresponds to that of $N(q,\omega_\ell)$. The fact that $N(q,\omega_\ell)\to 0$ as one lets $q\to 0$ by keeping ω_ℓ finite is the reflection of the quite general symmetry of the Hamiltonian that the total number of electrons in the system is conserved; i.e. the density fluctuation with the vanishing wave vector is nothing but that of the total number of electrons which is the conserved quantity and hence $N(0,\omega_\ell)=0$ if $\omega_\ell\neq 0$. On the other hand $N(q,0)\to 2N(0)$ as $q\to 0$, since the density correlation function in this limit represents the change of the total carrier number as the chemical potential is varied. In obtaining eq.(2.12) the weak dependences on q and ω_ℓ of the contributions from the region of $\varepsilon_n(\varepsilon_n+\omega_\ell)>0$ are ignored.

3.
INTERACTION EFFECTS IN NORMAL METALS

In clean systems the lowest order contributions in the mutual interaction are given by the well-known Hartree and Fock processes for the self-energy shown in Fig.7, where the wavy lines are interactions, whose Fourier transform is defined as $v(q,\omega_\ell)$. If the interaction is not singular in the sense that the q- and ω_ℓ-dependences of $v(q,\omega_\ell)$ are weak as in the case of instantaneous and local interaction $v(q,\omega_\ell)$=const, these processes do not result in any interesting contributions except those which can be incorporated into the shift of the chemical potential.

In the presence of randomness, however, these Hartree-Fock processes coupled with randomness yield important contributions. First, let us consider the Fock process, which is affected by randomness as shown in Fig.8(a) and (b), where broken and double-broken lines are diffusons and Cooperons, respectively. (There exist less singular contributions, which can be ignored.) The contribution from Fig.8(a) is evaluated as follows,

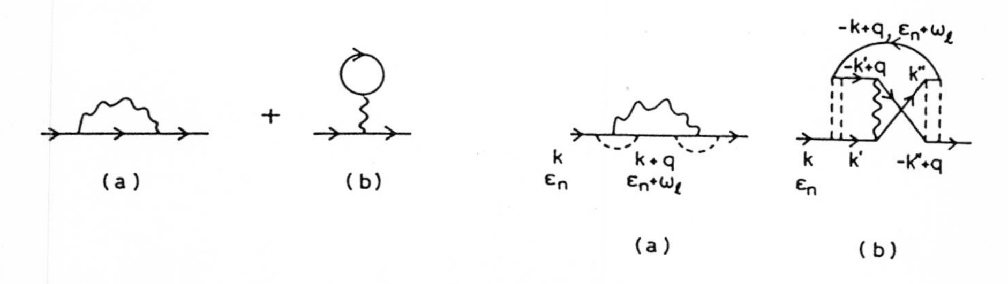

Fig. 7   *Fig. 8*

$$\bar{\Sigma}[\text{Fig. 8(a)}] \equiv \bar{\Sigma}_1 = -T\sum_{\omega_l}\sum_{q} v(q, \omega_l)\, \mathcal{G}(k+q, \varepsilon_n+\omega_l)[1-n_i v_0^2 X]^{-2}, \quad (3.1)$$

$$\simeq -\frac{T}{\tau^2}\sum_{\omega_l}{}'\sum_{q}{}' v(q,\omega_l)\, \mathcal{G}(k+q, \varepsilon_n+\omega_l)[Dq^2+|\omega_l|]^{-2}. \quad (3.2)$$

In deriving eq.(3.2) from eq.(3.1), only the singular contributions are retained (note that in the region of $\varepsilon_n(\varepsilon_n+\omega_\ell)>0$ we can approximate $X\sim 0$) and the summation over ω_ℓ in eq.(3.2) is in the region where $\varepsilon_n(\varepsilon_n+\omega_\ell)<0$ and $|\omega_\ell|<\tau^{-1}$. The integration over q is for the region of small q and is cut off by $q\le q_0\equiv(D\tau)^{-1/2}=\sqrt{d/2}\ \ell^{-1}\ll k_F$ if $\lambda<1$. Since q and $|\omega_\ell|$ are small, $\mathcal{G}(k+q,\varepsilon_n+\omega_\ell)$ for $|k|\simeq k_F$ and $\varepsilon_n\sim 0$ can be approximated as

$$\mathcal{G}(k+q, \varepsilon_n+\omega_l) \simeq -2i\tau\,\mathrm{sgn}(\varepsilon_n+\omega_l). \quad (3.3)$$

Hence eq.(3.2) obtains

$$\bar{\Sigma}_1 = -2i(\mathrm{sgn}\,\varepsilon_n)\frac{T}{\tau}\sum_{\omega_l}{}'\sum_{q}{}' v(q, \omega_l)/[Dq^2+|\omega_l|]^2. \quad (3.4)$$

As seen eq.(3.4) strongly depends on the q- and ω_ℓ- dependences of the mutual interaction and also on the dimensionality of the systems through the q-integrations.

If $v(q,\omega_\ell)$ can be considered as independent of q and ω_ℓ in the region of the summation and integration, $v(q,\omega_\ell)=v$, eq.(3.4) is evaluated for each dimensionality, d, as follows:

$$\bar{\Sigma}_1 = -i(\mathrm{sgn}\,\varepsilon_n)\frac{v}{2\pi D\tau}T\sum{}'\frac{1}{|\omega_l|}. \quad (d=2) \quad (3.5)$$

$$= -i(\mathrm{sgn}\,\varepsilon_n)\frac{v}{4\pi D^{3/2}\tau}T\sum{}'\frac{1}{\sqrt{|\omega_l|}}. \quad (d=3) \quad (3.6)$$

Here integrations over q have been extended to infinity since they are convergent. In the case of $\varepsilon_n>0$ (the case of $\varepsilon_n<0$ can be treated similarly) each summation over ω_ℓ in eqs.(3.5) and (3.6) is evaluated as follows; for d=2

$$\bar{\phi}_2(T) \equiv 2\pi T\sum_{\tau^{-1}<\omega_l<-\varepsilon_n}\frac{1}{-\omega_l} = \sum_{l=n+1}{}'\frac{1}{l}, \quad (3.7)$$

$$= \psi(n_0) - \psi(n+1), \quad (3.8)$$

$$\simeq \ln\frac{1}{2\pi\tau T} - \psi\left(\frac{\varepsilon_n}{2\pi T}+\frac{1}{2}\right), \quad (3.9)$$

where $\psi(z)$ is the di-gamma function[13] and n_0 is the cut-off parameter satisfying $\varepsilon_{n_0} \sim \tau^{-1}$ and in eq.(3.9) the asymptotic expression $\psi(z) \sim \ln z$ as $z \to \infty$ is employed. In the limit of $\varepsilon_n \to 0$ eq.(3.9) yields

$$\bar{\phi}_2(T) = \ln \frac{2e^{\gamma}}{\pi T \tau} = \ln(1.14/\tau T)\,, \tag{3.10}$$

where γ is the Euler constant and $\psi(1/2) = -\gamma - 2\ln 2$ is noted.

On the other hand for d=3 the summation over ω_ℓ is evaluated in the region of low temperature, i.e. $\tau T \to 0$, as follows:

$$\bar{\phi}_3(T) = C_3 2\pi T \sqrt{\tau} \sum_{-\tau < \omega_l < -\varepsilon_n} \frac{1}{\sqrt{\omega_l}} = C_3 \sqrt{2\pi T \tau} \sum_{l=n+1}' \frac{1}{\sqrt{l}}$$

$$= C_3 \left[2 + \sqrt{2\pi T \tau}\, \zeta\!\left(\frac{1}{2}, \frac{\varepsilon_n}{2\pi T} + \frac{1}{2} \right) \right], \tag{3.11}$$

where $C_3 = 3\sqrt{3}\pi^2/4$ and $\zeta(z,a)$ is the generalized zeta function, which has a particular value of $\zeta(1/2,1/2) = (\sqrt{2}-1)\zeta(1/2)$ with $\zeta(1/2) = -1.46$ corresponding to the case of $\varepsilon_n \to 0$.

By use of $\bar{\phi}_d(T)$, $\bar{\Sigma}_1$ is given as follows:

$$\bar{\Sigma}_1/\Sigma_0 = 2g_1 \lambda^{d-1} \bar{\phi}_d(T)\,, \tag{3.12}$$

where $\Sigma_0 = -(i/2\tau)\mathrm{sgn}(\varepsilon_n)$ is the self-energy correction introduced in eq.(2.1) and $g_1 = 2vN(0)$.

Hence the characteristic functions $\bar{\phi}_2(T)$ and $\bar{\phi}_3(T)$ have singular temperature dependences as schematically shown in Fig.9 ; especially $\phi_2(T)$ is logarithmically divergent. Note that these singular dependences result even from the mutual interaction which is not singular at all.

Similarly the process of Fig.8(b) is evaluated as follows:

$$\Sigma[\text{Fig.8 (b)}] \equiv \bar{\Sigma}_2$$

$$= -T \sum_{\omega_l}' \sum_{q}' \mathcal{G}(-k+q, \varepsilon_n+\omega_l)\, v(k'-k'', \omega_l)\, C(q, \omega_l)^2$$

$$\times \mathcal{G}(k', \varepsilon_n)\, \mathcal{G}(-k'+q, \varepsilon_n+\omega_l)\, \mathcal{G}(k'', \varepsilon_n+\omega_l)\, \mathcal{G}(-k''+q, \varepsilon_n)\,. \tag{3.13}$$

The momentum transfer associated with this process extends up to the large momentum of the order of $2k_F$ in contrast to the case of Σ_1, where it is

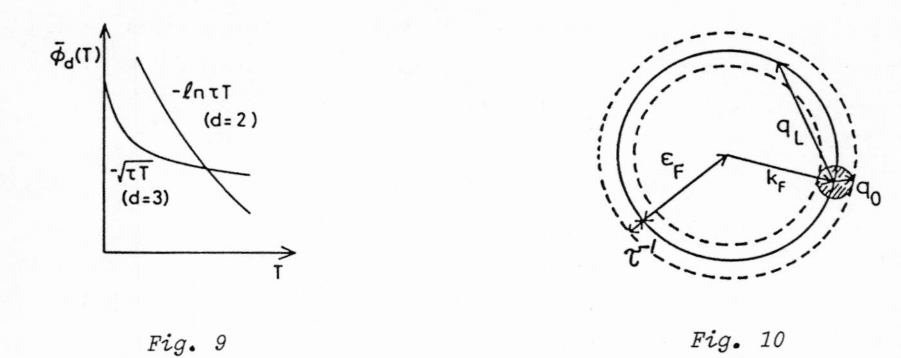

Fig. 9 *Fig. 10*

confined within the small momentum (as schematically shown in Fig.10), an hence $\bar{\Sigma}_2$ and $\bar{\Sigma}_1$ reflect different aspects of the interaction processes Unless the interaction has very singular q-dependence at relatively larg q, the contribution of eq.(3.13) will be roughly estimated by replacin $v(k'-k'',\omega_\ell)$ by its average over the scattering processes across the ferm surface,

$$2N(0)\langle v(\boldsymbol{k}'-\boldsymbol{k}'',\omega_l)\rangle \simeq 2N(0)\langle v(\boldsymbol{k}'-\boldsymbol{k}'',0)\rangle, \qquad (3.14)$$
$$\equiv F,$$

where < > implies the average over k' and k'' with $|k'|=|k''|=k_F$ and th dynamics of the interaction is ignored since it generally plays mino roles for large momentum transfer. By use of F thus defined eq.(3.14) i evaluated similarly to $\bar{\Sigma}_1$,

$$\bar{\Sigma}_2/\Sigma_0 = 2g_2\lambda^{d-1}\bar{\phi}_d(T) \qquad (3.15)$$

for each dimension, where g_2=F/2.

Besides the processes shown in Figs.8(a) and (b), there also exis processes shown as in Figs.11(a) and (b) contributing to the self-energ functions. These have contributions half but with different sign o Figs.8(a) and (b), respectively. Hence the quantum corrections to th Fock self-energy, Σ_1 and Σ_2, are given by (i=1 or 2)

$$\Sigma_i = g_i\lambda^{d-1}\bar{\phi}_d(T)\Sigma_0. \qquad (3.16)$$

As regards the Hartree process there exist processes shown a

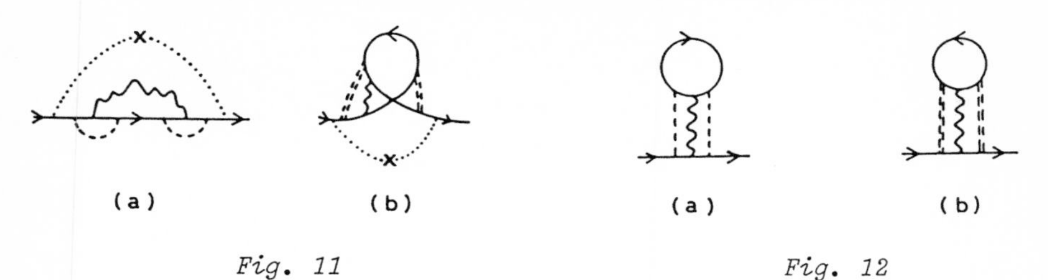

Fig. 11 *Fig. 12*

Figs.12(a) and (b), together with those similar to Figs.11(a) and (b) associated to Fig.8(a) and (b). The momentum transfer through the interaction in these processes is again large as in the case of Fig.8(b) and by approximating the interaction constant by F, eq.(3.14), we obtain (i=3 or 4)

$$\Sigma_i/\Sigma_0 = -2g_i\lambda^{d-1}\bar{\phi}_d(T)\,, \tag{3.17}$$

where g_3=F/2, g_4=F/2 in the present approximation of taking only the lowest order interaction effects into account and in eq.(3.17) contributions with impurity corrections similar to Figs.11(a) and (b) are included. The difference of the signs between eqs.(3.16) and (3.17) result from the that of direct and exchange processes of the interaction and the extra factor 2 in eq.(3.17) is due to the spin degeneracy of electrons interacting with incoming electrons.

Consequently, we obtain the quantum corrections to the self-energy of propagating electrons specified by momentum, k, and energy, ε_n, as follows for $|k|\to k_F$ and $\varepsilon_n\to 0$,

$$\Sigma \equiv \Sigma_1+\Sigma_2+\Sigma_3+\Sigma_4\,, \tag{3.18}$$

$$= g\lambda^{d-1}\bar{\phi}_d(T)\Sigma_0\,, \tag{3.19}$$

$$g = g_1+g_2-2(g_3+g_4)\,. \tag{3.20}$$

Although this self-energy correction is not directly observable, it turns out that there also exist similar quantum corrections, P', to physical quantities, P_0, which have similar dependences on λ and T, i.e.

$$P'/P_0 = g_P\lambda^{d-1}\phi_d(T)\,, \tag{3.21}$$

where g_P is some interaction constant and $\phi_d(T)$ here are slightly different from eq.(3.10) and eq.(3.11) and are given by

$$\phi_2(T) = 2\pi T \sum_{\omega_l>0}{}' \frac{1}{\omega_l} = \ln\frac{1}{2\pi\tau T} - \psi(1) ,$$

$$= \ln\frac{e^\gamma}{2\pi\tau T} = -\ln 3.5\tau T , \tag{3.22}$$

$$\phi_3(T) = C_3 2\pi T\sqrt{\tau} \sum_{\omega_l>0}{}' \frac{1}{\sqrt{\omega_l}}$$

$$= C_3\left[2+\sqrt{2\pi T\tau}\,\zeta\left(\frac{1}{2}\right)\right], \tag{3.23}$$

where ω_ℓ is summed up to τ^{-1} and $\tau T < 1$ is assumed.
For example corrections to the density of states at the Fermi energy, N', the conductivity, σ', and the spin susceptibility, χ', are given as follows,

$$N'/N_0 = -g\lambda^{d-1}\phi_d(T) , \tag{3.24}$$

$$\sigma'/\sigma_0 = \begin{cases} -g\lambda\phi_2(T), & d=2, \quad (3.25a) \\ -g\lambda^2\phi_3(T)/2, & d=3, \quad (3.25b) \end{cases}$$

$$\chi'/\chi_0 = \begin{cases} 2(g_3+g_4)\lambda\phi_2(T), & d=2, \quad (3.26a) \\ (g_3+g_4)\lambda^2\phi_3(T), & d=3, \quad (3.26b) \end{cases}$$

where $g=g_1+g_2-2(g_3+g_4)$.

The actual values of coupling constants, g_i, depend on the original interaction, $v(q,\omega_\ell)$. As stressed before g_i, (i=2,3 and 4) are not so sensitive to the details of $v(q,\omega_\ell)$ and the simple parametrization by F will be appropriate. In the lowest order of the interaction, $g_i=F/2$ for i=2,3 and 4. (If higher order effects of interactions are taken into account, these equations no longer hold.) On the other hand, the g_1 -process is very sensitive to the details of $v(q,\omega_\ell)$ in the region of small q and ω_ℓ because the diffusons couple to such a region exclusively in this process. In the case of dynamically screened Coulomb interaction $v(q,\omega_\ell)$ is given as follows

$$v(q, \omega_l) = v_B(q)/[1 - v_B(q)\pi(q, \omega_l)] , \tag{3.27}$$

where $v_B(q)$ is the bare Coulomb interaction, $v_B(q)=2\pi e^2/q$ (d=2), $4\pi e^2/q^2$ (d=3), and $\pi(q,\omega_\ell)$ is the polarization function given by use of the density-density correlation function, $N(q,\omega_\ell)$ of eq.(2.12). For small q and ω_ℓ

$$\pi(\boldsymbol{q},\omega_l) = -N(\boldsymbol{q},\omega_l)$$

$$= -N(0)\frac{Dq^2}{Dq^2+|\omega_l|} \equiv \pi_0(\boldsymbol{q},\omega_l)\,. \tag{3.28}$$

Equation (3.27) together with eq.(3.28) rsults in

$$v(\boldsymbol{q},\omega_l) = \frac{1}{2N(0)}\,\frac{Dq^2+|\omega_l|}{Dq^2}\,. \tag{3.29a}$$

For each dimension eq.(3.29a) will be valid approximation except when there exists a divergence associated with the q-integration in d=2. In such case q-integrations should be cut off by using more refined expression than eq.(3.29a), i.e.

$$v(\boldsymbol{q},\omega_l) = \frac{2\pi e^2}{q}\,\frac{Dq^2+|\omega_l|}{D\kappa q+|\omega_l|}\,. \tag{3.29b}$$

Here $\kappa=4\pi e^2N(0)$ is the inverse screening radius in d=2. The use of eq.(3.29b) instead of (3.29a) is essentially equivalent to use eq.(3.29a) but with the lower momentum cut off $q \geq q_s = 2\pi T/D\kappa = \pi\tau T a_B/\ell^2$ with $a_B=(me^2)^{-1}$ being the Bohr radius. Note that q_s is very small. By these particular features present in the dynamically screened interaction, we obtain for N'

$$g_1 = \begin{cases} 2\,, & d=3\,, \qquad (3.30a)\\ \ln(D\kappa^2/\tau T^2)\,, & d=2\,, \qquad (3.30b) \end{cases}$$

and for σ'

$$g_1 = \begin{cases} 2/3\,, & d=3\,, \qquad (3.31a)\\ 1\,, & d=2\,, \qquad (3.31b) \end{cases}$$

respectively.

Such corrections have been first recognized by Altshuler and Aronov[14] for N' in d=3, where they considered the g_1-process for the dynamically screened Coulomb interaction, $v(q,\omega_\ell)$, and by Altshuler, Aronov and Lee[8] (g_1- and g_3- processes for $v(q,\omega_\ell)$) and by Fukuyama[9] ($g_1 \sim g_4$ processes for local and instantaneous interaction) independently for σ' in d=2 and by Fukuyama[15] for χ' in d=2.

One important aspect of the interaction processes is that some are sensitive to the magnetic fields, i.e. g_3-processes between opposite spin direction, $g_3(\uparrow\downarrow)$, are suppressed by the spin Zeeman splitting, whereas g_2- and g_4-processes are by the orbital effects.[15-18]

Experimentally the remarkable ℓnT dependence of σ' has first been observed by Dolan and Osheroff[19] in metallic films, which is shown in Fig.13(a). Similar dependence has been observed repeatedly thence in various 2d systems and an example on Cu films[20] is also shown in Fig.13(b). On the other hand the $\sqrt{T}$ dependence in 3d systems has been identified in Si-P by Rosenbaum et al.,[21] as shown in Fig.13(c).

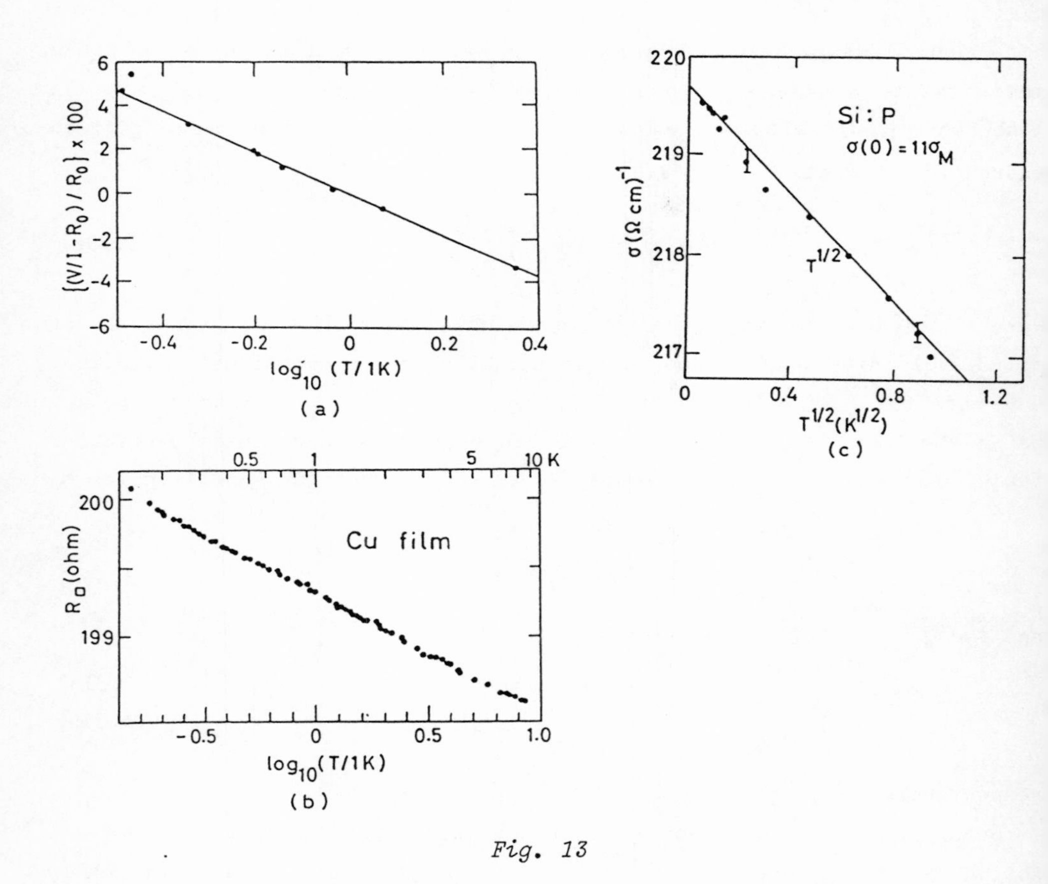

Fig. 13

4.
HIGHER ORDER INTERACTION EFFECTS

So far the combined effects of mutual interactions and the randomness are discussed in the lowest order both in λ and in interactions. Though

λ can be a genuine parameter in the sense that it can be varied by making the samples cleaner or dirtier, the values of interaction constants, g_i, can not be varied arbitrarily but they are rather intrinsic to each system. Moreover the actual values of g_i's are usually not so small and then it is necessary to consider the cases where λ is very small but g_i's are not. Such a problem can be treated by a theory lowest order in randomness but including the higher order contributions in the interaction constants. A theory based on the diagrammatical method,[23,24] will be explained in this section for the model of the dynamically screened Coulomb interaction. As has been discussed in Sec.2, the contributions lowest order in λ should result from processes with as many as $R(q,\omega_\ell)$ but with same momentum, q.

i) g_1-process

One can see that the g_1-process should consist of those of the type shown in Fig.14. Here the screening is affected by F as well as the interaction vertices. The former is known as exchange corrections to the screening in the problem of electron gas, while the latter is also known to be important. The interaction vertex, $\Lambda(q,\omega_\ell)$, is given as follows for $(\varepsilon_n+\omega_\ell)\cdot\varepsilon_n<0$

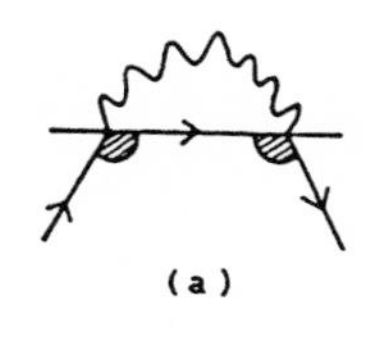

(a)

(b)

(c)

Fig. 14

$$\Lambda(q,\omega_l) = \left[1-\frac{E}{2}\frac{Dq^2}{Dq^2+|\omega_l|}\right]^{-1}\frac{1}{(Dq^2+|\omega_l|)\tau} = \frac{1}{(D'q^2+|\omega_l|)\tau}, \tag{4.1}$$

where $D'=D(1-F/2)$. Here it is noted that the wavy line in **Fig.14(c)** is the screened Coulomb interaction but that the effect of randomness can be ignored since the momentum transfer through the interaction is large. For such interaction processes with large momentum transfer the parametrization by F is employed for simplicity. The factor 1/2 is due to the fact that F is defined, as due to contributions from both spins whereas in the interaction vertex the spin index is fixed. The polarization function in the screening is also affected by F as

$$\pi(\boldsymbol{q},\omega_l)=\pi_0(\boldsymbol{q},\omega_l)\left[1-\frac{F}{4N(0)}\pi_0(\boldsymbol{q},\omega_l)\right]^{-1}$$

$$=-2N(0)\frac{Dq^2}{D'q^2+|\omega_l|}, \tag{4.2}$$

where $\pi_0(q,\omega_\ell)$ is defined by eq.(3.28) and then

$$2N(0)v(\boldsymbol{q},\omega_l)=\frac{D'q^2+|\omega_l|}{Dq^2}. \tag{4.3}$$

These result in the replacement of $N(0)v(q,\omega_\ell)$ by $g_1(q,\omega_\ell)$ defined as follows:

$$g_1(\boldsymbol{q},\omega_l)=\frac{(Dq^2+|\omega_l|)^2}{2Dq^2(D'q^2+|\omega_l|)}. \tag{4.4}$$

By use of this interaction we obtain g_1 in σ' as

$$g_1=\begin{cases}\dfrac{2}{F}\ln\dfrac{2}{2-F}, & d=2, \qquad (4.5)\\[2ex] \dfrac{8}{3F}\left(\dfrac{1}{\sqrt{1-\dfrac{F}{2}}}-1\right), & d=3. \qquad (4.6)\end{cases}$$

ii) g_3-process

The g_3-process should consist of processes shown as Fig.15, which can also be written as Fig.16(a) and (b) similar to the g_1-process. In this way the effective interaction characterizing this process can be introduced as in the g_1-process;

$$v_{\uparrow\uparrow}(\boldsymbol{q},\omega_l)=v_{+-}(\boldsymbol{q},\omega_l), \tag{4.7}$$

$$=-\frac{F}{2}\frac{Dq^2+|\omega_l|}{D'q^2+|\omega_l|}, \tag{4.8}$$

$$=-g_3(\boldsymbol{q},\omega_l), \tag{4.9}$$

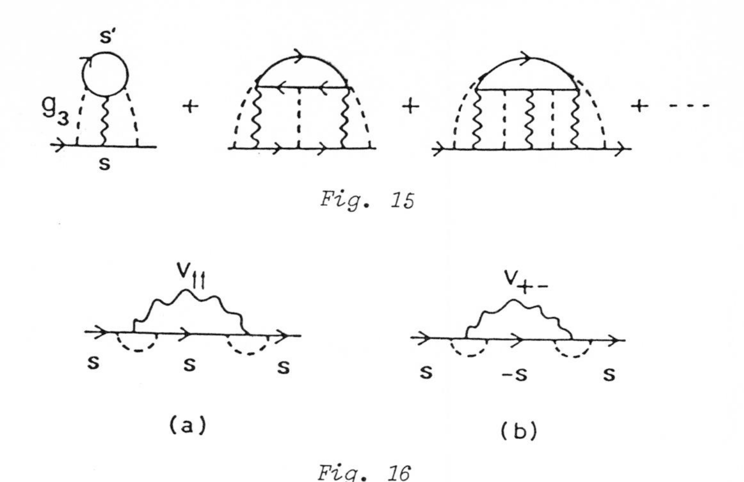

Fig. 15

Fig. 16

The minus sign in eq.(4.8) is due to the redefinition of the Hartree process to the effective Fock process. These effective interactions result in g_3 in σ' as

$$g_3(F)=\begin{cases} 2\left(\dfrac{2}{F}\ln\dfrac{2}{2-F}-1\right), & d=2, \quad (4.10) \\ \dfrac{8}{3F}\left(\dfrac{2}{\sqrt{1-\dfrac{F}{2}}}-2-\dfrac{F}{2}\right), & d=3. \quad (4.11) \end{cases}$$

The same results for σ' from g_1- and g_3-processes have also been obtained by different methods by Finkelstein[25] and Altshuler and Aronov.[26] (For Fermi liquid corrections to make a complete correspondence, see Ref. 24).

iii) g_2- and g_4- processes

Higher order contributions in g_2- and g_4- processes consist of processes as those shown in Fig.17(a) and (b). There exists a cancellation between the g_2-process and the g_4-process with same indices, s=s', and the net result is those of the g_4-process between opposite spins, s=-s', which is a familiar one in the theory of superconductivity.[27] This process is very sensitive to the actual type of interactions and their energy dependences. What is usually assumed in these discussions is the local (i.e. q-independent) interactions whose energy dependences are those shown in Fig.18. Here $-\lambda_{ph}$ is the dimensionless BCS interaction mediated by phonons, ω_D being the Debye frequency, and μ is the effective Coulomb interaction, which will correspond to F/2 for the Coulomb gas, F being

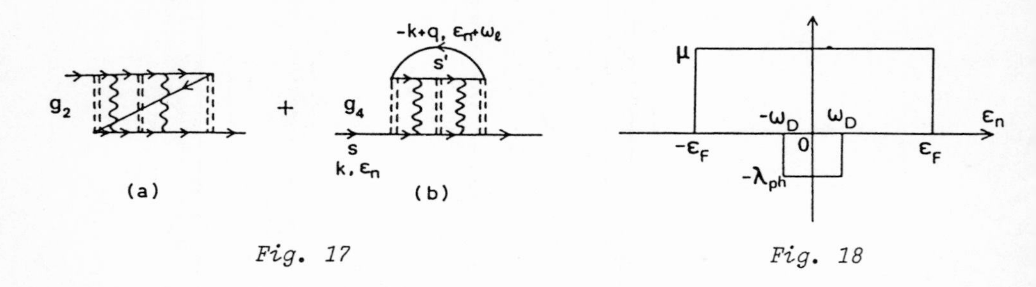

Fig. 17 *Fig. 18*

defined by eq.(3.14). For this model interaction the sum of g_2- and g_4 -processes is characterized by the effective interaction $g_4(q,\omega_\ell)$ defined for Fig.17 as

$$g_2 - 2g_4 = -g_4(q, \omega_l)\,, \tag{4.12}$$

$$g_4(q, \omega_l) = g_s \Big/ \left[1 + g_s\left\{\ln\frac{1.13\omega_D}{T} + \psi\left(\frac{1}{2}\right) - \psi\left(\frac{1}{2} + \frac{Dq^2 + |\omega_l|}{4\pi T}\right)\right\}\right], \tag{4.13}$$

where the effective interaction constant, g_s, is

$$g_s = -\lambda_{ph} + \mu^*\,, \tag{4.14}$$

$$\mu^* = \mu / [1 + \mu \ln \varepsilon_F/\omega_D]\,. \tag{4.15}$$

In eq.(4.15) μ^* is the renormalized Coulomb interaction effective in the low energy region, $|\varepsilon_n| < \omega_D$. This renormalization of Coulomb interaction, which is found by Morel and Anderson,[28] is due to the processes shown in Fig.19, where wavy lines are μ, and solid and thick solid lines are electron Green functions with energies $|\varepsilon_n| < \omega_D$ and $\omega_D < |\varepsilon_n| < \varepsilon_F$, respectively.

The onset of superconductivity is characterized by $g_4(0,0) = \infty$ as the temperature is lowered, which is possible if $g_s < 0$. Hence in the superconducting systems g_2- and g_4-processes are expected to play essential roles.

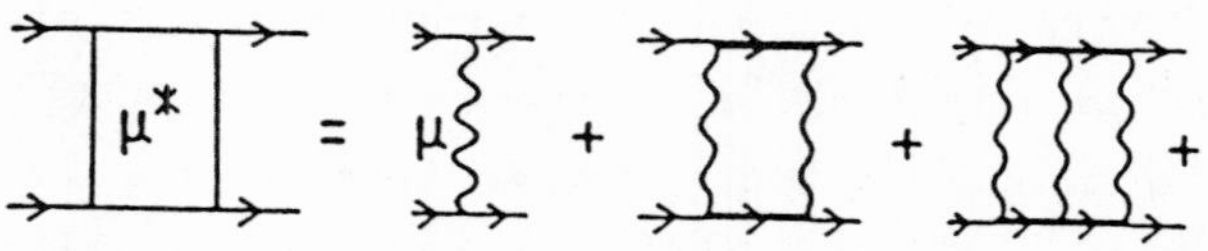

Fig. 19

On the other hand in the non-superconducting systems, i.e. if $g_s>0$, the effective interaction in g_2- and g_4-processes is reduced. This fact has been noted in the present context by Lee and Ramakrishnan,[17] and by Altshuler et al.[29]

Hence for electron gas with Coulomb interaction only contributions from g_1- and g_3-process are to be considered and then conductivity is governed by (g_1-2g_3). In the presence of magnetic field such that $g\mu_B H>T$ this prefactor should be (g_1-g_3) since $g_3(\uparrow\downarrow)$ is suppressed as discussed in Sec.4. It is then possible that the conductivity increases as the temperature is lowered in the absence of field, i.e. $g_1-2g_3<0$, but it decreases once $g\mu_B H>T$, $g_1-g_3>0$. Such has actually been observed,[30] e.g., in GeSb as shown in **Fig.20**. This experiment indicates the stronger temperature dependence than $\sqrt{T}$ in the absence of magnetic field, which is understood as due to the temperature renormalization of the interaction constant.[31]

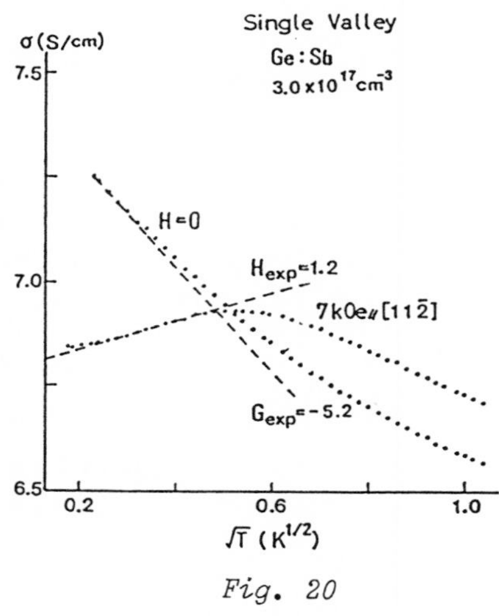

Fig. 20

5.
KONDO EFFECT IN IMPURE METALS

In the preceding section it is demonstrated that, even if the interaction is weakly dependent on energy variable, singular dependences result in various physical quantities. It will be natural to expect that if the interaction has a strong dynamics, i.e. strongly dependent on energy transfer, stronger singularities are possible. The Kondo effect[12] is such an example. The existence of $J^3 \ln\varepsilon_F/T$ (independent of d), even in clean systems is due to the presence of dynamics of a paramagnetic impurity. Here J is the exchange interaction between a spin and the conduction electrons. In this case $v(q,\omega_\ell)$ associated with a paramagnetic impurity is proportional to $|\omega_\ell|^{-1}$. By use of eq.(3.4) it will be easy to see that the Kondo effect in impure metals (i.e. $\lambda \neq 0$) has contributions of the order of $\lambda^{d-1}J^3(\tau T)^{d/2-2}$ in the self-energy function. This fact has been noted earlier for d=3,[32] and examined in detail recently in the present context.[33] It is now demonstrated that not only the self-energy but also the conductivity have same types of singular temperature dependences in contrast to the former claim[34] that there should not be in the conductivity. Recently the higher order effect in J[35] and the effects of mutual interactions between conduction electrons[36] have been examined . This subject deserves further experimental investigations, since there exist little after the pioneering work by Hasegawa.[37]

6.
EFFECTS OF RANDOMNESS ON SUPERCONDUCTIVITY

It had been a general belief that the superconductivity is insensitive to normal impurity, which is actually the case in the usual treatment of dirty superconductors.[38,39] However once the quantum effects associated with impurity scattering in the presence of mutual interaction are taken into account it is no longer the case. In superconductivity attractive interactions mediated by phonons and repulsive Coulomb interactions are competing. Usually the latter is represented by μ^*,[28] eq.(4.15), constant in energy and momentum transfer. The constancy of μ^* is valid in clean systems where the screening will be complete. On the other hand once electrons start to diffuse around the screening at small momenta and energies is no longer effective as seen in eqs.(3.28) and (3.29), and there results the large enhancement of the effective repulsive interactions. Although the attractive interaction is also enhanced, its effect is relatively small because it is associated mainly with large momentum. The net result is that the repulsive interaction is more enhanced leading to the suppression of the superconductivity.

Formally the onset of the superconductivity is characterized by the divergence of the propagator, $\mathcal{D}$, of the Cooper pair as defined by Fig.21,

$$\mathcal{D} = -g_s/[1+g_sK(T)], \tag{6.1}$$

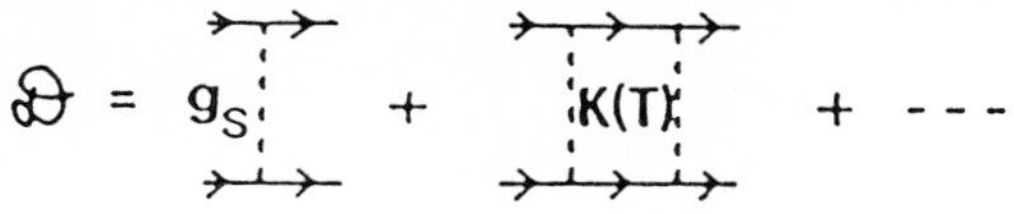

Θ = g_S + K(T) + - - -

Fig. 21

where g_s (<0) is given by eqn.(4.14) and K(T) is the pair polarization function, which is irreducible in the sense that it cannot be separated into two parts by cutting one g_s-line. In clean systems K(T) is given by

$$K(T)=2\pi T \sum_{\omega_D>\varepsilon_n>0} \frac{1}{\varepsilon_n},$$

$$\equiv K_0(T)\,. \tag{6.2}$$

Even in dirty systems K(T) is given by $K_0(T)$ if quantum corrections are not taken into account. This corresponds to the Anderson theorem. However the foregoing discussions would indicate that there exist quantum corrections, K'(T), to $K_0(T)$ and then the

$$\ln\frac{T_c}{T_{c0}}=K'(T_c)\,, \tag{6.3}$$

where T_{c0} is given by $1-g_s K_0(T_{c0})=0$. Detailed examinations on K'(T)[40-42] have revealed that there exist three different kinds of effects associated with this enhanced Coulomb interactions (i) the suppression of the one-particle density of state (ii) the suppression of the total attractive forces and (iii) the depairing effects due to the retardation of the interaction.

These theoretical predictions have been tested satisfactorily for metallic films, where the critical temperature, T_c, is seen to decrease in proportion to the inverse of the thickness, b, as seen in various experiments[43-45] shown in Fig.22. Here $R_\square=(\sigma b)^{-1}$ is the sheet resistance. More or less similar effects have been observed earlier[46] but causes had not been identified then unambiguously because of the less control of the sample quality than is now available. In these recent experiments it is confirmed that the conductivity is essentially unchanged if the thickness is varied and is independent of the substrate.

Quite recently[47] the case of thin wires has theoretically been

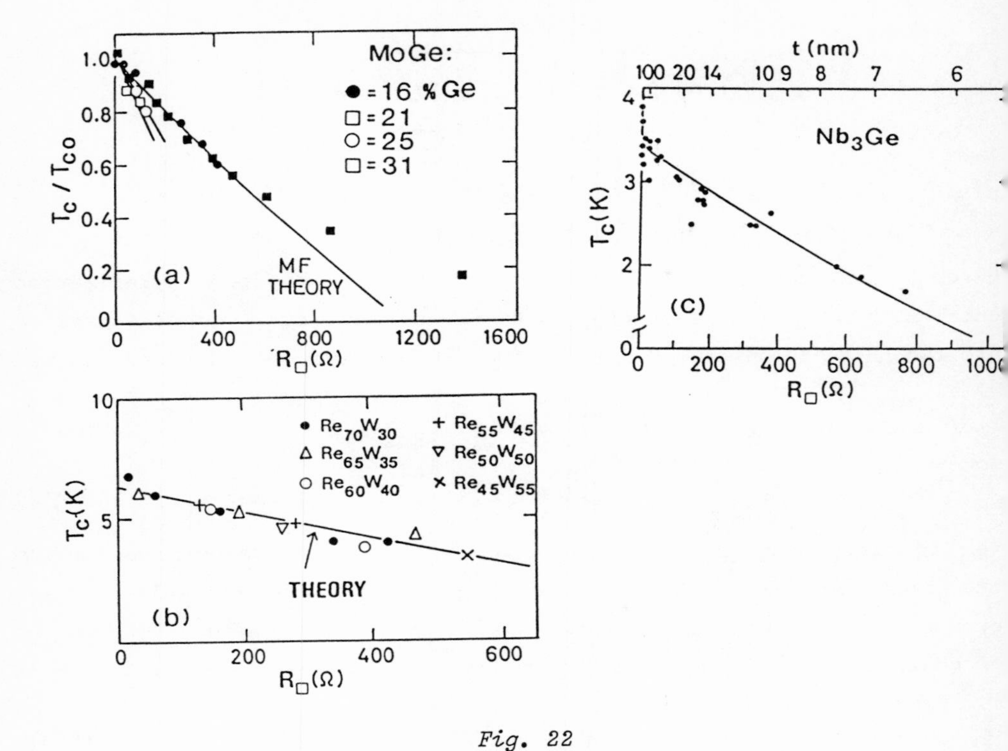

Fig. 22

examined where the present effects are naturally expected to be mor pronounced.

We note that the three-terminal Josephson device with MOS structure a schematically shown in Fig.23 suggested a while ago [48] and realized ver recently [49,50] is also interesting from the viewpoint of localiza tion. [51]

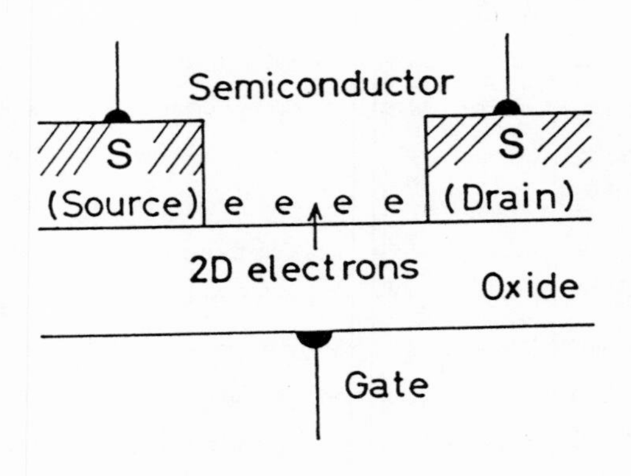

Fig. 23

7.
EFFECTS OF RANDOMNESS ON ITINERANT MAGNETISM

It will be natural to expect that the itinerant magnetism,[52] which is another typical example of phase transition driven by mutual interactions, is also affected by Anderson localization. Since it is due essentially to the repulsive interaction, effects can be more direct. This has been clarified for electrons with (long range) Coulomb interactions[53,54] relevant to doped semiconductors and for the Hubbard model[55] more applicable to such magnetic metals as transition elements and their alloys. As regards the spin susceptibility these different models predict essentially the same. For our explicit discussions we will examine the Hubbard model with the random potential on sites.

$$H=\sum_{k,s}\varepsilon(\boldsymbol{k})\,a^{+}_{ks}a_{ks}+U\sum_{i}n_{i\uparrow}n_{i\downarrow}+v_0\sum_{\{l\}}n_l\,, \tag{7.1}$$

where R_ℓ is the impurity site and $n_\ell=n_{\ell\uparrow}+n_{\ell\downarrow}$. We assume in the following that the band is far from half-filled and then the model is essentially the same as electron gas in the random potential together with short range (not long-range Coulomb) spin-dependent repulsive interaction. We will particularly be interested in the paramagnetic region of ferromagnetism. For this model the spin susceptibility in the paramagnetic phase, χ, is given by[56]

$$\bar{\chi}(q,\omega_l)=\frac{\chi_0(q,\omega_l)}{1-U\chi_0(q,\omega_l)}\,, \tag{7.2}$$

where χ_a and χ_b are irreducible parts of $\chi^{\uparrow\uparrow}$ and $\chi^{\uparrow\downarrow}$, respectively, i.e. sum of the processes which can not be separated into two parts by cutting one U-line. Here $\chi^{ss'}$ is defined in Fig.24. In RPA, χ_a and χ_b are approximated by $\chi_a=\chi_0(q,\omega_\ell)$, $\chi_b=0$. In RPA the spin fluctuations are independent for different q and ω_ℓ and called paramagnon.[57-59] The natural extention of the theory will be to take account of the mode-mode coupling between spin fluctuations. In these procedures the self-consistency will be important. Such self-consistent renormalization (SCR) scheme has been proposed some time ago[60,61] and elaborated by Moriya and collaborators[62]. This theory takes into account processes shown in

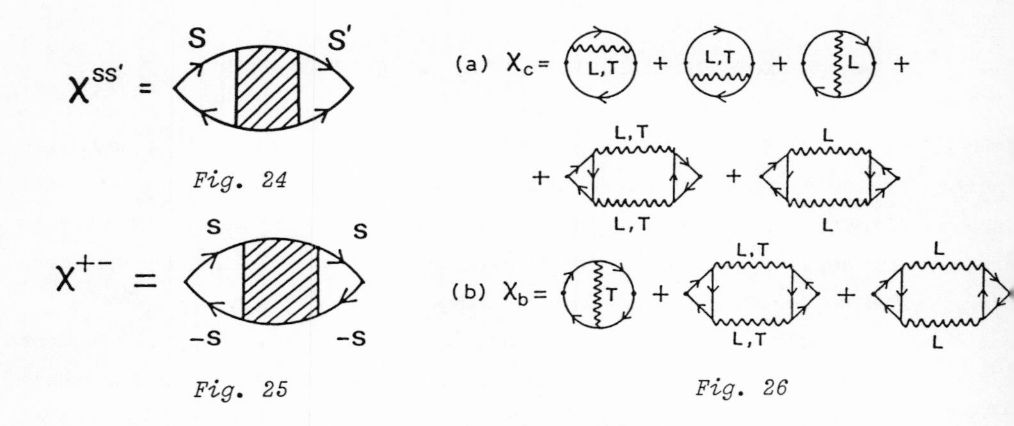

Fig. 24

Fig. 25

Fig. 26

Fig.26 for χ_a and χ_b, where $\chi_c=\chi_a-\chi_0$ and wavy lines are spin fluctuations. In this figure L and T indicates that these are effective interaction which are related to the longitudinal ($\chi^{ss'}$) and transverse (χ^{+-}) spi fluctuations, the latter being defined in Fig.25. These make self-consistent scheme for spin fluctuations, leading to the followin condition for the ferromagnetic instability

$$1-UN(0)+\gamma_0(0)+A(T_{c0}/\varepsilon_F)^{4/3}=1-UN(0)+\gamma_0(T_{c0}), \qquad (7.3$$

where $\gamma_0(0)$ is some constant of order unity weakly dependent on temper ature and represents the renormalization effect due to quantum fluctua tions, whereas the last term in eq.(7.3) is due to thermal fluctuation with A being a positive constant of order unity. In eq.(7.3) T_{c0} is th Curie temperature in clean systems.

As regards the effect of random impurity potential on spin fluctuation there already exist some investigations. As early as in 1960's th dynamical susceptibility has been discussed in the context of paramagnon in dirty metals.[63,64] These discussions were, however, within RPA, i.e $\chi_0(q,\omega_\ell)$ was properly replaced by that in dirty systems which is c diffusive character

$$\chi_0(\boldsymbol{q},\omega_l)=N(0)\frac{Dq^2}{Dq^2+|\omega_l|}. \qquad (7.4$$

This can be derived in a same way as for eq.(2.12). In this scheme th condition of the ferromagnetic instability is given by 1-UN(0)=0 and i

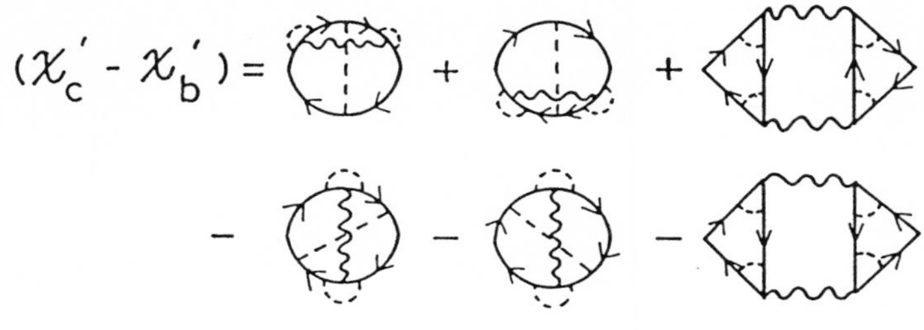

$$(\chi_c' - \chi_b') = \ldots$$

Fig. 27

ot affected by randomness. This corresponds to ignore the effects of nite λ.

The construction of the framework of weak itinerant ferromagnetism in he weakly localized regime is now straightforward;(55) this akes into account both finite λ and mode-mode coupling between spin uctuations. This can be achieved by taking into account the quantum orrections due to diffusions on the various processes in **Fig.26**. The uantum corrections to these, defined as χ_c' and χ_b', respectively, are ven by the processes shown in **Fig.27**. Note that only the transverse uctuations, χ^{+-}, are contributing to χ_c' and χ_b'. Calculations of these ocesses lead to

$$\chi = N(0) / \eta(T) \quad , \tag{7.5}$$

$$\eta(T) = 1 - u + \gamma_0(T) - \gamma'(T) \quad , \tag{7.6}$$

$$\gamma'(T) = G_s \lambda^{d-1} \phi_d(T) \quad , \tag{7.7}$$

ere $u=UN(0)$, $G_s=4u^2(1-u/2)$ (estimated for $\eta \to 0$) and $\phi_d(T)$ are essentially e same as those defined in eqs.(3.22) and (3.23). The present result dicates that in d=2 $\chi \to \infty$ at some finite temperature if exterpolated terally but more elaborate treatment(65) based on the renormalization oup treatment yields $\chi \sim T^{-\gamma}$, $\gamma=4/3$, once the temperature gets lower than me characteristic cross-over temperature as schematically shown in ig.28. This result is very remarkable: in d=2 metals in the presence of th randomness and repulsive forces become always magnetic in the limit low temperature. For the moment there is no clear experimental tests r this because of the experimental difficulty. In the case of d=3, eories indicate the existence of critical value of randomness in the etallic region for the divergence of χ at T=0, as schematically shown in

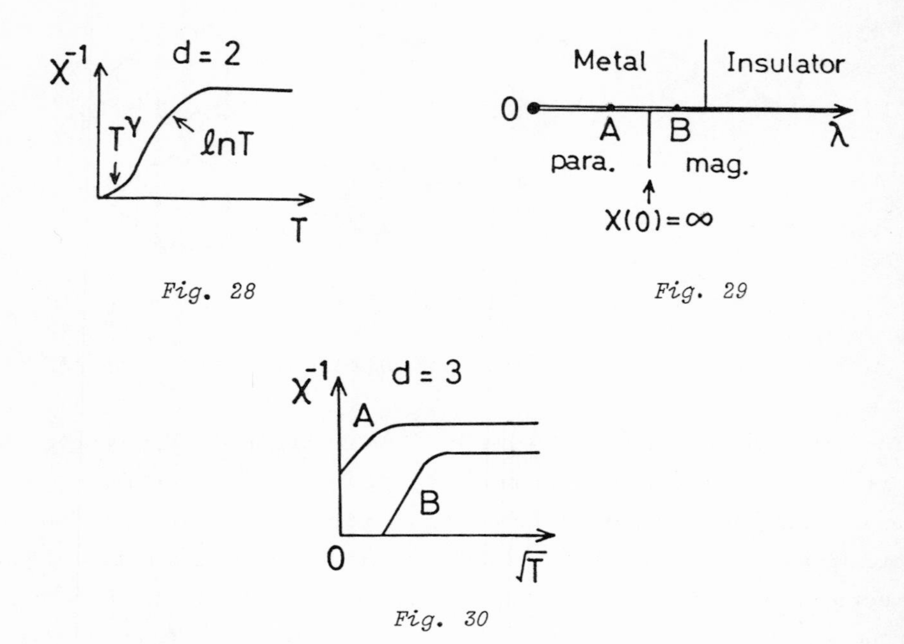

Fig. 28

Fig. 29

Fig. 30

Fig.29. In each region corresponding to the points A and B in Fig.29 t temperature dependences of χ^{-1} are expected as in Fig.30.

Quantitative aspects of these theoretical predictions have already be confirmed experimentally.(66-68)

8.
SUMMARY

In this lecture a brief review is given on the effect of mutu interaction in weakly localized regime, where the genuine perturbatior treatment of the randomness is valid. It is demonstrated that t interaction effects are enhanced by randomness leading to singul temperature dependences in various physical quantities dependent on t dimensionality. In normal metals ℓnT and $\sqrt{T}$ dependences have be predicted theoretically for conductivity and those have confirn experimentally. In the case of the Kondo effect $T^{d/2-2}$-dependence h been predicted for conductivity, whose stronger temperature dependence the result of the existence of dynamics of a paramagnetic impurity coupl to the diffuson processes.

These investigations on normal states have naturally resulted in the theoretical predictions of strong modification of the onset condition of two typical phase transitions driven by mutual interactions, superconductivity and magnetism, for which there already exist some experimental confirmations.

It will be concluded that the transport and magnetic measurements performed on the same sample and at low temperatures will further reveal the fascinating possibilities of interaction processes in itinerant electrons under the influence of random potential.

REFERENCES

. e.g. P. Nozieres, "Theory of Interacting Fermi Systems" (Benjamin, 1964).

. E. Abrahams, P.W. Anderson, D.C. Licciardello and T.V. Ramakrishnan, Phys. Rev. Lett. 42, 673 (1979).

. "Anderson Localization" Ed. Y. Nagaoka and H. Fukuyama, (Springer Verlag, 1982).

. P.A. Lee and T.V. Ramakrishnan, Rev. Mod. Phys. 57, 287 (1985).

. "Localization Interaction and Transport Phenomena" Ed. B. Kramer, G. Bergmann and Y. Bruynseraede (Springer Verlag, 1985).

. "Anderson Localization" Ed. Y. Nagaoka, Suppl. Prog. Theor. Phys. 85 (1985).

. P.W. Anderson, E. Abrahams and T.V. Ramakrishnan, Phys. Rev. Lett. 43, 718 (1979).

. B.L. Altshuler, A.G. Aronov and P.A. Lee, Phys. Rev. Lett. 44, 1288 (1980).

. H. Fukuyama, J. Phys. Soc. Jpn. 48, 2169 (1980).

). B.L. Altshuler and A.G. Aronov, "Electron-Electron Interaction in Disordered Systems" Ed. A.L. Efros and M. Pollak (North Holland, 1985) p.1.

. H. Fukuyama, p.155 of the same book as ref.10.

. J. Kondo, Prog. Theor. Phys. 32, 37 (1964).

. "Handbook of Mathematical Functions" Ed. M. Abramowitz and I.A. Stegun (Dover, 1964) p.253.

. B.L. Altshuler and A.G. Aronov, Solid State Commun. 30, 115 (1979).

15. H. Fukuyama, J. Phys. Soc. Jpn. **50**, 3407 (1981).
16. A. Kawabata, J. Phys. Soc. Jpn. **50**, 2416 (1981).
17. P.A. Lee and T.V. Ramakrishnan, Phys. Rev. **B26**, 4009 (1982).
18. Y. Isawa, K. Hoshino and H. Fukuyama, J. Phys. Soc. Jpn. **51**, 3262 (1981).
19. G.J. Dolan and P.P. Osheroff, Phys. Rev. Lett. **43**, 721 (1979).
20. S. Kobayashi, F. Komori, Y. Ootuka and W. Sasaki, J. Phys. Soc. Jpn. **49**, 1635 (1980).
21. T.F. Rosenbaum, R.F. Milligan, G.A. Thomas, P.A. Lee, T.V. Ramakrishnan, R.N. Bhatt, K. De Conde, H. Hess and T. Perry, Phys. Rev. Lett. **47**, 1758 (1980).
22. For experimental review, G. Bergmann, Phys. Rept. **107**, 1 (1984).
23. H. Fukuyama, Y. Isawa and H. Yasuhara, J. Phys. Soc. Jpn. **52**, 16 (1983).
24. Y. Isawa and H. Fukuyama, J. Phys. Soc. Jpn. **53**,1415 (1984).
25. A.M. Finkelstein, Sov. Phys. JETP **57**, 97 (1983).
26. B.L. Altshuler and A.G. Aronov, Solid State Commun. **46**, 429 (1983).
27. J.R. Schrieffer, "Theory of Superconductivity" (Benjamin, 1964).
28. P. Morel and P.W. Anderson, Phys. Rev. **125**, 1263 (1962).
29. B.L. Altshuler, A.G. Aronov, D.E. Khmelnitzkii and A.I. Larkin, Sov. Phys. JETP **54**, 4111 (1981).
30. Y. Ootuka, S. Kobayashi, S. Ikehata, W. Sasaki and J. Kondo, Solid State Commun. **30**, 169 (1979).
31. C. Castellani, C. Di Castro, H. Fukuyama, P.A. Lee and M. Ma, Phys. Rev.**B33**, 7277 (1986).
32. H.U. Everts and J. Keller, Z. Phys. **240**, 281 (1970).
33. F.J. Ohkawa, H. Fukuyama and K. Yosida, J. Phys. Soc. Jpn. **52**, 1701 (1983); F.J. Ohkawa and H. Fukuyama, ibid. **53**, 2640 (1984).
34. K.P. Bohnen and K.H. Fisher, J. Low. Temp. Phys. **12**, 559 (1973).
35. S. Suga, H. Kasai and A. Okiji, J. Phys. Soc. Jpn. **55**, #7 (1986).
36. H. Fukuyama, J. Phys. Soc. Jpn. **55**, #7 (1986).
37. R. Hasegawa, Phys. Rev. Lett. **28**, 1376 (1972).
38. P.W. Anderson, J. Phys. Chem. Solids **11**, 26 (1959).
39. L.P. Gorkov, Sov. Phys. JETP **10**, 998 (1960).
40. S. Maekawa and H. Fukuyama, Physica **107B**, 123 (1981); J. Phys. Soc. Jpn. **51**, 1380 (1982).

1. H. Ebisawa, S. Maekawa and H. Fukuyama, J. Phys. Soc. Jpn. **54**, 2257 (1985).
2. H. Fukuyama, Physica **126B**, 306 (1984); **135B**, 458 (1985).
3. J. Graybeal and M.R. Beasley, Phys. Rev. **B29**, 4167 (1984).
4. H.R. Raffy, R.B. Laibowitz, P. Chaudhari and S. Maekawa, Phys. Rev. **B28**, 6607 (1983).
5. P.H. Kes, C.C. Chi and C.C. Tsuei, Proc. Int. Conf. Localization and Transport Phenomena in Impure Metals, Ed. L. Schweitzer and B. Kramer (PTB, Braunschweig, 1984).
6. M. Stongin, R.S. Thompson, O.F. Kammerer and J.E. Crow, Phys. Rev. **B1**, 1078 (1970).
7. H. Ebisawa, H. Fukuyama and S. Maekawa, J. Phys. Soc. Jpn., **55** (1986).
8. T.D. Clark, Ph. D. Thesis (Univ. of London, 1971), T.D. Clark, R.J. Prance and A.D. Grassie, J. Appl. Phys. **51**, 2735 (1980).
9. H. Takayanagi and T. Kawakami, Phys. Rev. Lett. **54**, 2449 (1985).
0. T. Nishino, M. Miyake, Y. Harada and U. Kawabe, IEEE EDL-6, 297 (1985).
1. H. Fukuyama and S. Maekawa, J. Phys. Soc. Jpn. **55**, #6 (1986).
2. e.g. C. Herring, "Magnetism IV" Ed. T. Rado and H. Suhl (Academic Press, 1966).
3. A.M. Finkelstein, Z. Phys. **B56**, 189 (1984).
4. C. Castellani, C. Di Castro, P.A. Lee and M. Ma, Phys. Rev. **B30**, 527 (1984).
5. H. Fukuyama, J. Phys. Soc. Jpn. **54**, 2092 (1985); J. Mag. Magn. Mat. **54-57**, 1437 (1986).
6. A. Kawabata, J. Phys. **F4**, 1477 (1974).
7. N. Berk and J.R. Schrieffer, Phys. Rev. Lett. **17**, 433 (1966).
8. S. Doniach and S. Engelsberg, Phys. Rev. Lett. **17**, 750 (1966).
9. M.T. Beal-Monod, S.-K. Ma and D.R. Fredkin, Phys. Rev. Lett. **20**, 929 (1968).
0. K.K. Murata and S. Doniach, Phys. Rev. Lett. **29**, 285 (1972).
1. T. Moriya and A. Kawabata, J. Phys. Soc. Jpn. **34**, 639 (1973).
2. T. Moriya, Physica **91B**, 235 (1977); J. Magn. Magn. Mater. **14**, 1 (1979).
3. P. Fulde and A. Luther, Phys. Rev. **170**, 570 (1968).

64. S. Engelsberg, W.F. Brinkman and S. Doniach, Phys. Rev. Lett. 20, 1040, (1968).
65. C. Castellani, C. Di Castro, P.A. Lee, M. Ma, S. Sorella and E. Tabet Phys. Rev. B30, 1596 (1984); ibid B33, 6169 (1986).
66. M.A. Paalanen, A.E. Ruckenstein and G.A. Thomas, Phys. Rev. Lett. 54, 1295 (1985).
67. M.A. Paalanen, S. Sachdev, R.N. Bhatt and A.E. Ruckenstein, preprint.
68. S. Ikehata, M. Fujita and S. Kobayashi, private communications.

PHYSICS OF WEAK LOCALIZATION

Gerd Bergmann

University of Southern California

Disordered conductors show in their resistance deviations from Boltzmann-theory (weak localization) because the conduction electrons are coherently backscattered by the impurities and form an echo which decays as $1/t^{-d/2}$ with time (d is the dimension of the conductor). The intensity of the echo (integrated over time) can be easily measured by the resistance. A magnetic field suppresses the echo after a "flight"-time proportional to 1/H - and one can determine the inelastic lifetime, the spin-orbit and the magnetic scattering time of the conduction electrons in the pico-second range.

I. INTRODUCTION

Disordered metals, in particular thin films, show resistance anomalies which were theoretically not understood until a few years ago. In this talk we want to discuss one origin of the anomaly which is generally called weak localization. This effect has been first pointed out by Abrahams et al.(1) (see for example the review-articles 2-5). Weak localization exists in one, two and three dimensions as well but for an experimental investigation the two-dimensional case is the most favourable one. Here the correction of the resistance is of the order of 10^{-2} to

10^{-3} and can be easily measured with an accuracy of 1%. One can in particular investigate weak localization in two dimensions in a magnetic field perpendicular to the film (which is not possible in one dimension). The following discussion of weak localization applies to all dimensions but when we have to write equations we take first the two-dimensional cas and turn only later to the three dimensional one.

In disordered metals the conduction electrons are scattered by the impurities. If we consider the conduction electrons as plane waves then the scattered waves propagate in all directions. The usual Boltzmann-theory neglects interferences between the scattered partial waves and assumes that the momentum of the electron wave disappears exponentially after the elastic scattering time τ_0. The neglect of the interference is, however, not quite correct. There is a coherent superposition of the scattered electron wave which results in back-scattering of the electron wave and lasts as long as its coherence i not destroyed. This causes a correction to the conductance which is generally calculated in the Kubo-formalism by evaluating "Kubo-graphs". The most important correction has already been discussed by Langer and Neal[6] in 1966 and is shown in Fig.1a. Anderson et al.[7] and Gorkov et al.[8] showed that at low but finite temperature the conductance of a two dimensional electron system has a correction

$$\Delta L = -\Delta R/R^2 = L_{oo} \; \ell n(\tau_i/\tau_o); \; L_{oo} = e^2/(2\pi^2\hbar) \qquad (1$$

where τ_i is the inelastic lifetime. This correction is temperature dependent because the inelastic lifetime depends on the temperature (for example $1/\tau_i \propto T^P$).

II.

THE ECHO OF A SCATTERED CONDUCTION ELECTRON

We consider at the time t=0 an electron of momentum k which has the wave function exp[ikr]. The electron in state k is scattered after the time into a state k'_1, after $2\tau_o$ into the state k'_2 etc. There is a finite probability that the electron will be scattered into the vicinity of the state -k; for example after n scattering events. This scattering sequen (with the final state -k)

$$k \rightarrow k'_1 \rightarrow k'_2 \rightarrow .. \rightarrow k'_{n-1} \rightarrow k'_n = -k \tag{2a}$$

is drawn in Fig. 1b in k-space. The momentum transfers are $g_1, g_2 \cdots g_n$. There is an equal probability for the electron k to be scattered in n steps from the state k into -k via the sequence

$$k \rightarrow k''_1 \rightarrow k''_2 \rightarrow .. \rightarrow k''_{n-1} \rightarrow k''_n = -k \tag{2b}$$

where the momentum transfers are $g_n, g_{n-1}, \cdots g_1$. This complementary scattering series has the same changes of momentum in opposite sequence. If the final state is -k, then the intermediate states for both scattering processes lie symmetric to the origin. The important point is that the amplitude in the final state -k is the same for both scattering sequences.

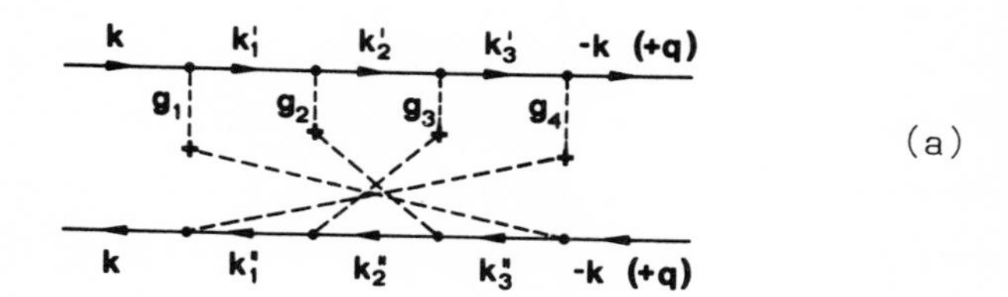

(a)

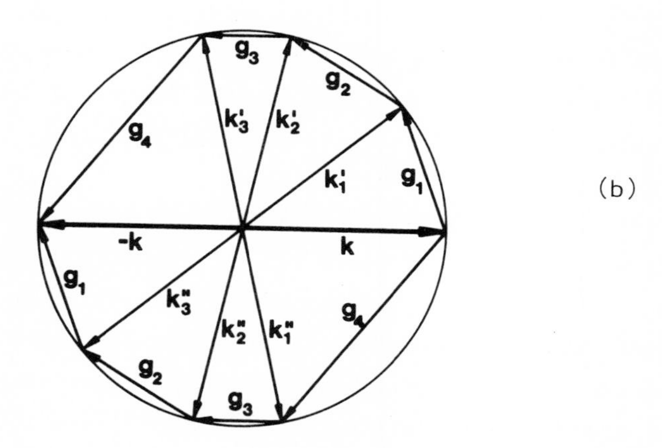

(b)

Fig. 1

Since the final amplitudes A' and A" are phase coherent and equal, A'=A"=A, the total intensity is $|A' + A"|^2 = |A'|^2 + |A"|^2 + A'^*A" + A'A$ $= 4|A|^2$. If the two amplitudes were not coherent then the total scattering intensity of the two complementary sequences would only be $2|A|^2$. This means that the scattering intensity into the state -k is by $2|A|^2$ larger than in the case of incoherent scattering. This additional scattering intensity exists only in the back-scattering direction.

At finite temperature the scattering processes are partially inelasti As a consequence the amplitudes A' and A" lose their phase coherence (after the time τ_i) and the coherent back-scattering disappears after τ_i The integrated momentum of the electron k decreases with increasing τ_i. In the following we treat this scattering semi-quantitatively.

After the elastic lifetime τ_o the electron k is scattered into a shel at the Fermi-surface which is assumed to contain Z intermediate states. The amplitude in the intermediate state k'_1 is $Z^{-1/2}e^{i\delta_1}$ where $e^{i\delta_1}$ is essentially given by $iV(g_1)/|V(g_1)|$. The intensity in the next intermediate state k'_2 at the time $2\tau_o$ is Z^{-2}. After n scattering processes the intensity in the final state -k is Z^{-n} and the amplitude $Z^{-n/2}\ e^{i\Sigma\ \delta_\nu}$. The second scattering series yields the same amplitude. The cross product or interference term is $A'^*A"+A'A"^*=2\ Z^{-n}$. Now we hav to sum over all possible intermediate states. This yields the factor 1/ Z^{n-1}. (1/2 occurs because the two complimentary series appear twice in the sum). Therefore the coherent additional back-scattering intensity i Z^{-1}. It is independent of the number of intermediate scattering states and equal to the scattering intensity from k into k'_1. This intensity i of course, completely calculated in evaluating the diagram with the appropriate rules. However, one can easily estimate this intensity in a rather direct and less formal manner.

For the calculation of Z in two dimensions we consider the scattering from the state k into the state k'_1. This state is an intermediate stat for the scattering sequence which does not have to conserve the energy (sometimes called virtual scattering process). Since the elastic lifeti is τ_o the intermediate state can lie within $\pi\hbar/\tau_o$ of the Fermi energy. This corresponds to a smearing of the Fermi-sphere by π/ℓ (ℓ=mean free path of the conduction electrons). Therefore the available area in k-space is $2\pi k_F*\pi/\ell=2\pi^2k_F/\ell$ and Z is obtained by multiplying by the density of states in k-space, i.e. $(2\pi)^{-2}$.

The coherent back-scattering is not restricted to the exact state -k; one has a small spot around the state -k in momentum space which contributes. We calculate the coherent back-scattering intensity into the state -k+q which is reached after n scattering processes with the transfer of momentum g_i where $\Sigma\, g_i = -2k_F + q$. The sum of the momenta of the initial and final states is +q. The same applies for each pair of scattering states in fig. 1b which lie opposite to the centre, i.e. $q = k'_1 + k''_{n-1} = k'_2 + k''_{n-2} = \ldots$ The corresponding intermediate states differ not only in momentum but also in their kinetic energy (which need not be conserved). The energy difference is $\hbar q{*}v_F$ and since the phase rotates with $Et/\hbar$ one obtains during the time τ_o a phase difference between the two complementary waves which is $q{*}v_F\tau_o$. The important fact is that the different intermediate states have independent directions of momentum. Therefore the phase differences are independent in sign and value. This means that only the square of the phase shifts adds. Therefore after the n scattering processes one obtains phase differences between the complementary waves whose width is

$$[\Delta\phi]^2 = n\overline{(q{*}v_F^2)} = n\,\frac{1}{2}\,(v_F q)^2 = nD\tau_o \qquad (3)$$

In two dimensions the average over $(v_F{*}q)^2$ is $(v_F q)^2/2$ (and in three dimensions $(v_F q)^2/3$ but the diffusion constant D absorbs the factor of dimension). The neighbouring states of -k contribute less to the coherent back-scattering because they lose the phase coherence with increasing n and q. Their contribution is proportional to $\exp[-Dq^2 t]$ since $t = n\tau_o$. The area of the spot for the coherent back-scattering is obtained by integration over q. In two dimensions this yields $\pi/(Dt)$. This corresponds to about $\pi(Dn\tau_o)^{-1}/(2\pi)^2$ states, i.e. their number shrinks with time. Therefore the portion of coherent back-scattering is given by

$$\begin{aligned} I_{coh} &= [\pi/(Dt)]/[2\pi^2 k_F/\ell] \\ &= \tau_o/(\pi k_F \ell t) = \hbar/(2\pi E_F t) \end{aligned} \qquad (4)$$

In the presence of an external electrical field the conduction electrons contribute to the current. However, the echo, i.e., the coherent back-scattering reduces the current and therefore the conductance. A pulse of an electrical field generates a short current (for the time τ_o) in the direction of the electric field and then a reversed current which decays as 1/t. The dc conductance is obtained by integrating the moment over time. For the normal contribution this yields $k\tau_o$ and for the echo $[k\tau_o/(\pi k_F \ell)] \ln(\tau_i/\tau_o)$. Therefore the electron in the state k contribut to momentum

$$k\tau_o[1-1/(\pi k_F \ell) \ln(\tau_i/\tau_o)] \tag{5}$$

The contribution of the electron k to the current is reduced by the fact in the brackets and the conductance is decreased by the same factor.

$$\begin{aligned} L &= (ne^2\tau_o/m)*[1-1/(\pi k_F \ell) \ln(\tau_i/\tau_o)] \\ &= (ne^2\tau_o/m) - e^2/(2\pi^2\hbar)*\ln(\tau_i/\tau_o) \end{aligned} \tag{6}$$

with $n = 2\pi k_F{}^2/(2\pi)^2$. This correction to the conductance was introduced by Anderson et al. and Gorkov et al.

The important consequence of the above consideration is that the conduction electrons perform a typical interference experiment. The (incoming) wave k is split into two complementary waves k'_1 and k''_1. Th two waves propagate individually, experience changes in phase, spin orientation, etc. and are finally unified in the state -k where they interfere. The intensity of the interference is simply measured by the resistance. In the situation which has been discussed above the interference is constructive in the time interval from τ_o to τ_i.

III.
TIME OF FLIGHT EXPERIMENT BY A MAGNETIC FIELD

One of the interesting possibilities for an interference experiment is t shift the relative phase of the two interfering waves. For charged particles this can be easily done by an external magnetic field. In a magnetic field the phase coherence of the two partial waves is weakened destroyed. In real space the two partial waves propagate on a closed lo in opposite direction. When the two partial waves surround the area F containing the magnetic flux Φ, then the relative change of the two phases is $(2e/\hbar)\Phi$. The factor of 2 arises because the two partial waves surround the area twice.

Altshuler et al.[9] suggested performing such an "interference experiment" with a cylindrical film in a magnetic field parallel to the cylinder axis. Then the magnetic phase shift between the complementary waves is always a multiple of $2e\Phi/\hbar$ (Φ=flux in the area of the cylinder). Sharvin and Sharvin[10] showed in a beautiful experiment that the resistance of a hollow Mg cylinder oscillates with a flux period of $\Phi=h/(2e)$. This is shown in Fig. 2. However, for a thin film in a

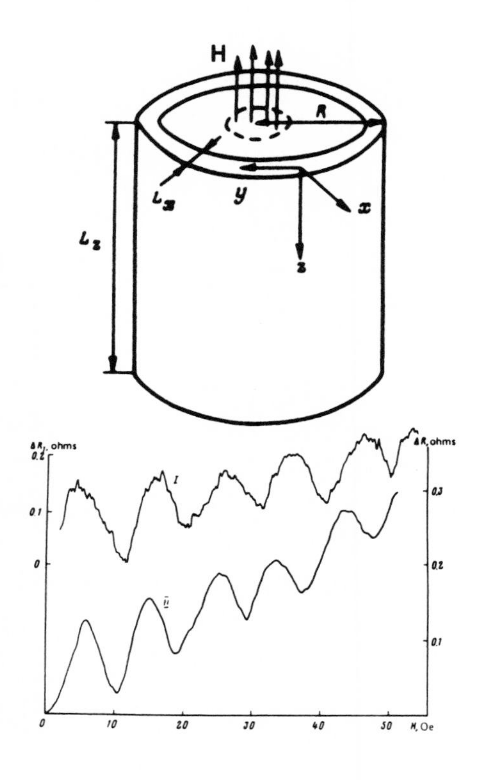

Fig. 2

perpendicular magnetic field the pairs of partial waves enclose areas between -2Dt and 2Dt. When the largest phase shift exceeds 1, the interference is constructive and destructive as well and the average cancels. This happens roughly after the time $t_H=\hbar/(4eDH)$. This means essentially that the conductance correction in the field H i.e. $\Delta L(H)$ yields the coherent back-scattering intensity integrated from τ_o to T_H

$$\Delta L(H) \propto \int_{\tau_o}^{t_H} I_{coh} dt \propto -L_{oo} \ \ell n(t_H/\tau_o) \quad (7)$$

This means that the magnetic field allows a time of flight experiment. I a magnetic field H' is applied the contribution of coherent back scattering is integrated in the time interval between τ_o and $t_{H'}=\hbar/(4eDH')$. If one reduces the field from the value H' to the value H" and measures the change of resistance this yields the contribution of the coherent back-scattering in the time interval $t_{H'}$ and $t_{H''}$. Since the magnetic field introduces a time t_H into the electron system all characteristic times τ_n of the electrons can be expressed in terms of magnetic fields H_n

$$\tau_n \Longleftrightarrow H_n$$

where $\tau_n * H_n = \hbar/(4eD)$. In a thin film this is given by $\hbar e\rho N/4$ which is of the order of 10^{-12} to 10^{-13} Ts (ρ=resistivity of the film and N=density o electron states for both spin directions).

The exact formula for the magneto-resistance by Hikami et al.[11] is

$$\Delta L_{wl}/L_{oo} = (3/2)f_2(H_T/H) - (1/2)f_2(H_S/H) \quad (8a$$

where $f_2(x) = \ell n(x)+\Psi(1/2+1/x)$, Ψ is the digamma function and H is the applied field. The H_S and H_T are defined in the following manner

$$H_S = 2*H_s+H_i$$

$$(8b$$

$$H_T = 4/3*H_{so}+2/3*H_s+H_i$$

where H_i corresponds to the inelastic life-time τ_i, H_{so} to the spin-orbit coupling time τ_{so} and H_s to the magnetic scattering time τ_s.

Magneto-resistance measurements on thin films have been performed by several groups (see for example ref. 3). For avoiding the influence of spin-orbit coupling the magneto-resistance experiment must be performed with a very light metal because spin orbit coupling causes severe complications. In Fig. 3 the magneto-resistance of a Mg film is plotted as a function of the applied magnetic field. The units of the field are

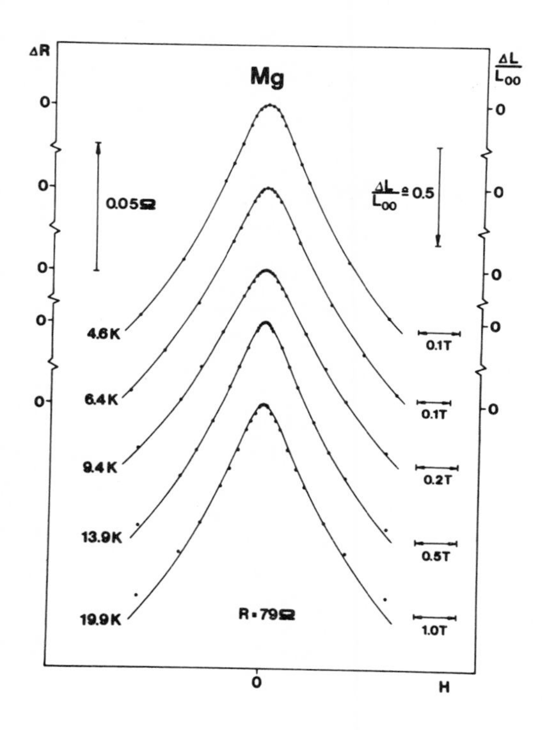

Fig. 3

shown on the right side of the curves. The Mg is quench-condensed at helium temperature, because the quench condensation yields homogeneous films with high resistances. The agreement between the experimental points and the theory is very good (the spin-orbit coupling of Mg is already incorporated; see below). The experimental result proves the destructive influence of a magnetic field on weak localization. It measures the area in which the coherent electronic state exists as a function of temperature and allows the quantitative determination of coherent scattering time τ_i. The temperature dependence follows a T^{-2} law for Mg.

IV.
SPIN-ORBIT SCATTERING

One of the most interesting questions in weak localization is the influence of spin-orbit scattering. Hikami et al.[11] predicted, in the presence of strong spin-orbit scattering, a logarithmic decrease of the resistance with decreasing temperature. As a consequence the magneto-resistance should change sign as well. This prediction is contrary to the picture of localization. The prediction by Hikami et al. could be experimentally confirmed. For this purpose a thin Mg-film has been prepared in an ultra high vacuum. After the measurement the Mg-film has been covered with the strong spin-orbit coupler Au. This causes a significant change of the magneto-resistance. In Fig. 4 the magneto-

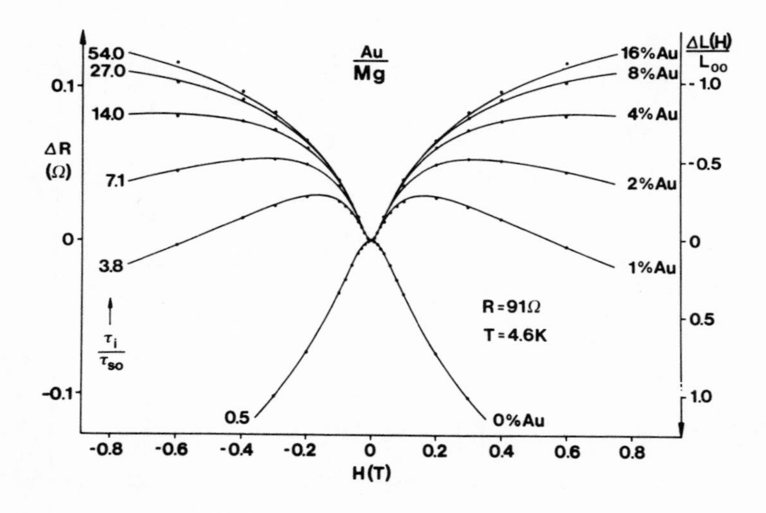

Fig. 4

resistance of the Mg-film at 4.5K is plotted for increasing coverage with Au. The numbers on the right of the curves give the Au-coverage in % of a mono-layer. The points represent the experimental results, whereas the full curves are calculated with the theory of Hikami et al. The adjustable parameter is the spin-orbit scattering time which decreases with increasing Au coverage (this experiment also yields the spin-orbit scattering of the pure Mg-film). Obviously weak localization provides a new and very sensitive method to measure the spin-orbit scattering directly i.e. with a substructure and not only by a broadening of a resonance.

For other metal films where the nuclear charge is higher than in Mg one finds even in the pure case the substructure caused by spin-orbit scattering. For Au-films the spin-orbit scattering is so strong that it completely dominates the magneto-resistance. The natural question is, why does weak localization change to weak anti-localization in the presence of spin-orbit scattering?

V.
INTERFERENCE OF ROTATED SPINS

It is a consequence of quantum theory and proved by a rather sophisticated neutron experiment that spin 1/2 particles have to be rotated by 4π to transfer the spin function into itself. A rotation by 2π reverses the sign of the spin state. Weak anti-localization gives another experimental proof of this fact. In the presence of spin-orbit scattering the spin of the conduction electron is not preserved. During the whole scattering series (') the spin orientation diffuses into a final state σ' which can be obtained by a rotation T of the original spin state σ. ($\sigma'=T\sigma$). It is straight forward to show that the finite spin state of the complementary scattering series (") is $\sigma''=T^{-1}\sigma$. Without the spin rotation the interference of the two partial waves is constructive (in the absence of an external field). In the presence of spin-orbit scattering the interference becomes destructive if the relative rotation of σ' and σ'' is 2π. It can be shown that for strong spin-orbit scattering the destructive part exceeds the constructive one[3]. This means that the back-scattering is reduced below the statistical one. This corresponds to an echo in the forward direction and an decrease of the resistance. The magneto-resistance curve in Fig. 4 for 1% Au on top of Mg can be interpreted as follows. In a high magnetic field where $t_H<\tau_{so}$ the spin states of the complementary states are almost unchanged and one obtains the usual negative magneto-resistance. For $t_H>\tau_{so}$ (and $t_H<\tau_i$) the interference is destructive and shows the opposite sign. For $t_H\approx\tau_{so}$ it changes sign. The resistance maximum in a finite field corresponds to a relative rotation of σ' and σ'' by the angle $\pi/2$ (in an average).

VI.
MAGNETIC IMPURITIES

The weak localization allows us to study the properties of magnetic impurities because the magnetic scattering brings the two complementary waves out of phase and the magneto-resistance yields the magnetic scattering time as a function of temperature. This effect is frequently an undesired disturbance in the investigation of so-called "pure" films.

One can study the influence of Fe on the magneto-resistance of Au. First a thin Au-film has been investigated and its inelastic lifetime (and spin-orbit scattering time) have been determined. Afterwards the Au-film has been covered with about 1/1000 layer of Fe. Since only the magnetic scattering time i.e. the corresponding field H_s is changed the evaluation

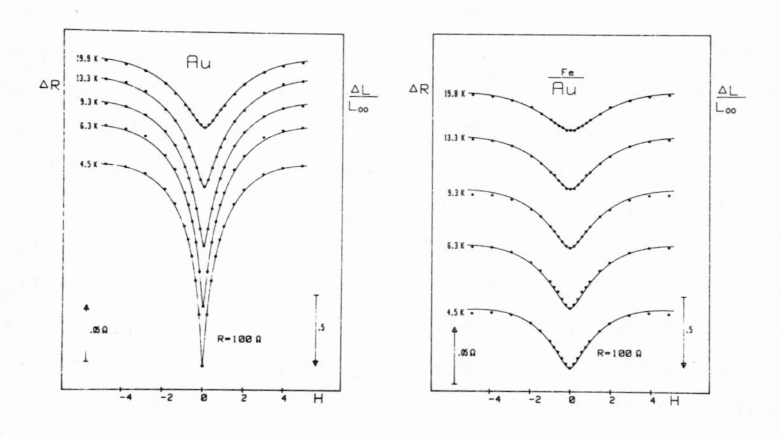

Fig. 5

allows an unambiguous determination of H_s to τ_s respectively as a function of temperature. Fig. 5 shows the magneto-resistance curves of the pure Au and the Au/Fe film for different temperatures. The theoretical curves for the Au/Fe sandwich are obtained with the same values of H_i and H_{so} as for the pure Au film and an additional magnetic scattering which corresponds to a τ_s = 2.9 ps.

VII.
TEMPERATURE DEPENDENCE

The first prediction of the theory of weak localization was the logarithmic anomaly of the resistance with temperature. Such a resistance anomaly has been observed experimentally by several authors shortly after the theory was worked out. However, at about the same time Altshuler et al.[12] showed that the dynamic modification of the electron-electron interaction in disordered metals yields in two dimensions a very similar anomaly of the resistance. First the two theoretical predictions have been considered as alternatives. However, after the first magneto-resistance measurements it became clear that both effects coexist. On the other hand the spin-orbit scattering should modify or even invert the contribution of weak localization.

If one considers a two-dimensional system in which the spin-orbit scattering is changed from 0 to infinity while the temperature dependent

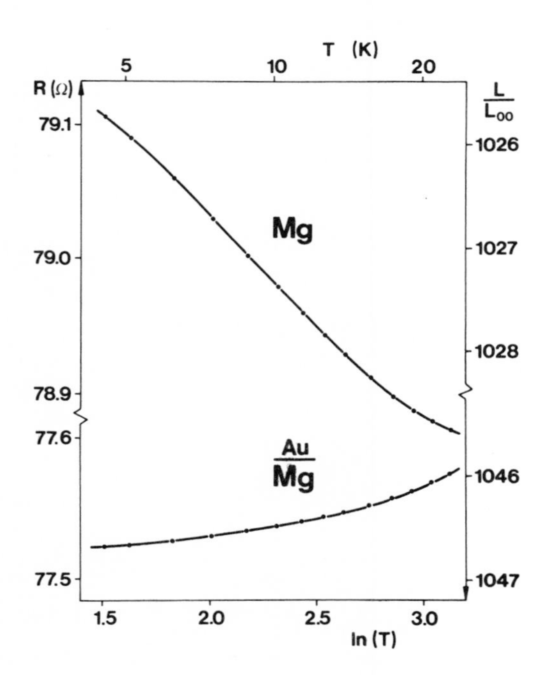

Fig. 6

inelastic lifetime is not altered then the temperature dependent correction of the conductance ΔL_{wl} changes sign and reduces by a factor of 1/2. Experimentally this manipulation of the spin-orbit scattering can be achieved by superimposing Mg with a fraction of a monolayer Au. One can measure the temperature dependence of the pure Mg-film and the Mg covered with 1/4 layer of Au. The temperature dependence of the resistance is shown in Fig. 6. Obviously one finds the strong influence of the spin-orbit scattering. From the evaluation of the magneto-resistance measurements we know the temperature dependence of $H_i(T)$. Together with the two values of H_{so} we can calculate the temperature dependence of the resistance which is caused by weak localization. This is, of course not the only contribution to the temperature dependence of the resistance. In addition we have the Altshuler contribution caused by the electron-electron interaction and (at higher temperature) the thermal part. But when we take the difference in the resistance between the Mg and the MgAu film then the other contributions should cancel and we can check whether the temperature dependence of weak localization also obeys the theory. There is, indeed, a good agreement for the difference between experiment and theory.

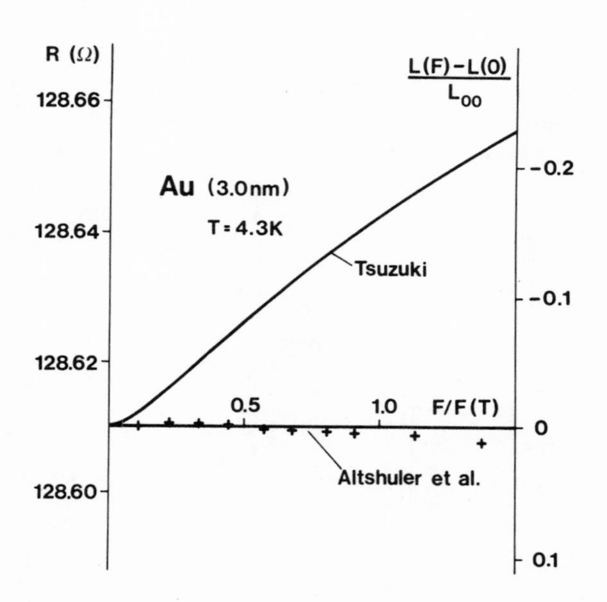

Fig. 7

VIII.
INFLUENCE OF AN ELECTRICAL FIELD

While the effect of a magnetic field on weak localization is indisputable and experimentally proved the influence of an electric field on weak localization was quite controversial. The experiment confirms the theory by Altshuler and Aronov[13]. Fig.7 shows the resistance of an Au-film 30 A thick as a function of the (normalized) electric field. The resistance is with high accuracy independent of the electric field in a field range where Tsuziki[14] and Kaveh et al[15] predicted a field dependence.

IX.
SUPERCONDUCTORS

The difficulty in disordered films of superconductors is that the magneto-resistance (above the superconducting transition temperature T_c) is composed of several different terms: a) the Aslamazov-Larkin[16] fluctuations (AL), b) the Maki-Thompson[17,18] fluctuations (MT), c) weak localization (WL), d) the Coulomb-contribution in the particle-hole channel (CPH), e) the Coulomb contribution in the particle-particle channel (CPP). For quench condensed Aℓ which is a weak coupling superconductor one finds that only the contributions of weak localization and the Maki-Thompson fluctuation are important. By comparing the magneto-resistance of a pure Aℓ-film with the magneto-resistance of the same Aℓ-film after superposition with .25 atola of Au* one can determine the contribution of weak localization and the Maki-Thompson fluctuation independently. The experimental results show a high consistency with the theory.

X.
THE THREE DIMENSIONAL CASE

As we discussed before the physics of weak localization is the same in three dimensions as an in two. However, the relative weight of the quantum correction is much smaller. The echo decays as $1/t^{3/2}$ with time and this makes the essential difference because the integrated contribution does not diverge as in two dimensions but saturates.

* *1 atola = 1 atomic layer*

Therefore the relative correction is often by a factor 100 smaller than in two dimensions. Despite this difficulty there has been theoretical as well as some experimental work on this subject (see for further references (19)). According to Altshuler et al.[20] The magneto-resistance in three dimensions is given by

$$\frac{\Delta\rho}{\rho^2} = \frac{e^2}{2\pi^2\hbar}\left(\frac{eH}{\hbar}\right)^{1/2}\left[\frac{1}{2} f_3\left(\frac{H_S}{H}\right) - \frac{3}{2} f_3\left(\frac{H_T}{H}\right)\right] \tag{9}$$

where

$$f_3(x) \approx \begin{cases} x^{3/2}/48 & \text{for } x \ll 1 \\ .605 & \text{for } x \gg 1 \end{cases}$$

Fig. 8 shows the magneto-resistance of amorphous $Cu_{50}Y_{50}$ as a function of $H^{1/2}$ from ref. (10). The sample has an intermediate spin-orbit scattering strength. The full curves are the theoretical results. These authors

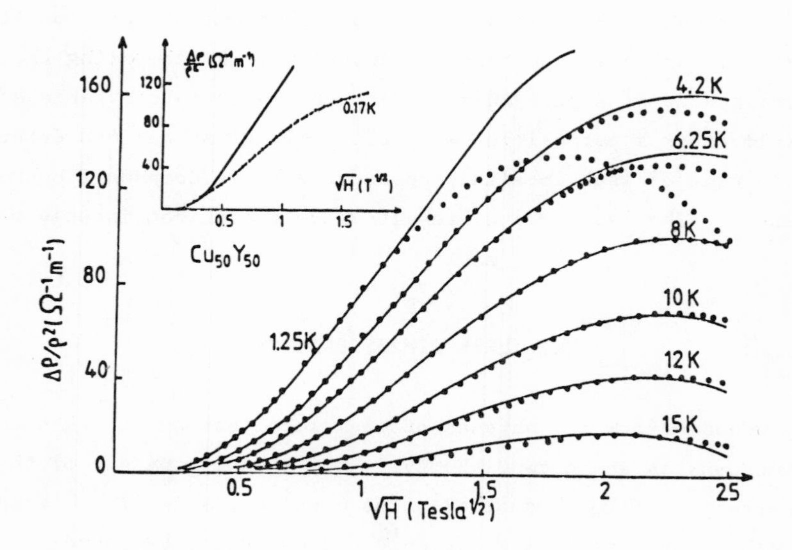

Fig. 8

found a surprisingly good agreement between the experimentally determined strength of the spin-orbit scattering H_{so} and their theoretical calculation. Weak localization proves to be also in three dimensions a method which allows the determination of basic parameters of the solid state.

XI.
COULOMB-INTERACTION

Resistance anomalies of thin films have been observed quite a few times in the past. For example Buckel and Hilsch(21) found already 25 years ago an increasing resistance with decreasing temperature in disordered Bi-films. Most of these anomalies can now be explained by the Coulomb-anomaly of the resistance. As we mention below Altshuler et al.(22) predicted almost in coincidence with the theory of weak localization another very similar resistance-anomaly as a function of temperature. The mechanism which causes the Coulomb anomaly is considerably more difficult than weak localization. The origin of this anomaly is the fact that the electron-electron interaction in disordered metals is retarded. A sudden change of the charge distribution in a disordered metal cannot be screened immediately. Since the electrons can only propagate by diffusion they need time to screen the charge distribution. Since the diffusion is particularly slow over large distances one finds a small screening for small q, i.e. a large effective electron-electron interaction. The result by Altshuler et al. for the effective Coulomb-interaction in disordered electron-systems is (for small q)

$$V_{sc} = \frac{1}{N}\,\frac{Dq^2 - i\omega}{Dq^2} \tag{10}$$

(N = density of states). The screening of a static potential is, however, hardly changed by impurities.

The physics of the electron-electron interaction in disordered electron systems is a large field. I refer the reader to two review articles on this field by Fukuyama(4) and Altshuler and Aronov(23). In the present talk I do not want to treat this field in any detail. However, an experimentalist is confronted with the fact that his measured temperature dependence of the resistance is not only determined by weak localization but also by Coulomb interaction. Here we discuss only the resistance anomaly and the Hall anomaly which is closely related to the former and not influenced by weak localization. Many of the interesting consequence of the electron-electron interaction on density of states, thermo-power, tunneling resistance etc. are not discussed here.

1) Resistance anomaly

Altshuler et al.[22] and Fukuyama[24] evaluated a Kubo-graph which contained the dynamic Coulomb interaction and consisted of the Fock and the Hartree term. They obtained a correction for the conductance

$$\Delta L(T) = -\Delta R/R_o^2 = -L_{oo}\,(1-F)\,\ell n(T) \qquad (11.a$$

Here F was originally defined as a screening factor.

$$F = [K/(2k_F)]^2 * \ell n[1+(2k_F/K)^2] \qquad (11.b$$

where K^{-1} is the screening length in three dimensions $K^2=Ne^2/\epsilon_o$. (N=density of states at the Fermi-surface). However, Finkelstein[25] showed recently that the perturbation theory does not treat the Coulomb-interaction consistently and F had to be redefined. We will trea it as an adjustable parameter as it is in the experiment. The temperatur dependence of the resistance is hardly influenced by a magnetic field. According to Fukuyama[26] the spin-orbit scattering modified the anomaly slightly due to the Hartree-contribution but again Finkelstein's results require a recalculation.

According to Lee and Ramakrishnan[27] the Hartree part of the Coulomb-anomaly shows a (small) magneto-resistance. They obtained for th change of the conductance in a magnetic field

$$[L(H)-L(0)]/L_{oo} = \begin{cases} (F/2)\,0.084\,h^2 & \text{for } h \ll 1 \\ F/2\,\ell n[h/1.3] & \text{for } h \gg 1 \end{cases} \qquad (12$$

where $h=g\mu_B H/(k_B T)$.

The relation is only derived for vanishing spin-orbit scattering. For strong spin-orbit scattering $(1/\tau_{so} > 1/\tau_T$ the magneto-resistance vanishes.

There are also Kubo-graphs which contain a combination of weak localization and electron-electron interaction. According to Altshuler et al.[20] one has to include the Coulomb-interaction repeatedly. In a normal conducting metal the contribution of these "particle-particle-diagrams" can be neglected.

Experimentally the Coulomb-anomaly and weak localization are superimposed. One may separate the two contributions by applying a large magnetic field. The field suppresses the temperature dependence of the weak localization and only the Coulomb anomaly remains. Fig. 9 shows the

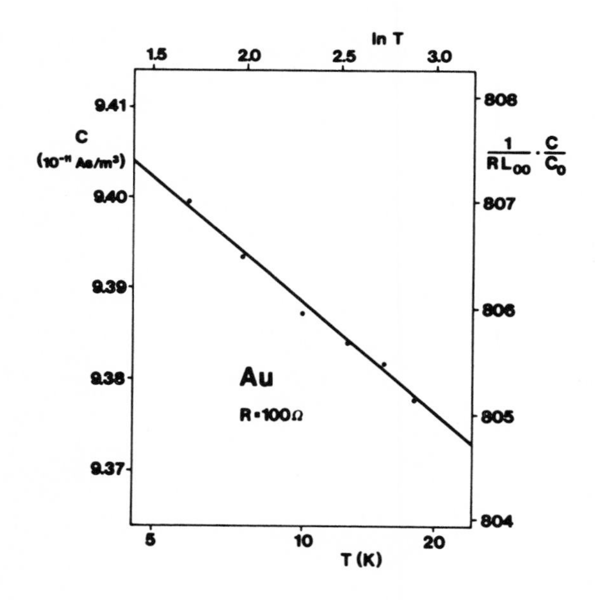

Fig. 9

experimental result for thin Au film. The Coulomb-anomaly is hardly changed by a change of the spin-orbit scattering[19]. While Mg and Mg/Au have very different temperature dependences of the resistance in zero field the influence of the Au in a field of 7 T is negligible. The adjustable parameter F is of the order of 0.2.

2) Hall-effect

While the theory of weak localization predicts no change of the Hall-constant (29) Altshuler et al.[13] showed that the Coulomb-interaction causes a clear change of the Hall-constant C with decreasing temperature. They obtained

$$\Delta C_C(T)/C_o = 2\Delta R_C(T)/R_o \qquad (13)$$

(the index C stands for the correction caused by the Coulomb interaction). As a consequence of equation (11.a) and (13) one comes to the following:

$$\frac{\Delta C(T)}{C_o \; R_o L_{oo}} = 2(1-F) \; \ell n(T) \qquad (14)$$

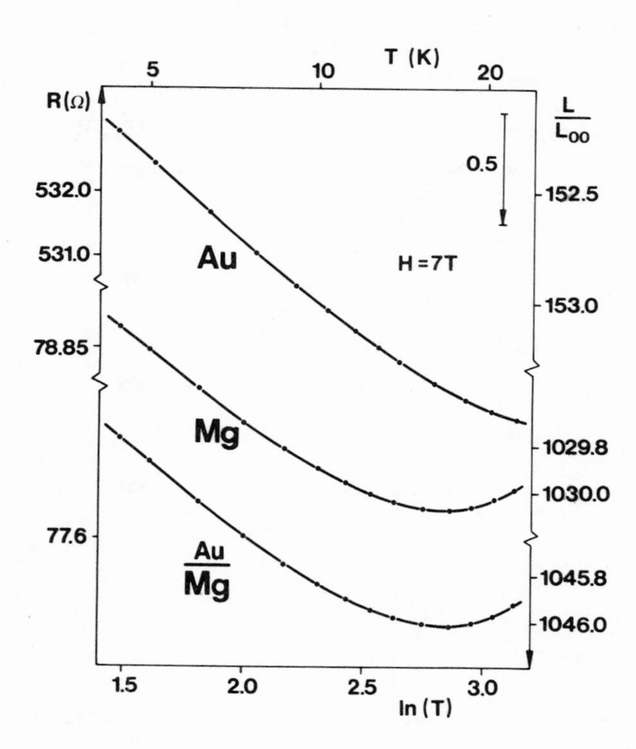

Fig. 10

The experimental difficulty for thin metallic films is that for reasonable film resistances the anomalies have a relative magnitude of 10^{-2} to 10^{-3}.

The Hall-effect requires a measurement of a Hall-angle of about 10^{-3} with an accuracy of about 10^{-4} which is rather tedious. Measurements of the temperature dependent Hall-constant have been performed on electron inversion layers. For thin Au-films the author[30] found an increase of the Hall-constant with decreasing temperature as shown in Fig. 10. The value of the adjusted parameter F from the Hall-effect measurement was 0.25.

REFERENCES

1. E. Abrahams, P.W. Anderson, D.C. Licciardello and T.V. Ramakrishnan, Phys. Rev. Lett. 42, 673 (1979).
2. B.L. Altshuler, A.G. Aronov, D.E. Khmelnitskii and A.I. Larkin, "Quantum Theory of Solids" Ed. I.M. Lifshits (MIR Publishers, Moscow, 1982).
3. G. Bergmann, Physics Reports 107, 1 (1984).
4. H. Fukuyama, to be published in Modern Problems in Condensed Sciences.
5. P.A. Lee and T.V. Ramakrishnan, Rev. Mod. Phys. 57, 287 (1985).
6. J.S. Langer and T. Neal. Phys. Rev. Lett 16, 984 (1966).
7. P.W. Anderson, E. Abrahams and T.V. Ramakrishnan, Phys. Rev. Lett 43, 718 (1979).
8. L.P. Gorkov, A.I. Larkin and D.E. Khmelnitzkii, JETP Lett 30, 228, (1979).
9. B.L. Altshuler, A.G. Aronov and B.Z. Zpivak, JETP Lett. 33, 94.
10. D.Y. Sharvin and Y.V. Sharvin, JETP Lett. 34, 272 (1981), Pis'ma Zh. Eksp. Theor. Fiz. 34, 285 (1981).
11. S. Hikami, A.I. Larkin and Y. Nagaoka, Prog. Theor. Phys. 63, 707 (1980).
12. B.L. Altshuler, D. Khmelnitzkii, A.I. Larkin and P.A. Lee, Phys. Rev. B22, 5142 (1980).
13. B.L. Altshuler and A.G. Aronov, JETP Lett. 30, 482 (1979).
14. T. Tsuziki, Physica. 107B, 679 (1981).
15. M. Kaveh, M.J. Uren, R.A. Davies and M. Pepper. J. Phys. C. Solid State Phys. 14, 413 (1981).
16. L.G. Aslamazov and A.I. Larkin. Phys. Lett 26A, 238 (1968).
17. K. Maki. Prog. Theor. Phys. 40, 193 (1968).
18. R.S. Thompson. Phys. Rev. B 1, 327 (1970).
19. J.B. Bieri, A. Fert, G. Creuzet and A. Schuhl, to be published.
20. B.L. Altshuler, A.G. Aronov, A.I. Larkin and D. Khmelnitzkii, Sov. Phys. JETP 54, 411 (1981), Zh. Eksp. Theor. Fiz. 81, 768 (1981).
21. W. Buckel and R. Hilsch, Z. Physik 138, 109 (1954).
22. B.L. Altshuler, A.G. Aronov and P.A. Lee, Phys. Rev. Lett. 44, 1288 (1980).

23. B.L. Altshuler and A.G. Aronov, to be published in Modern Problems in Condensed Sciences.
24. H. Fukuyama, J. Phys. Soc. Jpn. 50, 3407 (1981).
25. A.M. Finkelstein, Zh. Eksp. Theor. Fiz. 84, 168 (1983), Sov. Phys. JETP 57, 97 (1983).
26. H. Fukuyama, J. Phys. Soc. Jpn. 50, 1105 (1982).
27. P.A. Lee and T.V. Ramakrishnan, Phys. Rev. B26, 4009 (1982).
28. G. Bergmann, Phys. Rev. B28, 515 (1983).
29. H. Fukuyama, J. Phys. Soc. Jpn. 49, 644 (1980).
30. G. Bergmann, Solid State Commun. 49, 775 (1984).

MEASUREMENTS NEAR THE METAL NON-METAL CRITICAL POINT

Gordon A. Thomas

Department of Physics, Harvard University, Cambridge, Mass.
and
AT&T Bell Laboratories, Murray Hill, New Jersey

1.
INTRODUCTION

Several recent results present interesting new information about the metal non-metal transition in disordered systems. First, a large optical nonlinearity found in doped semiconductors has probed the local density fluctuations. Second, similar materials in the metallic state have been driven insulating by large magnetic fields and unexpected critical exponents have been found. Third, enhancements of the spin susceptibility have been found in disordered metallic materials along with indications that the spin diffusion is slower than that of the electrons. Fourth, a correlated electronic instability may have been observed at large magnetic fields and low temperatures.

1.1 Outline

We begin with a selective review of key results to set the context for a discussion of new results. We shall organize the discussion around attempts to resolve the difference between theory and experiment regarding the critical exponent of the metal-insulator transition. The class of systems in question are those in which a Fermi sea of electrons attempts to diffuse through a random potential. The exponent describes the electrical conductivity or the dielectric constant of these materials as the relative concentration of electrons is changed. In the limit of zero temperature, a phase transition occurs between the metallic and insulating states. The Coulomb interactions among the electrons along with the disorder produce an environment that is significantly different from our expectations in nearly ordered systems.

A major uncertainty is that essentially all of the quantitative theoretical analyses have been carried out perturbatively and must be extrapolated into the region of large effects in order to describe the particularly interesting region near the phase transition.

1.2 General Context

An important motivation for the continued intensive study of the metal-insulator transition in disordered systems[1-14] is that most theories[15-56] extrapolate to an estimate of 1 for the critical exponent - a result significantly different from the first measured value[57] of 1/2. Experimentally, the exponent 1 is found in many materials[58-77], while the value 1/2 has been confirmed in a number of measurements[78-96] including those closest[86] to the critical point. The experimental systems fall into these two categories perhaps because of varying strengths[6] of the Coulomb interactions[10,50-56,97-105] and the resultant spin scattering[6,13,46,106-112]. If so, there may be a cross-over[112] to a universal exponent 1/2 at the critical point.

2.

THEORETICAL BACKGROUND

Following Anderson's demonstration[113] that a non-interacting electron could be localized in a random potential at zero temperature, Mott presented an exposition of the physics of localization[1] and suggested[114,115] that it might occur discontinuously. Specifically, the dielectric constant would increase up to a critical value and then jump discontinuously to infinity. At the same point, the electrical conductivity would jump to a minimum value and then would increase smoothly at higher densities. In this theory, the critical exponent would not exist at zero temperature.

(There would presumably be a second order critical point at finite temperature.) All experiments on quantum[57-96] (as opposed to percolative[116,117]) systems have found the transition to be very sharp, and many have been interpreted in terms of a minimum metallic conductivity[1,2,29-32,114,115,118-126].

In contrast, a class of theories has been formulated which assume that the generalized electrical conductance varies continuously from an atomic to a macroscopic length scale[127-130] and consequently that the metal-insulator transition is continuous. In this case the exponent exists and is a key variable describing the transition. Such a scaling theory of localization was first worked out for non-interacting electrons without spin-flip or spin-orbit scattering[15]. This work and subsequent extensions of it in the same class[15-49] have evaluated the electron scattering without Coulomb interactions under the assumption that the scattering is weak. Mott and Kaveh[29-32,131-136] have analysed these perturbation theories in the context of a possible discontinuity and conclude that the continuity assumed in the scaling theories is physically reasonable (except possibly in the presence of an applied magnetic field). The critical region has also been discussed using a self-consistent current-relaxation model, and again continuous behaviour is assumed to be reasonable[137,138].

Most of the scattering processes considered, some including the Coulomb interactions approximately[50-56,97-105], suggest a critical exponent 1. An exception has been found in the case of strong spin-flip scattering without interactions, where the calculation (first carried out within the non-linear sigma model[11,21-24]) extrapolates to an exponen 1/2. These theoretical results can be summarized in over simplified form as follows:

<u>Theoretical estimates of the exponent</u>

expnt=1	scaling	local.		non-int.	[15]
1	perturbation	local.		non-int.	[26-33]
1/2	n.lin. sigma	local.+	spins	non-int.	[21-24]
1	disorder		spins+	int.	[50-56]

Let us comment on these four theoretical cases. The scaling theory of localization for non-interacting electrons to which we refer can be described in terms of a beta function[15]. This function is illustrated in Figure 1 for the cases where the critical exponent is 1 or 1/2. Beta is defined as follows:

$$beta=[d(g)/dL]*[L/g],$$

where g is the dimensionless conductance; i.e., the conductivity times the sample length L normalized to e^2/h. The construction of a single curve assumes that the way the conductance changes with length is uniquely determined by the value of g.

2.1 Short length scales and far infrared.

Short length scales can be visualized in terms of finite frequencies[16], f, according to

$$L=(D/f)^{1/2},$$

where D is the electron diffusion coefficient. The scaling theories calculate g at a short L (or a finite f) and then scale the result to infinite L (or f=0). For an insulating sample of a doped semiconductor, for instance, the far-infrared conductivity[138-140] tends to zero rapidly as f goes to 0, as illustrated in Fig. 2a. For a metallic sample[6,140] (of Si:P, or of GaAs:Si as shown in Fig. 2b), the values of ln(g) at finite f are larger than in the insulator. As f decreases, the conductivity varies slowly toward a finite value. Thus, when it is multiplied by L, the resulting conductance has a positive derivative with respect to L and beta is positive. In the limit of zero f, the conductivity goes to a constant and beta goes to 1.

2.2 Scattering with Spins

The scaling theories of localization without interactions or spins are illustrated in Fig. 1 by the upper solid curve. The first order contribution to this deviation is linear and the second and third order contributions are zero[11]. To investigate the metal-insulator transition, the solid curve must be extrapolated to beta=0. Because contributions above third order are not known, this extrapolation is uncontrolled. Nevertheless, we take this extrapolation as indication of the value of

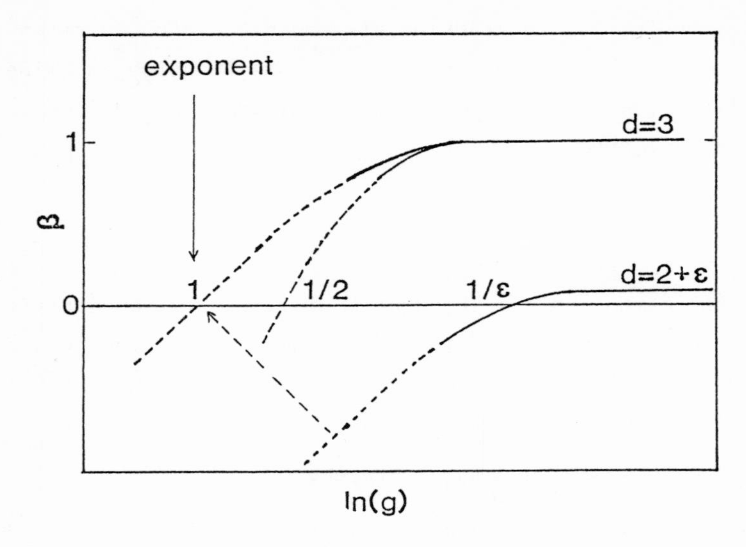

Fig. 1. Theoretical beta functions versus the log of the conductivity [redrawn based on Refs. 6, 15].

the exponent. The exponent 1 is obtained from the inverse of the slope of beta at the transition.

The non-interacting theory of localization with strong spin scattering calculated using the non-linear sigma model[21-24], is also illustrated in Fig. 1. The first order correction becomes zero, the second order correction takes on a finite value, and the third order correction remains zero. When this form is extrapolated to beta=0, as shown by the lower curve, its steeper slope leads to an exponent=1/2. (A note of caution regarding the extrapolation of these non-interacting theories arises from the result[11,21-24] that the deviation from beta=1 is upward [antilocalization] in the case of spin-orbit scattering. It is not clear how this result can be extrapolated sensibly.)

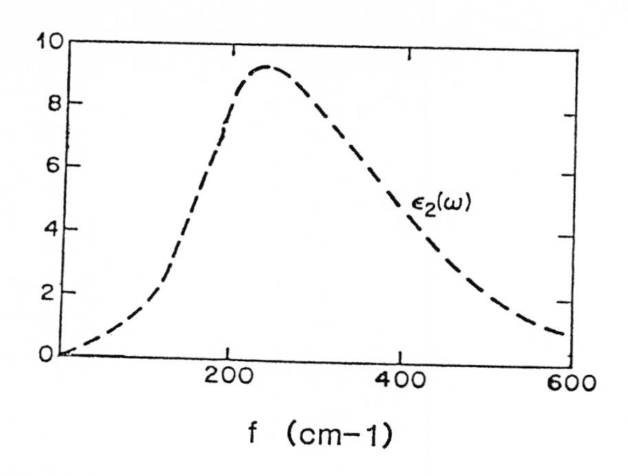

Fig. 2a. Frequency dependent conductivity and dielectric susceptibility in insulating Si:P [Ref. 6].

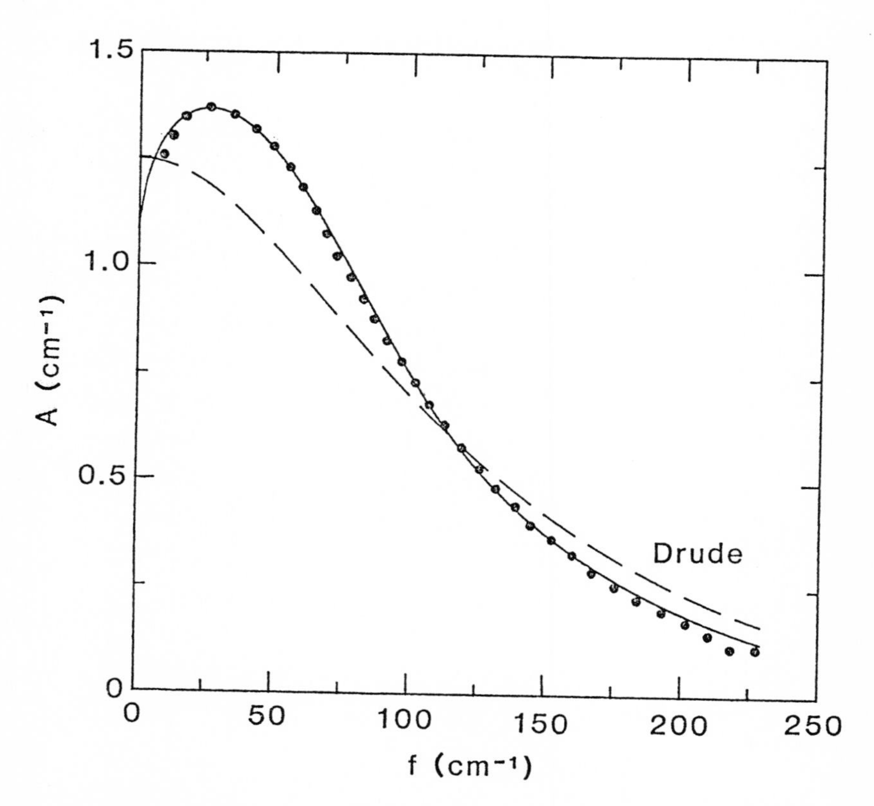

Fig. 2b. Frequency dependent conductivity in metallic GaAs:Si compared to Drude behaviour [Ref. 140].

3.

TRANSPORT NEAR THE TRANSITION

We shall illustrate the perturbation theories by considering the conductivity as a function of carrier density as illustrated in Figure (3). The figure shows results[6,33,40,57] for Si:P. The values of conductivity at zero temperature must be obtained from extrapolations to T=0 K of measurements at low temperatures (here, of order 10mK, or 10^{-4} of the Fermi temperature of 100K). The metallic state is defined by a non-zero intercept of this extrapolation. The values of density were obtained from the concentration of P atoms in the Si crystals. Each P donates its one valence electron to the electronic system (shown schematically in Fig. 3B), while the P ions form the random potential in which the electrons move and interact. Here, the metallic state occurs at densities above $n_c=3.75\times10^{18}/cm^3$. (At this point, the donors would be 65A apart if they were ordered in a cubic lattice.)

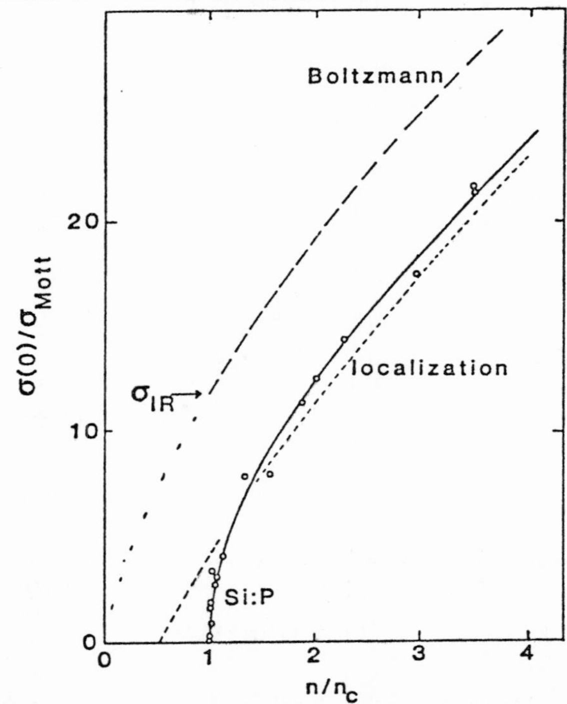

Fig. 3a. Linear plot of the conductivity as a function of carrier density [Refs. 6, 33].

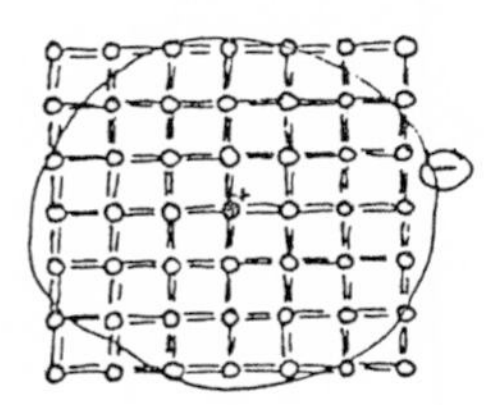

Fig. 3b. Schematic of Si:P.

The usual Boltzmann conductivity[1,2] is shown as the long dashes in Figure 3a. Here, the electrons in their Fermi distribution are imagined to scatter weakly with a mean free path l so that the conductivity is $sigma_B=ne^2l/(hk_F)$. (Including screening, in the Born approximation, l is roughly constant[33,48] in metallic Si:P near n_c, and the decrease in the conductivity arises from the decreasing density.) This picture becomes unreasonable below the Ioffe-Regel conductivity[146], at which the Fermi wave length becomes equal to l

$$sigma_{IR}=ne^2/(hk_F) .$$

Mott[114,115] estimated that the characteristic conductivity for metallic conduction would be less than this because of a reduction in the density of states due to disorder. He estimated this reduction to be a constant (determined to within a factor of 2) so that the Mott conductivity can be written

$$sigma_M=1300/a ,$$

for sigma in $(ohm\text{-}cm)^{-1}$ and the inter-electron spacing, a, in Angstroms. For example, in Si:P, $a=(n_c)^{-1/3}=64A$, and $sigma_M=20\ (Ohm\text{-}cm)^{-1}$, whereas $sigma_{IR}=130\ (Ohm\text{-}cm)^{-1}$. As can be seen in Figure 3, the density at which the measured conductivity reaches $sigma_{IR}$ is relatively far from n_c compared to n at $sigma_M$. However, even with weak scattering and without electron-electron interactions, coherent back-scattering tends to localize the electrons and reduces the conductivity below the Boltzmann curve.

The magnitude of this correction has been calculated[26-33] in a perturbation expansion in $1/k_F1$, which gives

$$sigma=sigma_B[1-3/(k_F1)^2] ,$$

as illustrated by the long dashes in the figure for the parameters[4,5] of Si:P. Extensions of this formula have been discussed by Kaveh and Mott[131-136]. This theory extrapolates to a metal-insulator transition with critical behaviour of the following form[26-33]:

$$sigma=sigma_B(n/n_c-1)^1,$$

i.e., exponent 1.

Significant recent theoretical advances in our understanding of the Coulomb interactions have been obtained by neglecting a full treatment of localization and including spins or a large magnetic field[50-56]. This new theory, based on analysis by Finkelstein[56,105], extrapolates to an exponent=1 as does the non-interacting theory.

3.1 Experimental exponent values.

The critical exponent provides a simple way of comparing the theoretical and experimental behaviours. The solid line fit[57] to the experimental points in Figure 3 gives exponent=1/2. The Mott characteristic conductivity[1] is used for normalization in the figure. A number of careful experiments have confirmed this exponent:

Experiments indicating exponent=1/2
(one electron/ scattering site)

Si:P	AT&T Bell	1980-84[57,79-88]
Ge:Sb	Tokyo	1981[59]
Si:P	Tokyo	1982[81]
Si:P	Cornell	1984[89]
Si:As	Cornell	1984[89]
Si:P+As	Cornell	1984[89]
Si:Sb	Cambridge	1984[91]
Ge:As	Harvard	1986[90]

(dielectric susceptibility:)

Si:P	AT&T Bell	1980,83[57,78]
Si:P	Rochester	1984[93-96]
Si:As	Rochester	1984[93-96]

The confirmation of this critical behaviour to as close to the transition as 0.1% in relative density has come from measurements[82-87] very near the transition under uniaxial stress, and from studies of the dielectric susceptibility, as illustrated in Figure 4. The right axis and solid points show the square of the conductivity as a function of relative density, in agreement with those in Figure 3. The left hand axis

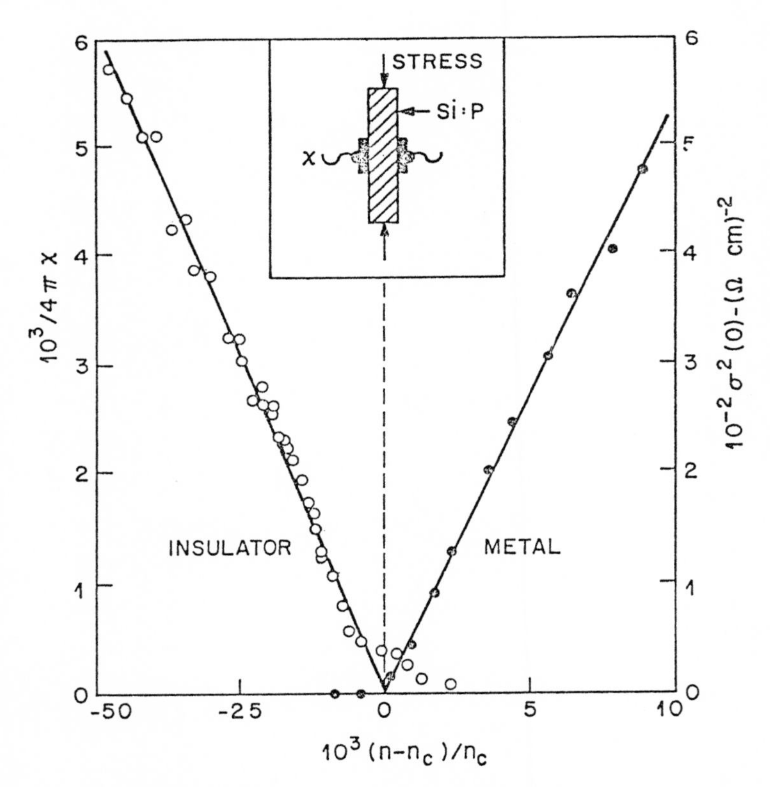

Fig. 4. Dielectric susceptibility and conductivity very near the transition in Si:P, both indicating exponent 1/2 [Refs. 85, 86].

and open circles show the inverse of the dielectric susceptibility and indicate a linear critical approach to a dielectric catastrophe. Recent unpublished results[93] confirm these results in the insulator. Simple dimensional considerations suggest[57] (and calculations[41,138,147] indicate) that this dependence should give twice the characteristic exponent.

There are, however, a range of systems in which the observed exponent is 1, an example of which[59] is shown in Figure 5. The data are plotted with the same axes as Figure 3, with the open circles again the results for Si:P, which are in agreement with the results for Ge:Sb. The solid circles are measurements for special Ge:Sb samples in which additional scattering centres were added beyond one per electron, by putting in additional, equal amounts of Sb and B. These two impurities exchange an electron, and contribute charged scattering centres, but no carriers at low T, as illustrated schematically in Figure 5b. This addition of compensating impurities increases the scattering and thus reduces the conductivity at fixed carrier density n (by reducing l), as shown in the figure. The exponent appears to change to 1 in the region where data is available.

Experiments indicating exponent=1
(several scattering sites/elec.)

Ge:Sb(comp.)	Tokyo	1982[59]
a-Si:Nb	AT&T Bell	1983,84[60,61]
a-Si:Al	"	1984[61]
a-Si:Au	Tokyo	1984[65-67]
GaAs:n	Tohoku	1984[68,69]
InSb:n	"	1984[68,69]
a-Ge:Au	Illinois	1984[71-73]
a-Si:Mo	Stanford	1984[74]
A-Kr:Bi	Ruhr	1984[75,76]
InO_x	AT&T Bell	1984[77]
	(same + large field):	
Gd_3S_4:v	IBM	1983,84[62-64]
InSb:n	London	1984[70]

The exponent 1 is also found in compound semiconductors which are doped at a series of concentrations and unavoidably compensated, such as GaAs:n studied by Morita et al[68,69]. Here, n indicates unspecified impurities that donate electrons to the system. As noted

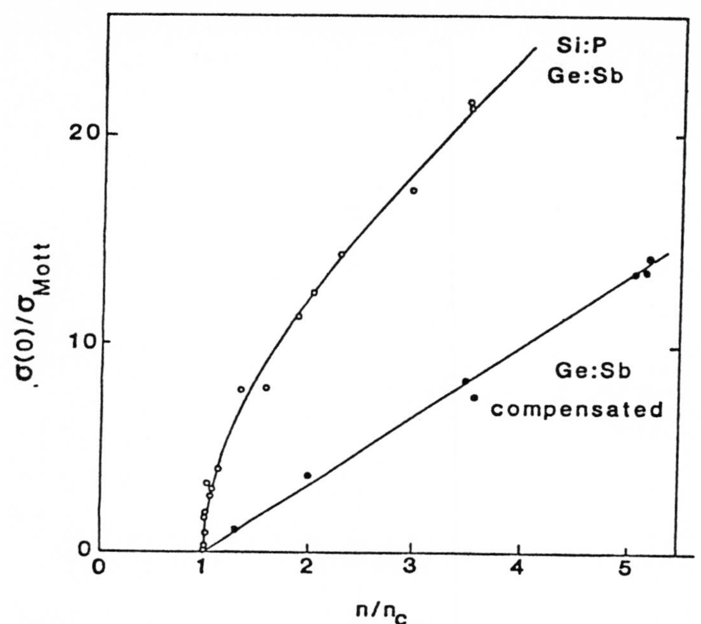

Fig. 5a. Indication of the effect of adding many additional scattering centers on the zero-T conductivity in dirty metals. The open circles show a series of samples of Si:P which, when scaled by Mott's characteristic conductivity [Ref. 1], are equivalent to those for Ge:Sb [Refs. 6,59]. The solid line indicates a critical exponent of 1/2. Theoretically, such an exponent might be produced by non-interacting electrons with strong spin flip scattering, as suggested by Hikami [Ref. 11]. In contrast, the solid circles indicate the behavior for samples to which ~30% additional Sb impurities compensated by B have been added, so that the scattering by centers that do not donate an electron at low temperatures is increased at constant electron density. Here, the solid line corresponds to an exponent 1, similar to that seen in Figs. 6-9.

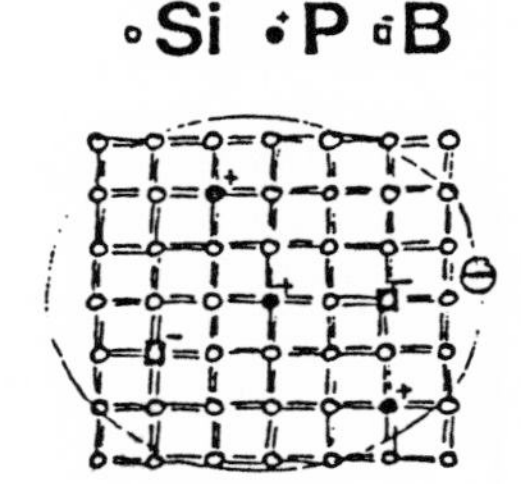

Fig. 5b. Schematic of compensated Si.

above, other results give the same exponent in a variety of systems where there are several scattering centres per electron. However, none of the results giving exponent 1 resolve the region within a few percent of the critical point.

Measurements with particular low scatter in the data were summarized recently[61], in addition to those[65,66] illustrated in Figures 6-8 and others[71-76], on amorphous alloys of a semiconductor and a metal. These amorphous alloys are illustrated schematically in Figure 6B. The critical concentration of the metal component is x~10%, with an average spacing between the metal atoms of a few lattice spacings. The open circles in Figure 6 indicate that a moderate field (5 Tesla) does not seem to affect the exponent.

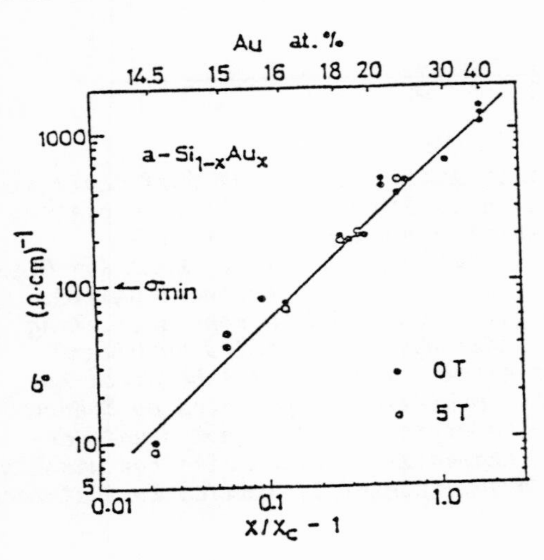

Fig. 6a. *Evidence for exponent 1, both at zero and large magnetic field, in samples of aSi:Au. The zero-T conductivities (in zero and 5 Tesla fields) and the superconducting transition temperatures in a log-log plot for a series of samples of amorphous Si:Au alloys by Nishida, Furubayashi, Yamaguchi, Tshimoto, and Morigaki [Ref. 65].*

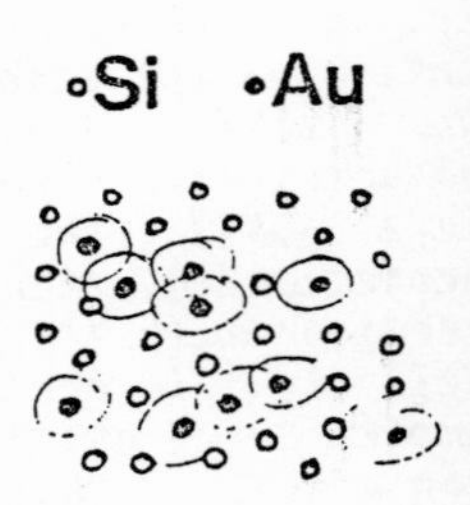

Fig. 6b. *Schematic of an amorphous alloy*

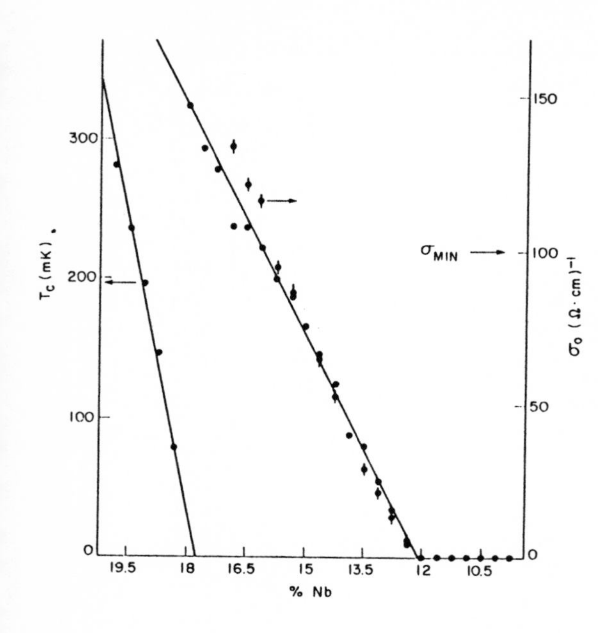

Fig. 7. The same quantities on a linear scale (with the concentration axis reversed) for aSi:Nb by Bishop, Dynes and Spencer [Ref. 61].

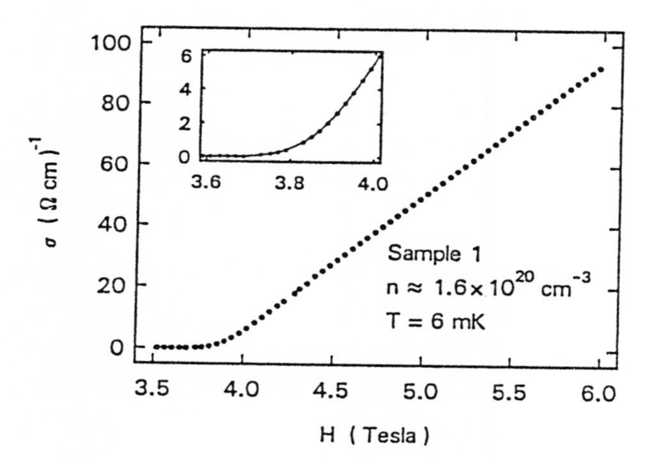

Fig. 8a. Example of behaviour with exponent 1 in Gd_3S_4:v where the metal-insulator transition is achieved by decreasing the disorder with an applied magnetic field. Washburn, Webb, von Molnar, Holtzberg, Flouquet, and Remenyi [Ref. 63] held the sample at T=6mK for these measurements, and found that the transition occurred at H_c=3.9T in this sample with a carrier density, produced by vacancies, of $1.6 \times 10^{20} \cdot cm^{-3}$. Without extrapolating to T=0K, these measurements indicate a rounding near the transition, as shown in the inset, which corresponds to a relative density of ~3%.

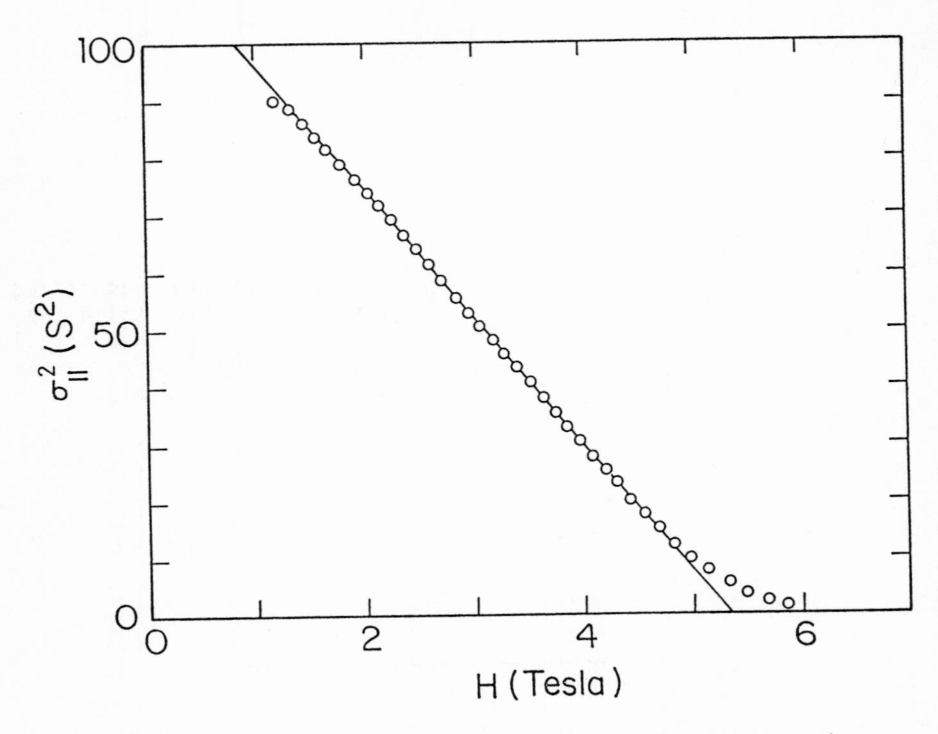

Fig. 8b. Observed suppression of the conductivity by magnetic field in Ge:As [Ref. 90].

3.2 Exponents in Large H

Another case in which data are available is that in which a sample near the transition is driven through it by a large applied magnetic field as discussed by Washburn et al[62-64] for Gd_3S_4:v, Burns et al[90] for Ge:As, and by Mansfield et al[70] for InSb:n. Here, v indicates vacancies which free carriers in the system. Figure 8a shows the results of the IBM group, in which the applied field reduces the degree of disorder, turning an insulator into a metal. In Figure 8b the reverse tuning is shown in Ge:As, where the shrinkage of the wave functions by the field drives a metallic sample insulating. This material is uncompensated and, as noted above, has a critical exponent of 1/2 indicating that these relatively high fields do not supress the mechanism responsible for this exponent. In Figure 9, the results for InSb are shown as a function of T for a series of fields. In this case, the field primarily shrinks the electron's wave function, turning a metal into an insulator.

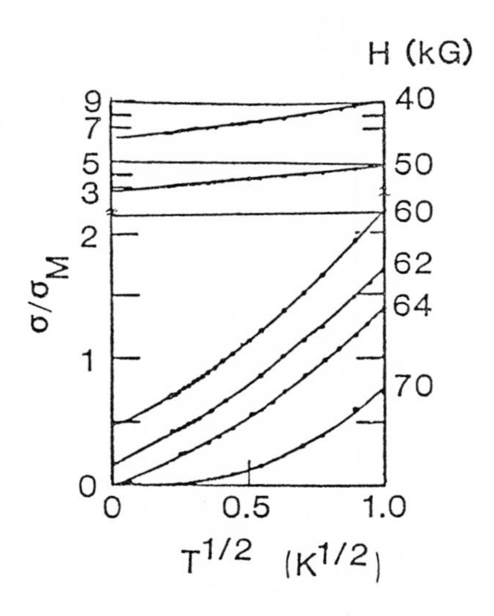

Fig. 9. Example of anomalous diffusion below Mott's characteristic conductivity in a metallic sample of InSb:n [Ref. 70].

Other measurements[124-126] on InSb:n and InP:n as a function of field appear to us to give results qualitatively similar to the studies cited above, but they have been interpreted as indicating a discontinuous approach to the transition. If these data are replotted, they appear consistent with a critical exponent 1. Several other experimental results do not fit easily into the categories discussed above. A critical exponent between 1/2 and 1 is found[149] in Ge:n when the transition is approached by varying compensation. A similar value is derived[150] from studies of variable range hopping in three samples of Ge:As. An exponent 1/2 with broadening near n_c, possibly due to percolation, is found [151,152] in a-Xe:Hg, and less clear results[153,154] are found in aSi:Cr. Generally, these results and those above in the two tables do not give the precise values 1/2 or 1, but we believe them to be consistent with these two classes, with the possibility of a cross-over, and with some rounding near n_c.

3.3 Anomalous Diffusion.

Mott's characteristic conductivity sigma$_M$ appears to be a good indicator of the conductivity region in which anomalous diffusion is found. This anomalous diffusion was first analysed as a correction to a metallic, T=0K conductivity of the form A*T**1/2, with A positive, in studies[80] of Si:P. Further studies[63,64,70,86,155] have found that, near n_c, A increases as the transition is approached, as illustrated in Figures 9 and 10. This behaviour is consistent with theories of Coulomb interactions[50,80,86,97-104] under the conditions present near n_c, and also without interactions for some types of inelastic scattering[47]. Although this positive derivative is qualitatively characteristic of hopping conduction[156-162], as discussed extensively by Mott[1,2] for an insulator, the extrapolation to zero temperature using the $T^{1/2}$

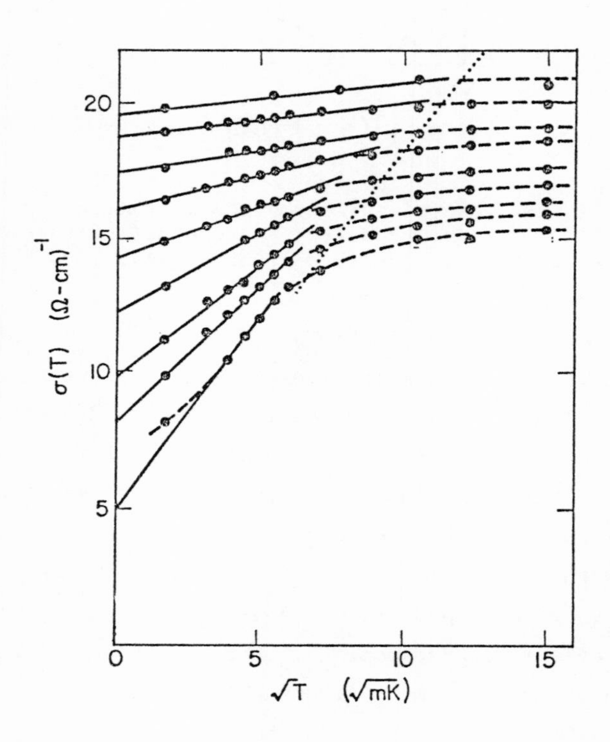

Fig. 10. Theory of localization, based on the non-interacting scaling theory of Abrahams, Anderson, Licciardello, and Ramakrishnan [Ref. 15] and on the formulation by Berggren [Ref. 48] as evaluated by Bhatt and Ramakrishnan [Ref. 33] is indicated by the solid line and is compared with three amorphous alloy systems. These systems are analogous to the compensated semiconductors shown in Fig. 5 and to the random vacancies illustrated in Fig. 8. This behaviour with exponent 1 is quantitatively consistent with non-interacting theory provided that the spin flip scattering isweak. However, theories of Coulomb interactions in a random potential, without localization, but with either a strong field or spins, as evaluated by Castellani et al. [Refs. 52,53], also give an exponent 1.

functional form indicates finite, and therefore metallic, values below Mott's characteristic conductivity in many materials[5-9,57-64,70,71]. The perturbation theories of Coulomb interactions[97-104] indicate that the increasing value of A corresponds physically to a decreasing diffusion temperature[9,86], i.e. to a greater sensitivity of the electrons' motion to T as they become more nearly localized. This process is a metallic analogue of electron assisted variable range hopping[156-162].

4.

INTRINSIC, RESONANT SPINS IN DISORDERED METALS.

With strong, short-range interactions, spins that act as if they were localized can occur within a metal: resonant spin states. Mott[1] has presented a clear discussion of one way in which this can happen, based on the Brinkman-Rice analysis[108,163,164] of Gutzwiller's approximate solution[165] to the Hubbard Hamiltonian[166,167]. Within this picture, an electron's effective mass can be greatly enhanced if its neighbouring sites are occupied by other electrons to which it feels a strong repulsion. The spin diffusion is slowed compared to the charge diffusion[6,13,110-112] by a small Stoner factor, and additionally by an enhanced effective mass, which can diverge and produce an insulating state in a half filled band. However, intrinsic clusters of spins (which are not present in periodic systems) may play a more important role[163,168] in all random systems, and may account for the spin scattering as well as a variety of effects[4-8,169-175] indicative of quasi-localized, resonant spins.

Well into the insulating state, the spectrum has been explained quantitatively[174,176,177] in terms of clusters of electrons, with the energies shown to be strongly dependent on the Coulomb forces. Qualitatively similar behaviour must occur near the transition, but with very large clusters occurring with increasing probability.

The specific heat[81,175] in Si:P, illustrated in Figure 11, shows a sharp upturn in the effective density of states C/T at low T. The roughly linear slope at high T, in this plot as a function of T**2, is partly due to the phonons. However, part of the variation in C/T above 1K is due to the quasi-static electron spins, whose contribution is probably slowly temperature dependent. The values of C/T in Figure 11 are normalized to the free electron density of states, gamma, which is about 0.03mJ/K for this sample. In Mott's terms[1], both this enhanced density of states for all excitations and the reduced one for current carriers (Fig. 2), are indications of strong correlations in an impurity band[6].

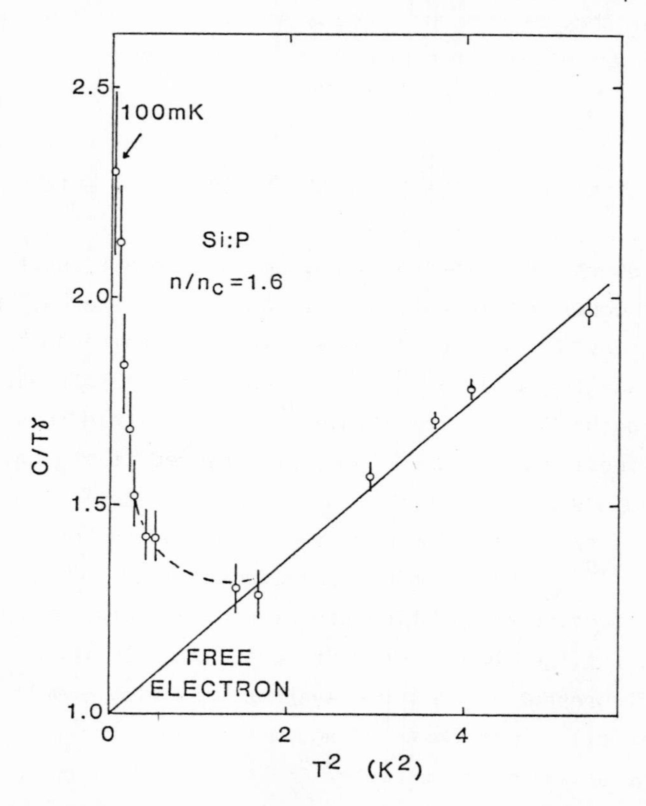

Fig. 11. Anomalous specific heat at low temperature in metallic Si:P [Ref. 7].

A related strong enhancement of the magnetic susceptibility occurs[173] at low temperatures in Si:B as illustrated in Figure 12a. These results confirm earlier studies[1,107,171] showing smaller effects. In the insulating state, studies of the magnetic susceptibility[174] show that clusters of spins with a wide distribution of antiferromagnetic interaction strengths account quantitatively[176,177] for the suppression of the susceptibility below the Curie curve. Qualitatively similar behaviour occurs near the transition, but with larger clusters involved.

In going to the metallic side, recent results by Paalanen et al[178] illustrated in Figure 12b show that there is a significant enhancement of the finite T susceptibility, and a remaining question as to whether its zero T value becomes finite. Certainly these results confirm anomalous spin effects[179,180], and the same experiments confirm that the spin diffusion is significantly slower than the charge diffusion.

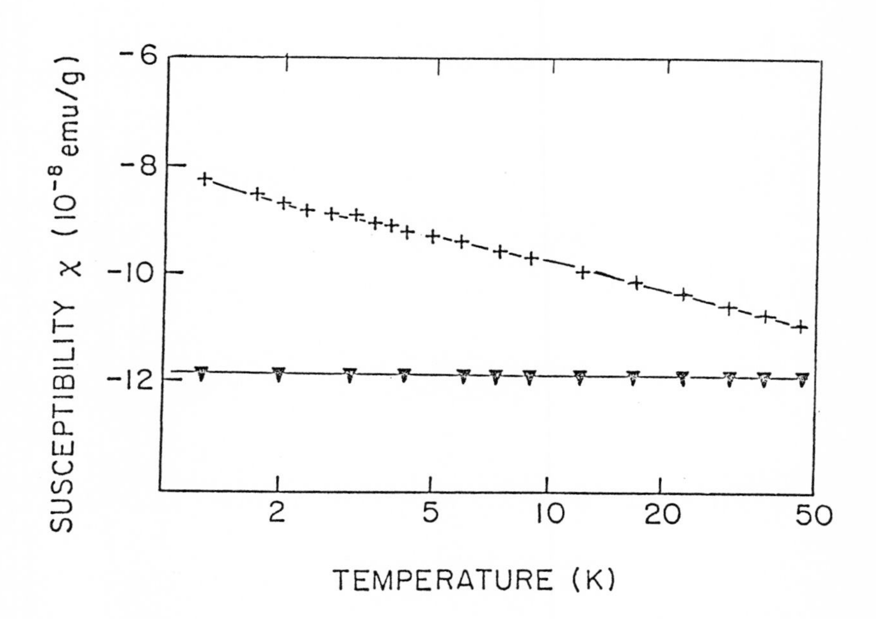

Fig. 12a. Evidence of enhancement of the magnetic susceptibility in a sample of metallic Si:B by Sarachik, He, Li, and Brooks [Ref. 173]. The solid triangles illustrate the behaviour of a metallic sample with carrier (hole) density $n=52\times10^{18}$ cm^{-3}, far above the metal-insulator transition which occurs at $n_c=3.6$ in the same units. For this clean sample, the expected Pauli susceptibility is observed. However, for a sample (+ symbols) with $n=4.9$, relatively near n_c, there is a substantial enhancement of the susceptibility that is suggestive of quasi-static, interacting spins, but which is theoretically unexplained (curves are visual guides).

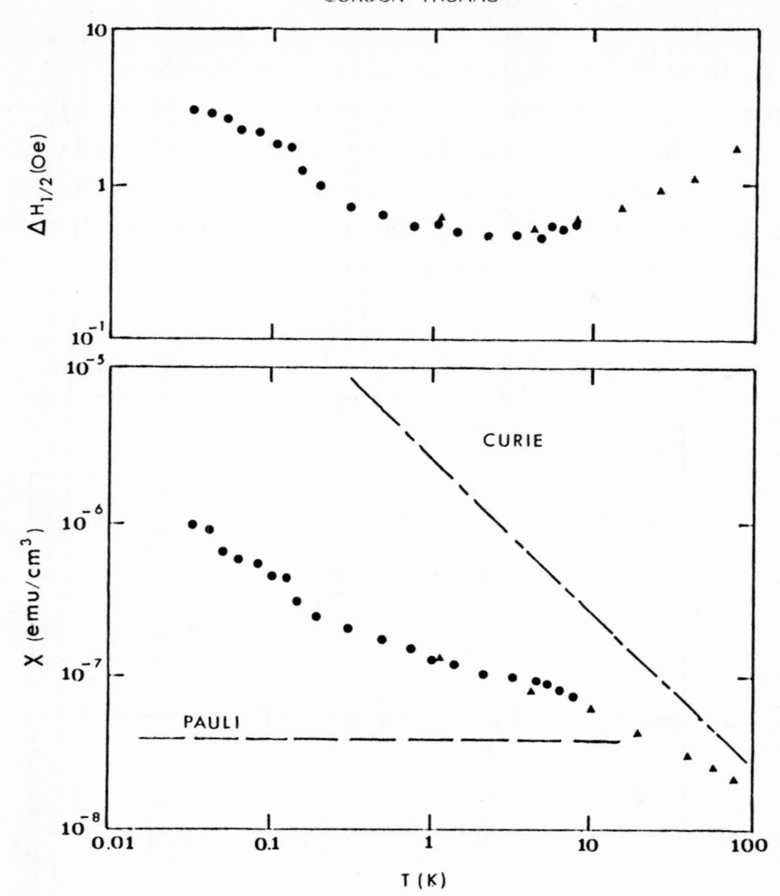

Fig. 12b. More dramatic results at lower temperatures in Si:P from measurements of electron spin resonance [Ref. 178].

A further indication of the presence of electron spins with quasi-localized character in the metallic state is the ^{29}Si nuclear spin relaxation[107,110-112,172], illustrated in Figure 13. In samples near n_c, these spins have been found[6,110-112] to relax up to almost 10^4 times faster, as a result of their interaction with their environment, than they would in the presence of free electrons[181]. The temperature dependence of this enhanced rate relative to the Korringa rate[181] can be analysed[110-112] as approximately T**1/2, which indicates a distinct difference from non-interacting spins (either delocalized, as in the Korringa model, or localized). If the analysis is restricted to T<100mK, the rate also fits a constant plus a linear term in T, as illustrated in Figure 13. This behaviour can be thought of as a sum of contributions[168] from insulating (constant) and enhanced Korringa (linear) contributions. The rate gets even faster as the frequency (and applied magnetic field) are decreased. In this case the dependence is roughly linear, perhaps because of the shifting of the distribution of Zeeman levels or a slowing of the spin fluctuations.

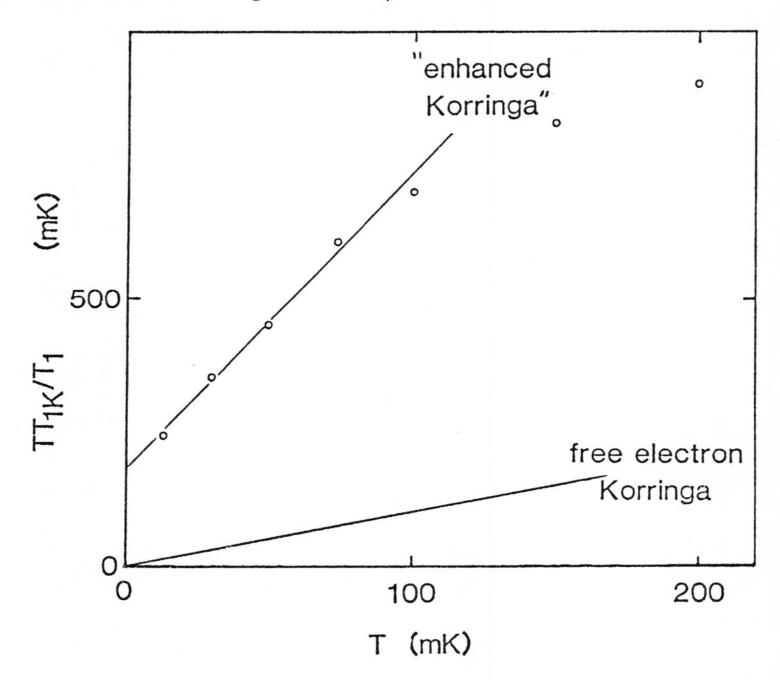

Fig. 13. Evidence of quasi-static electron spins is seen in the enhancement of the NMR spin-lattice relaxation rate in barely metallic Si:P by Paalanen, Thomas, and Ruckenstein [Ref. 8]. The sample has been tuned to a relative density 1% above the critical point (here, $n_c = 3.7 \times 10^{18} cm^{-3}$) using uniaxial stress. The open circles indicate the relaxation rate, T_1^{-1}, of Si^{29} nuclear spins relative to the Korringa rate, T_{1K}^{-1}, for a metal at this electron density. Although the Korringa rate is field independent, the observed rate tends to diverge as 1/H.

5.

NON-LINEAR OPTICS

New results analysing the third order optical non-linearity may yield[182] new information about the local electron density variation in disordered systems. The conditions appear to be favourable in doped Si to probe this variation, defined as

$$n' = [n(r_i)-n], \tag{5.1}$$

where $n(r_i)$ is the local density at the Fermi energy at the impurity position r_i. Of particular interest are the relative values of n' at high and low temperatures, because at high T the electrons spread out while at low T they pile up in the random potential responsible for localization.

This temperature derivative of n' can be probed by laser heating of the electrons with a simultaneous measurement of the ratio, K, of the third order non-linearity to the absorption coefficient.

$$K=K_oL/L_o, \tag{5.2}$$

where K_o is a normalization that can be estimated from a combination of measurement and theory for Si:P to be $K_o=10^{-10}$ esu-cm. The density variation expected for isolated donor electrons is $L_o=20n_c$. The interesting density variation parameter, L, has a dominant contribution which arises from the change induced in the dielectric constant by the laser heating.

$$L=[dn'/dT]E_o/c_v, \tag{5.3}$$

where the characteristic energy E_o is the isolated donor binding energy and c_v is the electronic specific heat.

Figure 14 shows the variation of K as a function of electron density over a region of n that spans n_c. The upper data set was observed with the sample immersed in liquid He at T=2K and the lower set at room temperature.

The principal result is the large L peaking near n_c at about 10^2L_o. An estimate of L within the random phase approximation indicates values only of order L_o because this approximation neglects the diagrams in perturbation theory causing diffusion and localization. The experiment indicates a substantial change in the wave function character as the Fermi energy passes through the mobility edge. The enhancement between the high and low T data shown suggests the effect would be even larger if the experiment could be carried out at lower electron temperatures. In the current work this temperature may be about 20K. The peak position also shifts with T as shown in a way suggesting that in the limit of zero T the peak might occur at n_c.

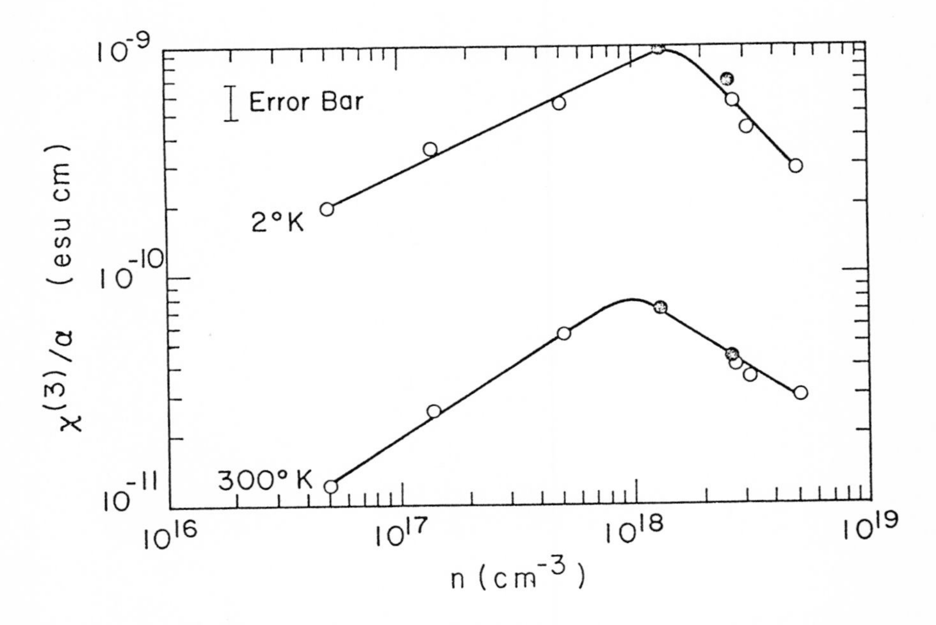

Fig. 14. The relative optical non-linearity of Si:P as a function of electron density indicating the enhancement of the variation in the local electronic density that peaks near the critical density.

6.

CORRELATED ELECTRONIC INSTABILITIES

Speculative and exciting results suggesting the formation of charge density waves in graphite[183] and a Wigner crystal in HgCdTe[184] near the metal non-metal critical point have been observed in large magnetic fields at low temperatures. New findings in this same regime have recently come to light[185] suggesting the formation of a correlated electronic instability oriented along the applied magnetic field in Ge:As. These findings are illustrated in Figure 15 where an anisotropy larger than 10^3 is seen in the resistivity along, compared to across the applied field. The poorer transport along the field appears to be the opposite of what is expected from single particle theories of magnetic freeze-out, in which the electron wave functions are shrunk by the field in the transverse direction. The strong effects of the random potential probably force a predominately disordered state rather than the usual charge density waves or Wigner crystal, and may supress the critical temperature of the instability to 0 so that the results may represent a cross-over between the disordered metal and the anisotropic, correlated insulator.

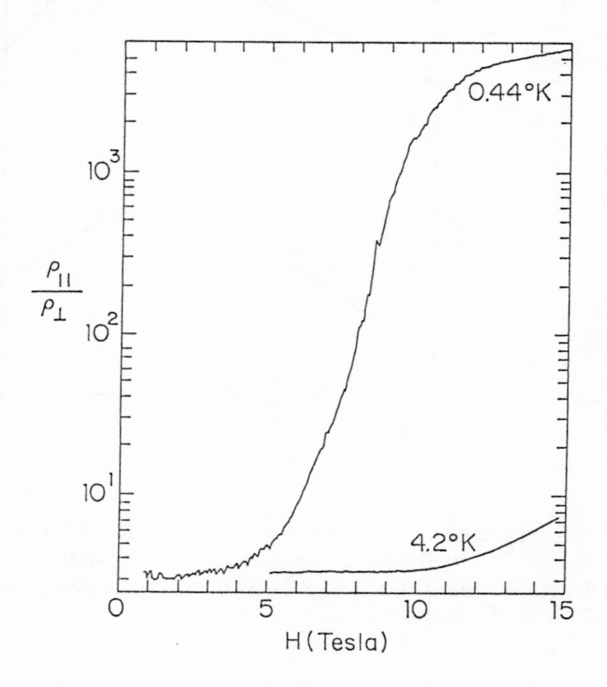

Fig. 15. Anisotropy in the transport of Ge:As produced by large magnetic fields at low temperatures, suggesting the formation of a correlated electronic state [Ref. 185].

7.

CONCLUSION

The discussion above expands upon a 1985 review[186] the conclusions of which need little revision for a description of the situation now. In the disordered, metallic systems discussed above, we imagine the electrons scattering strongly from the random potential, from the other electrons' collective Coulomb field, and from spin states. The range of applicability of the exponent 1/2 may be wider in materials where there is a particularly strong influence of combined localization and Coulomb interactions, possibly via effects due to spin states. This speculation is based on the possiblity that, although the disordered Coulomb interactions produce these intrinsic spins, the exponent can be described within a non-interacting model. Helpful discussions and important collaborative interactions leading to this paper have come particularly from E. Abrahams, P.W. Anderson, R.N. Bhatt, A. Kawabata, S. Kobayashi, P.A. Lee, T.M. Rice, T.F. Rosenbaum, A.E. Ruckenstein, W. Sasaki, and others as indicated in the references.

REFERENCES

Abbreviations:

SSE: Proceedings of the International Conference on Heavy Doping and the Metal-Insulator Transition in Semiconductors, P.T. Landsberg, Ed., Santa Cruz, CA, 1984. Solid State Electronics 28. 1-216 (1985).

ICPS17: Proceedings of the 17th International Conference on the Physics of Semiconductors, San Fransisco, CA, 1984, in press.

LITPIM: Proceedings of the Conference on Localization and Interactions in the Transport Properties of Impure Metals, Braunschweig, FRG, 1984, Y. Bruynseraede, Ed., in Topics in Solid State Physics (Springer Verlag, 1985).

1. N.F. Mott, "The Metal-Insulator Transition",(Taylor and Francis, London, 1974).
2. N.F. Mott and E.A. Davis, "Electronic Processes in Non-Crystalline Materials", (Oxford, 1979).
3. N.F. Mott, in "The Metal-non-Metal Transition in Disordered Systems", L.R. Friedman and D.P. Tunstall, eds., (SUSSP, Edinburgh, 1978); and other articles therein.
4. M.H. Alexander and D.F. Holcomb, Rev. Mod. Phys. 40, 815 (1968).

5. R.F. Milligan, T.F. Rosenbaum, R.N. Bhatt, and G.A. Thomas, in "Electron-Electron Interactions in Disordered Systems", A.L. Efros, M. Pollak, eds., (Elsevier Science, Amsterdam, 1985), p. 231.
6. G.A. Thomas and M.A. Paalanen, LITPIM.
7. R.F. Milligan, in "Metal-Non-Metal Transition: An Important Facet of the Chemistry and Physics of Condensed Matter", P.P. Edwards and C.N.R. Rao, eds., (Taylor and Francis, London, 1985), in press.
8. R.F. Milligan and G.A. Thomas, in Annual Review of Physical Chemistry, 1984, in press.
9. G.A. Thomas, Phil. Mag. B50, 169 (1984) [Issue contains proceedings of a symposium on the metal-insulator transition honouring Prof. N.F. Mott, Bar Ilan Univ., Nov. 1983, organized by Prof. M. Kaveh.].
10. P.A. Lee and T.V. Ramakrishnan, Reviews of Modern Physics, April (1985); Phys. Rev. B26, 4009 (1982).
11 S. Hikami, in "Anderson Localization", ed. by Y. Nagaoka and H. Fukuyama (Springer Verlag, 1982), p.15. [This book is the proceedings of a conference held in Kyoto.]
12. N.F. Mott, Proc. Cambridge Philos. Soc., 32, 281 (1936); Proc. Phys. Soc. London, A62, 416 (1949); Phil. Mag. B37, 377 (1978).
13. P.W. Anderson, SSE; LITPIM.
14. G.A. Thomas, "Proc. 16th Int. Conf. Phys. Semiconductors", M. Averous, Ed., (North Holland, Amsterdam, 1983), p. 81.
15. E. Abrahams, P.W. Anderson, D.C. Licciardello, and T.V. Ramakrishnan Phys. Rev. Letters 42, 673 (1979).
16. L.P. Gorkov, A.I. Larkin, and Khmelnitskii, Sov. Phys. JETP Letters 30, 228 (1979).
17. F.J. Wegner, A. Phys. B25, 327 (1976).
18. F.J. Wegner, 35, 207 (1979).
19. F.J. Wegner, 44, 9 (1981).
20. Y. Imry, Phys. Rev. Lett 44, 469 (1980).
21. S. Hikami, Phys. Rev. B24, 2671 (1981).
22. S. Hikami, A.I. Larkin, and Y. Nagaoka, Prog. Theor. Phys. 63, 707 (1980).
23. S. Hikami, Prog. Theor. Phys. 64, 1466 (1980).
24. S. Hikami, Phys. Rev. B24, 2671 (1981).

25. K.B. Efetov, A.I. Larkin, and K.E. Khmelnitskii, Sov. Phys. JETP 52, 568 (1980).
26. K.F. Berggren, J. Phys C 15, L45 (1982).
27. D. Vollhardt and P. Wolfle, Phys. Rev. Lett. 45, 842 (1980).
28. A. McKinnon and B. Kramer, ibid., 47, 1546 (1981).
29. N.F. Mott, SSE.
30. N.F. Mott and M. Kaveh, J. Phys. C, 15, L45 (1982).
31. N.F. Mott, Philos. Mag. B 44, 256 (1981).
32. N.F. Mott, Proc. Roy. Soc. London A382, 1 (1982).
33. R.N. Bhatt and T.V. Ramakrishnan, Phys. Rev. B28, 6091 (1983).
34. H. Fukuyama, SSE.
35. H. Fukuyama, LITPIM.
36. H. Fukuyama, J. Phys. Soc. Japan 48, 2169 (1980).
37. H. Fukuyama, 50, 3407 (1981).
38. H. Fukuyama, 50, 3652 (1981).
39. H. Fukuyama, 51, 1105 (1982).
40. R.N. Bhatt, SSE.
41. A. Kawabata, SSE.
42. A. Kawabata, Solid State Comm. 34, 431 (1980).
43. A. Kawabata, J. Phys. Soc. Japan, 49, Suppl. A, 375 (1980).
44. A. Kawabata, ibid., 53, 318 (1984).
45. A. Kawabata, ibid., 53,1429 (1984).
46. E. Abrahams, LITPIM.
47. E. Abrahams, P.W. Anderson, and T.V. Ramakrishnan, Phil. Mag. B 42, 827 (1980).
48. K.F. Berggren, Phil. Mag. 30, 1 (1974).
49. B.E. Sernelius and K.F. Berggren, Phil. Mag. 43, 115 (1981).
50. W.L. McMillan, Phys. Rev. B24, 2739 (1981).
51. C. Castellani, C. DiCastro, P.A. Lee, M. Ma, and S. Sorella, Phys. Rev. B30, 1596 (1985).
52. C. Castellani, SSE.
53. C. Castellani, C. DiCastro, P.A. Lee, M. Ma, S. Sorella, and E. Tabet. Phys. Rev. B33, in press.
54. P.A. Lee, SSE.; Rev. Mod. Phys., in press.
55. A.M. Finkelstein, JETP Lett. 37, 517 (1983).
56. A.M. Finkelstein, Sov. Phys. JETP 57, 97 (1983).

57. T.F. Rosenbaum, K. Andres, G.A. Thomas, R.N. Bhatt, Phys. Rev. Lett. 45, 1723 (1980).
58. G.A. Thomas, T.F. Rosenbaum, and R.N. Bhatt, ibid., 46, 1435 (1981).
59. G.A. Thomas, Y. Ootuka, S. Katsumoto, S. Kobayashi, and W. Sasaki, Phys. Rev. B25, 4288 (1982).
60. G. Hertel, D.J. Bishop, E.G. Spencer, J.M. Rowell, and R.C. Dynes, Phys. Rev. Letters 50, 743 (1983).
61. D.J. Bishop, R.C. Dynes, and E.G. Spencer, SSE; LITPIM.
62. S. von Molnar, A. Briggs, J. Flouquet, G. Remenyi, Phys. Rev. Lett. 51, 706 (1983).
63. S. Washburn, R.A. Webb, S. von Molnar, F. Holtzberg, J. Flouquet and G. Remenyi, Phys. Rev. B30, 6224 (1985).
64. S. von Molnar, J. Flouquet, F. Holtzberg, and G. Remenyi, SSE.
65. N. Nishida, T. Furubayashi, M. Yamaguchi, H. Ishimoto and K. Morigaki, SSE.
66. M. Yamaguchi, N. Nishida, T. Furubayashi, K. Morigaki, H. Ishimoto, and K. Ono, Physica (b), 118, 694 (1983).
67. N. Nishida, T. Furubayashi, M. Yamaguchi, K. Morigaki, Y. Miura, Y. Takano, H. Ishimoto, and S. Ogawa, preprint.
68. S. Morita, SSE; Y. Isawa, T. Fukase, S. Ishida, Y. Koike, Y. Takeuti, N. Mikoshiba, Phys. Rev. B25, 5570 (1982).
69. S. Morita, N. Mikoshiba, Y. Koike, T. Fukase, M. Kitagawa, and S. Ishida, LITPIM.
70. R. Mansfield, M. Abdul-Gader and P. Fozooni, SSE.
71. J.M. Mochel, private comm.
72. W.L. McMillan and J. Mochel, Phys, Rev. Lett. 46, 556 (1981).
73. B.W. Dodson, W.L. McMillan, J.M. Mochel, and R.C. Dynes, Phys. Rev. Letters 46, 46 (1981).
74. T. Yoshizumi and T. Geballe, SSE.
75. H. Micklitz. LITPIM.
76. R. Ludwig and H. Micklitz, Solid State Comm., 50, 861 (1984).
77. A.T. Fiore and A.F. Hebard, Phys. Rev. Lett. 52, 2057 (1984).
78. M. Capizzi, G.A. Thomas, F. DeRosa, R.N. Bhatt, and T.M. Rice, Phys. Rev. Lett. 44, 1019 (1980).
79. H.F. Hess, K. DeConde, T.F. Rosenbaum, and G.A. Thomas, Phys. Rev. B25, 5578 (1982).

80. T.F. Rosenbaum, K. Andres, G.A. Thomas, P.A. Lee, Phys. Rev. Lett. 46, 568 (1981); T.F. Rosenbaum, R.F. Milligan, G.A. Thomas, P.A. Lee, T.V. Ramakrishnan, R.N. Bhatt, H. Hess, K. DeConde, T. Perry, Phys. Rev. Lett. 47, 1758 (1981).
81. G.A. Thomas, A. Kawabata, Y. Ootuka, S. Katsumoto, S. Kobayashi, and W. Sasaki, Phys. Rev. B24 4886 (1981).
82. M.A. Paalanen, T.F. Rosenbaum, G.A. Thomas, and R.N. Bhatt, Phys. Rev. Lett., 48, 1284 (1982).
83. R.N. Bhatt, Phys. Rev. B24, 3630 (1981).
84. R.N. Bhatt, Phys. Rev. B26, 1082 (1982).
85. M.A. Paalanen, T.F. Rosenbaum, G.A. Thomas and R.N. Bhatt, Phys. Rev. Letters 51, 1896 (1983).
86. G.A. Thomas, M. Paalanen and T.F. Rosenbaum Phys. Rev. B27, 3897 (1983).
87. G.A. Thomas, J. Phys. Chem. 88, 3749 (1983) (Proc. of Colloque Weyl, Asilomar, CA, 1983).
88. T.F. Rosenbaum, R.F. Milligan, M.A. Paalanen, G.A. Thomas, R.N. Bhatt, and W. Lin, Phys. Rev. B27, 7509 (1983).
89. P.F. Newman and D.F. Holcomb, Phys. Rev. B28, 638 (1983); P.F. Newman and D.F. Holcomb, Phys. Rev. Lett. 51, 2144 (1983).
90. M. Burns, G.A. Thomas, P. Hopkins and R.M. Westervelt, preprint (1986).
91. A.P. Long and M. Pepper, SSE.
92. M. Pepper and A.P. Long, LITPIM.
93. T.G. Castner, SSE.
94. T.G. Castner, N.K. Lee, G.S. Cieloszyk and G.L. Salinger, Phys. Rev. Lett. 34, 1627 (1975).
95. T.G. Castner, J. Low Temp. Phys. 38, 447 (1980); Philos. Mag. B42, 837 (1980).
96. H.S. Tan and T.G. Castner, Phys. Rev. B23, 3983 (1982).
97. B.L. Altshuler and A.G. Aronov, JETP Lett. 27, 662 (1978).
98. B.L. Altshuler and A.G. Aronov, Sov. Phys. JETP 50, 968 (1979).
99. B.L. Altshuler and A.G. Aronov, Solid State Comm. 30, 115 (1979).
100. B.L. Altshuler and A.G. Aronov, Solid State Comm. 38, 11 (1981).
101. B.L. Altshuler, A.G. Aronov, and D.E. Khmelnitskii, J. Phys. C,15, 7367 (1982).
102. B.L. Altshuler, K. Khmelnitzkii, A.I. Larkin, and P.A. Lee, Phys. Rev. B22, 5142 (1980).

103. B.L. Altshuler, A.G. Aronov, and P.A. Lee, Phys. Rev. Lett. 44, 1288 (1980).
104. B.L. Altshuler, A.G. Aronov, and P.A. Lee, Phys. Rev. B22, 5142 (1980).
105. A.M. Finkelstein, Z. Physik B 56, 189 (1984).
106. Y. Ootuka, S. Kobayashi, S. Ikehata, W. Sasaki, and J. Kondo, Solid State Comm. 30, 169 (1979).
107. W. Sasaki, SSE.; W. Sasaki, Phil. Mag. B42, 725 (1980).
108. W.F. Brinkman and T.M. Rice, Phys. Rev. B2, 4302 (1970).
109. Y. Ootuka, S. Katsumoto, S. Kobayashi, and W. Sasaki, SSE.
110. M.A. Paalanen, A.E. Ruckenstein, and G.A. Thomas, SSE.
111. A.E. Ruckenstein, M.A. Paalanen and G.A. Thomas, ICPS17.
112. G.A. Thomas, Comments on Condensed Matter Physics, in press.
113. P.W. Anderson, Phys. Rev. 109, 1492 (1958).
114. N.F. Mott, Philos. Mag. 26, 1015 (1972).
115. N.F. Mott, Adv. Phys. 21, 785 (1972).
116. I. Webman, J. Jortner, M.H. Cohen, Phys. Rev. B8, 2885 (1975).
117. S. Kirkpatrick, Rev. Mod. Phys. 45, 574 (1973).
118. N.F. Mott and M. Kaveh, J. Phys. C 14, L177 (1981).
119. N.F. Mott and M. Kaveh, ibid., 14, L183 (1981).
120. N.F. Mott and M. Kaveh, ibid., 14, L649 (1981).
121. H. Fritzsche, in L.R. Friedman and D.P. Tunstall, eds., "The Metal-non-Metal Transition in Disordered Systems", (SUSSP, Edinburgh, 1978), p.183. [Including a summary of some experimental results interpreted in terms of a minimum metallic conductivity.]
122. P.P. Edwards, M.J. Sienko, Phys. Rev. B17, 2575 (1978).
123. P.P. Edwards and M.J. Sienko, Int. Revs. of Phys. Chem., 3, 83 (1983).
124. A.P. Long and M. Pepper, J. Phys. C17, L425 (1984).
125. A.P. Long and M. Pepper, J. Phys. C17, 3391 (1984).
126. G. Biskupski, H. Dubois, J.L. Wojkiewicz, A. Briggs, and G. Remenyi, J. Phys. C17, L411 (1984).
127. D.J. Thouless, J. Phys. C8, 1803 (1975).
128. F.J. Wegner, Z. Phys. B25, 327 (1976).
129. D.J. Thouless, in "Ill Condensed Matter", R. Balian, R. Maynard, and G. Thoulouse, eds., (North Holland, Amsterdam, 1979), p.31.

130. D.J. Thouless, Phys. Rev. Lett. 39, 1167 (1979).
131. M. Kaveh, Phil. Mag., B50, (1984).
132. N.F. Mott and M. Kaveh, Phil. Mag. B47, L17 (1983).
133. N.F. Mott and M. Kaveh, Phil. Mag. B47, 577 (1983).
134. N.F. Mott, Phil. Mag. B43, 942 (1981).
135. N.F. Mott, Phil. Mag. B44, 256 (1981); ibid., 49, L75 (1984).
136. M. Kaveh and N.F. Mott, J. Phys. C15, L697 (1982); ibid., L707 (1982).
137. W. Götze, J. Phys. C 12, 1279 (1979).
138. D. Belitz, A. Gold and W. Götze, Z. Phys. B44, 273 (1981).
139. G.A. Thomas, M. Capizzi, F. DeRosa, R.N. Bhatt, and T.M. Rice, Phys. Rev. B 23, 5472 (1981).
140. M. Capizzi, G.A. Thomas, F. DeRosa, R.N. Bhatt, and T.M. Rice, Solid State Comm. 31, 611 (1979); H.K. Ng, M. Capizzi, G.A. Thomas and R.N. Bhatt, Phys. Rev. B33 , 7329 (1986).
141. R.N. Bhatt, and P.A. Lee, J. Appl. Phys. 52, 1703 (1981).
142. R.N. Bhatt, and P.A. Lee, Phys. Rev. Lett. 48, 344 (1982).
143. R.N. Bhatt, Phys. Rev. Lett. 48, 707 (1982).
144. See, e.g., A.M. Davidson, P. Knowles, P.A. Makado, R.A. Stradling, Z. Wasilewski, and S. Porowski, Solid State Sciences 24, 84 (1980).
145. J.B. Mock and G.A. Thomas, unpublished.
146. A.F. Ioffe and A.R. Regel, Prog. in Semiconductors, 4, 237 (1960).
147. Y. Imry, Y. Gefen, and D.J. Bergmann, Phys. Rev. B26, 3436 (1982).
148. R.C. Dynes and J.P. Garno, Phys. Rev. Lett. 46, 137 (1981).
149. A.G. Zabrodsky and K.N. Zinov'eva, Zh.E.T.P. 86, 727 (1984).
150. A.N. Ionov, I.S. Shlimak, and M.N. Matveev, Solid State Comm. 47, 763 (1983).
151. K. Epstein, E.D. Dahlberg, and A.M. Goldman, Phys. Rev. Lett. 43, 1889 (1979).
152. K. Epstein, A.M. Goldman, and A.M. Kadin, Phys. Rev. B27, 6685 (1983).
153. A. Mobius, H. Vinzelberg, C. Gladun, A. Heinrich, D. Elefant, J. Schumann, and G. Zies, J. Phys. C, in press.
154. A. Mobius, J. Phys. C, in press.
155. G.A. Thomas, Y. Ootuka, S. Katsumoto, S. Kobayashi, and W. Sasaki, Phys. Rev. B26, 2113 (1982).

156. M. Pollak and T.H. Geballe, Phys. Rev. 122, 1742 (1961).
157. I.G. Auston and N.F. Mott, Adv. Phys. 18, 41 (1969).
158. A.L. Efros and B.I. Shklovskii, J. Phys. C8, L49 (1975).
159. A.L. Efros, J. Phys. C9, 2021 (1976).
160. A.L. Efros, Phil. Mag. B43, 829 (1981).
161. B.I. Shklovskii and A.L. Efros, Sov. Phys. JETP 54, 218 (1981).
162. T. Takimori and H. Kamimura, J. Phys. C16, 5167 (1983).
163. T.M. Rice, SSE.
164. T.M. Rice and W.F. Brinkman, Phys. Rev. B5, 4350 (1972).
165. M.C. Gutzwiller, Phys. Rev. 137, A1726 (1965).
166. J. Hubbard, Proc. Roy. Soc. (London) A276, 238 (1963).
167. J. Hubbard, Proc. Roy. Soc. (London) A277, 238 (1964).
168. A. Gan and P.A. Lee, Phys. Rev. B33, 3595 (1986).
169. R.K. Sundfors and D.F. Holcomb, Phys. Rev. 136, 810 (1964).
170. S. Kobayashi, Y. Fukugawa, S. Ikehata, and W. Sasaki, J. Phys. Soc. Japan 45, 1276 (1978).
171. R.E. Walstedt, R.B. Kummer, S. Geschwind, V. Narayanamarti and G.E. Devlin, J. Appl. Phys. 50, 1700 (1979).
172. S. Ikehata, T.Ema, S. Kobayashi and W. Sasaki, J. Phys. Soc. Japan, 50, 3655 (1981).
173. M.P. Sarachik, D.R. He, W. Li, and J.S. Brooks, preprint.
174. K. Andres, R.N. Bhatt, P. Goalwin, T.M. Rice, and R.E. Walstedt, Phys. Rev. B24, 244 (1981).
175. N. Kobayashi, S. Ikehata, S.Kobayashi, and W. Sasaki, Solid State Comm. 24, 67 (1977).
176. R.N. Bhatt and T.M. Rice, Phil. Mag. B42, 859 (1980).
177. R.N. Bhatt and P.A. Lee, Phys. Rev. Lett. 48, 344 (1982).
178. M.A. Paalanen, D. Sachdev, and R.N. Bhatt, preprint (1986).
179. D. Sachdev and R.N. Bhatt, Preprint (1986).
180. A. Ruckenstein and G. Kotliar, preprint (1986).
181. J. Korringa, Physica (Utrecht) 16, 601 (1950).
182. P.A. Wolff, S.Y. Yuen, and G.A. Thomas, Phys. Rev. Lett., in press (1986).
183. G. Timp, P.D. Dresselhaus, T.C. Chieu, G. Dresselhaus, and Y. Iye Phys. Rev. B28, 7393 (1983).
184. T.F. Rosenbaum, S.B. Field, D.A. Nelson, and P.B. Littlewood, Phys. Rev. Lett. 54, 241 (1985).

185. M. Burns, P. Hopkins, B.I. Halperin, G.A. Thomas, and R.M. Westervelt, preprint (1986).

186. G.A. Thomas, Phil. Mag. B52, 479 (1985).

THE QUANTIZED HALL EFFECT *

The Nobel Foundation 1986

by KLAUS v. KLITZING,

Max-Planck-Institut für Festkörperforschung, D-7000 Stuttgart 80

1. *Introduction*

Semiconductor research and the Nobel Prize in physics seem to be contradictory since one may come to the conclusion that such a complicated system like a semiconuctor is not useful for very fundamental discoveries. Indeed, most of the experimental data in solid state physics are analyzed on the basis of simplified theories and very often the properties of a semiconductor device is described by empirical formulas since the microscopic details are too complicated. Up to 1980 nobody expected that there exists an effect like the Quantized Hall Effect which depends exclusively on fundamental constants and is not affected by irregularities in the semiconductor like impurities or interface effects.

The discovery of the Quantized Hall Effect (QHE) was the result of systematic measurements on silicon field effect transistors—the most important device in microelectronics. Such devices are not only important for applications but also for basic research. The pioneering work by Fowler, Fang, Howard and Stiles [1] has shown that new quantum phenomena become visible if the electrons of a conductor are confined within a typical length of 10 nm. Their discoveries opened the field of two-dimensional electron systems which since 1975 is the subject of a conference series [2]. It has been demonstrated that this field is important for the description of nearly all optical and electrical properties of microelectronic devices. A two-dimensional electron gas is absolutely necessary for the observation of the Quantized Hall Effect and the realization and properties of such a system will be discussed in chapter 2. In addition to the quantum phenomena connected with the confinement of electrons within a two-dimensional layer, another quantization—the Landau quantization of the electron motion in a strong magnetic field—is essential for the interpretation of the Quantized Hall Effect (chapter 3). Some experimental results will be summarized in chapter 4 and the application of the QHE in metrology is the subject of chapter 5.

2 *Two-Dimensional Electron Gas*

The fundamental properties of the QHE is a consequence of the fact, that the energy spectrum of the electronic system used for the experiments consists of a *discrete* energy spectrum. Normally, the energy E of mobile electrons in a

*The Nobel Foundation's permission to include the 1986 Nobel Lecture is gratefully acknowledged.

semiconductor is quasi continuous and can be compared with the kinetic energy of free electrons with wave vector k but with an effective mass m*

$$E = \frac{\hbar^2}{2m^*}(k_x^2 + k_y^2 + k_z^2) \tag{1}$$

If the energy for the motion in one direction (usually z-direction) is fixed one obtains a quasi two-dimensional electron gas (2DEG) and a strong magnetic field perpendicular to the two-dimensional plane will lead—as discussed later—to a fully quantized energy spectrum which is necessary for the observation of the QHE.

A two-dimensional electron gas can be realized at the surface of a semiconductor like silicon or galliumarsenide where the surface is usually in contact with a material which acts as an insulator (SiO_2 for silicon field effect transistors and e.g. $Al_xGa_{1-x}As$ for heterostructures). Typical cross-sections of such devices are shown in Fig 1. Electrons are confined close to the surface of the semiconductor by an electrostatic field F_z normal to the interface, originating from positive charges (see Fig. 1) which causes a drop in the electron potential towards the surface.

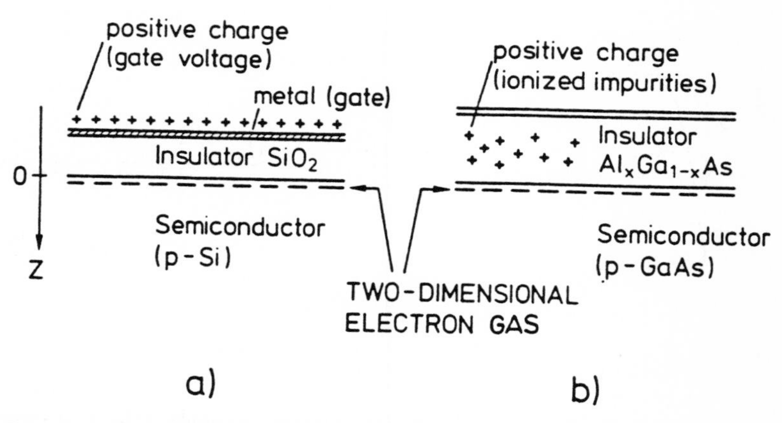

Fig. 1. A two-dimensional electron gas (2DEG) can be formed at the semiconductor surface if the electrons are fixed close to the surface by an external electric field. Silicon MOSFETs (a) and GaAs-$Al_xGa_{1-x}As$ heterostructures (b) are typical structures used for the realization of a 2DEG.

If the width of this potential well is small compared to the de Broglie wavelength of the electrons, the energy of the carriers are grouped in so-called electric subbands E_i corresponding to quantized levels for the motion in z-direction, the direction normal to the surface. In lowest approximation, the electronic subbands can be estimated by calculating the energy eigenvalues of an electron in a triangular potential with an infinite barrier at the surface ($z=0$) and a constant electric field F_s for $z \geqslant 0$ which keeps the electrons close to the surface. The result of such calculations can be approximated by the equation

$$E_j = \left(\frac{\hbar^2}{2m^*}\right)^{1/3} \cdot \left(\frac{3}{2}\pi e F_s\right)^{2/3} \cdot \left(j + \frac{3}{4}\right)^{2/3} \tag{2}$$

$j = 0, 1, 2\ldots$

In some materials, like silicon, different effective masses m^* and $m^{*\prime}$ may be present which leads to different series E_j and E_j'.

Equation 2 must be incorrect if the energy levels E_j are occupied with electrons since the electric field F_s will be screened by the electronic charge.

For a more quantitative calculation of the energies of the electric subbands it is necessary to solve the Schrödinger equation for the actual potential V_z' which changes with the distribution of the electrons in the inversion layer. Typical results of such calculation for both silicon MOSFETs and GaAs-heterostructures are shown in Fig. 2 [3,4]. Usually, the electron concentration of the two-dimensional system is fixed for a heterostructure (Fig. 1b) but can be varied in a MOSFET by changing the gate voltage.

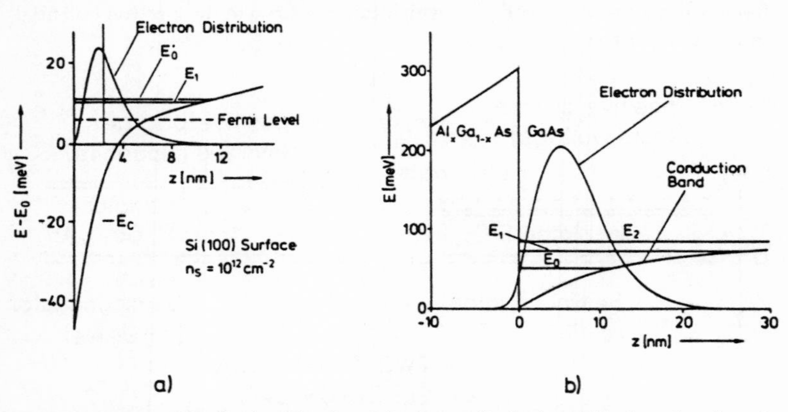

Fig. 2. Calculations of the electric subbands and the electron distribution within the surface channel of a silicon MOSFET (a) and a GaAs-$Al_xGa_{1-x}As$ heterostructure [3, 4].

Experimentally, the separation between electric subbands which is of the order of 10 meV, can be measured by analyzing the resonance absorption of electromagnetic waves with a polarization of the electric field perpendicular to the interface [5].

At low temperatures ($T<4$ K) and small carrier densities for the 2DEG (Fermi energy E_F relative to the lowest electric subbands E_0 small compared with the subband separation E_1-E_0) only the lowest electric subband is occupied with electrons (electric quantum limit) which leads to a strictly two-dimensional electron gas with an energy spectrum

$$E = E_0 + \frac{\hbar^2 k_{\parallel}^2}{2m^*} \tag{3}$$

$k_{\parallel}$: wavevector within the two-dimensional plane.

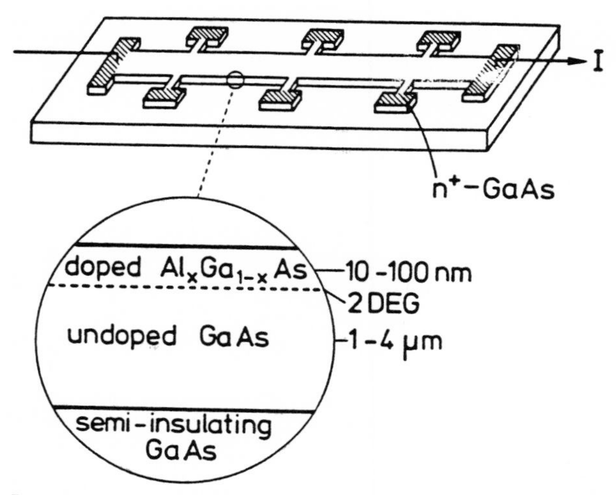

Fig. 3. Typical shape and cross-section of a GaAs-$Al_xGa_{1-x}As$ heterostructure used for Hall effect measurements.

For electrical measurements on a 2DEG, heavily doped n^+-contacts at the semiconductor surface are used as current contacts and potential probes. The shape of a typical sample used for QHE-experiments (GaAs-heterostructure) is shown in Fig. 3. The electrical current is flowing through the surface channel since the fully depleted $Al_xGa_{1-x}As$ acts as an insulator (the same is true for the SiO_2 of a MOSFET) and the p-type semiconductor is electrically separated from the 2DEG by a p-n junction. It should be noted that the sample shown in Fig. 3 is basically identical with new devices which may be important for the next computer generation [6]. Measurements related to the Quantized Hall Effect which include an analysis and characterization of the 2DEG are therefore important for the development of devices, too.

3. *Quantum Transport of a 2DEG in Strong Magnetic Fields*

A strong magnetic field B with a component B_z normal to the interface causes the electrons in the two-dimensional layer to move in cyclotron orbits parallel to the surface. As a consequence of the orbital quantization the energy levels of the 2DEG can be written schematically in the form

$$E_n = E_o(n+1/2)h\omega_c + s \cdot g \cdot \mu_B \cdot B \qquad (4)$$
$$n = 0,1,2,\ldots$$

with the cyclotron energy $h\omega_c = heB/m^*$, the spin quantum number $s = \pm 1/2$ the Landé factor g and the Bohr magneton μ_B ·

The wave function of a 2DEG in a strong magnetic field may be written in a form where the y-coordinate y_0 of the center of the cyclotron orbit is a good quatum number [7].

$$\psi = e^{ikx}\Phi_n(y-y_o) \qquad (5)$$

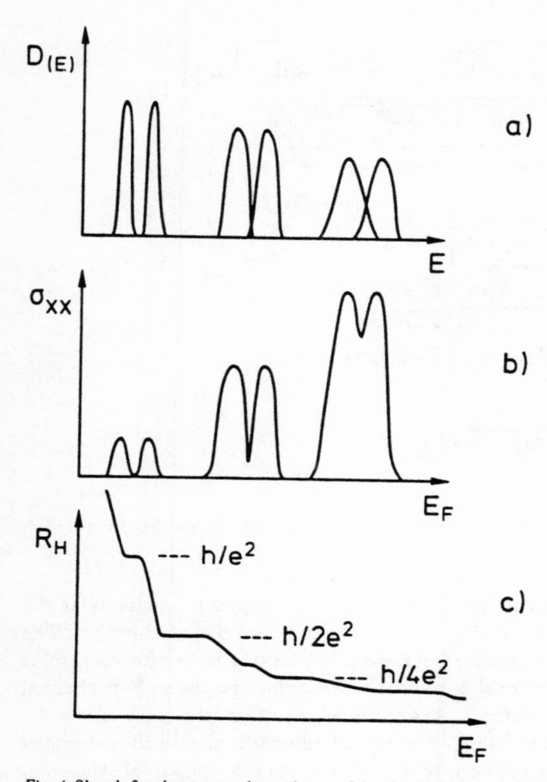

Fig. 4. Sketch for the energy dependence of the density of states (a), conductivity σ_{xx} (b), and Hall resistance R_H (c) at a fixed magnetic field.

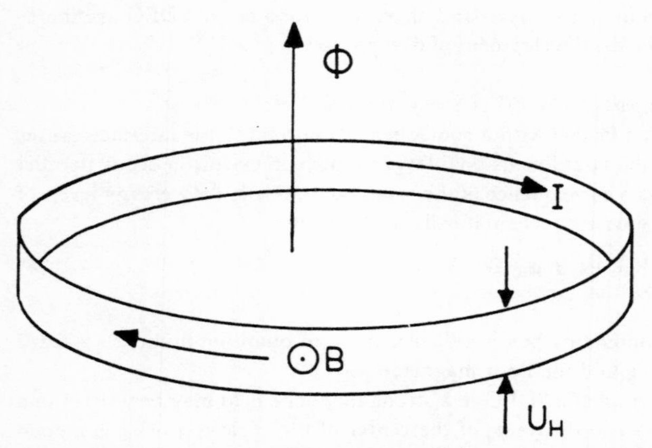

Fig. 5. Model of a two-dimensional metallic loop used for the derivation of the quantized Hall resistance.

where Φ_n is the solution of the harmonic-oscillator equation

$$\frac{1}{2m^*}\left[p_y^2 + (eB)^2y^2\right] \phi_n = (n + ½)\,\hbar\omega_c\,\phi_n \tag{6}$$

and y_0 is related to k by

$$y_0 = hk/eb \tag{7}$$

The degeneracy factor for each Landau level is given by the number of center coordinates y_0 within the sample. For a given device with the dimension $L_x \cdot L_y$, the center coordinates y_0 are separated by the amount

$$\Delta y_0 = \frac{\hbar}{eB}\Delta k = \frac{\hbar}{eB}\frac{2\pi}{L_x} = \frac{h}{eBL_x} \tag{8}$$

so that the degeneracy factor $N_0 = L_y/\Delta y_o$ is identical with $N_0 = L_xL_yeB/h$ the number of flux quanta within the sample. The degeneracy factor per unit area is therefore:

$$N = \frac{N_0}{L_xL_y} = \frac{eB}{h} \tag{9}$$

It should be noted that this degeneracy factor for each Landau level is independent of semiconductor parameters like effective mass.

In a more general way one can show [8] that the commutator for the center coordinates of the cyclotron orbit $[x_o, y_o] = ih/eB$ is finite which is equivalent to the result that each state occupies in real space the area $F_0 = h/eB$ corresponding to the area of a flux quantum.

The classical expression for the Hall voltage U_H of a 2DEG with a surface carrier density n_s is

$$U_H = \frac{B}{n_s \cdot e}\, I \tag{10}$$

where I is the current through the sample. A calculation of the Hall resistance $R_H = U_H/I$ under the condition that i energy levels are fully occupied ($n_s = iN$), leads to the expression for the quantized Hall resistance

$$R_H = \frac{B}{iN \cdot e} = \frac{h}{ie^2} \tag{11}$$

$$i = 1, 2, 3\ldots$$

A quantized Hall resistance is always expected if the carrier density n_s and the magnetic field B are adjusted in such a way that the filling factor i of the energy levels (Eq. 4)

$$i = \frac{n_s}{eB/h} \tag{12}$$

is an integer.

Under this condition the conductivity σ_{xx} (current flow in the direction of the electric field) becomes zero since the electrons are moving like free particles exclusively perpendicular to the electric field and no diffusion (originating from scattering) in the direction of the electric field is possible. Within the self-consistent Born approximation [9] the discrete energy spectrum broadens as shown in Fig. 4a. This theory predicts, that the conductivity σ_{xx} is mainly proportional to the square of the density of states at the Fermi energy E_f which leads to a vanishing conductivity σ_{xx} in the quantum Hall regime and quantized plateaus in the Hall resistance R_H (Fig. 4c).

The simple one electron picture for the Hall effect of an ideal two-dimensional system in a strong magnetic field leads already to the correct value for the quantized Hall resistance (Eq. 11) at integer filling factors of the Landau levels. However, a microscopic interpretation of the QHE has to include the influence of the finite size of the sample, the finite temperature, the electron-electron interaction, impurities and the finite current density (including the inhomogenious current distribution within the sample) on the experimental result. Up to now, no corrections to the value h/ie^2 of the quantized Hall resistance are predicted if the conductivity σ_{xx} is zero. Experimentally, σ_{xx} is never exactly zero in the quantum Hall regime (see chapter 4) but becomes unmeasurably small at high magetic fields and low temperatures. A quantitative theory of the QHE has to include an analysis of the longitudinal conducitivity σ_{xx} under real experimental conditions and a large number of publications are discussing the dependence of the conductivity on the temperature, magnetic field, current density, sample size etc. The fact that the value of the quantized Hall resistance seems to be exactly correct for $\sigma_{xx} = 0$ has led to the conclusion that the knowledge of microscopic details of the device is not necessary for a calculation of the quantized value. Consequently Laughlin [10] tried to deduce the result in a more general way from gauge invariances. He considered the situation shown in Fig. 5. A ribbon of two-dimensional system is bent into a loop and pierced everywhere by a magnetic field B normal to its surface. A voltage drop U_H is applied between the two edges of the ring. Under the condition of vanishing conductivity σ_{xx} (no energy dissipation), energy is conserved and one can write Faraday's law of induction in a form which relates the current I in the loop to the adiabatic derivative of the total energy of the system E with respect to the magnetic flux Φ threading the loop

$$I = \frac{\partial E}{\partial \phi} \tag{13}$$

If the flux is varied by a flux quantum $\Phi_0 = h/e$, the wavefunction enclosing the flux must change by a phase factor 2π corresponding to a transistion of a state with wavevector k into its neighbourstate $k + (2\pi)/(L_x)$ where L_x is the circumference of the ring. The total change in energy corresponds to a transport of states from one edge to the other with

$$\Delta E = i \cdot e \cdot U_H \tag{14}$$

The integer i corresponds to the number of filled Landau levels if the free electron model is used, but can be in principle any positive or negative integer number.

From Eq. (13) the relation between the dissipationless Hall current and the Hall voltage can be deduced

$$I = i \cdot e \cdot U_H/\phi_0 = i \frac{e^2}{h} \cdot U_H \tag{15}$$

which leads to the quantized Hall resistance $R_H = \frac{h}{ie^2}$.

In this picture the main reason for the Hall quantization is the flux quantization h/e and the quantization of charge into elementary charges e. In analogy, the fractional quantum Hall effect, which will not be discussed in this paper, is interpreted on the basis of elementary excitations of quasiparticles with a charge $\frac{e}{3}, \frac{e}{5}, \frac{e}{7}$ etc.

The simple theory predicts that the ratio between the carrier density and the magnetic field has to be adjusted with very high precision in order to get exactly integer filling factors (Eq. 12) and therefore quantized values for the Hall resistance. Fortunately, the Hall quantization is observed not only at special magnetic field values but in a wide magnetic field range, so that an accurate fixing of the magnetic field or the carrier density for high precision measurements of the quantized resistance value is not necessary. Experimental data of such Hall plateaus are shown in the next chapter and it is believed that localized states are responsible for the observed stabilization of the Hall resistance at certain quantized values.

After the discovery of the QHE a large number of theoretical paper were published discussing the influence of localized states on the Hall effect [11−14] and these calculations demonstrate that the Hall plateaus can be explained if localized states in the tails of the Landau levels are assumed. Theoretical investigations have shown that a mobility edge exists in the tails of Landau levels separating extended states from localized states [15−18]. The mobility edges are located close to the center of a Landau level for long-range potential fluctuations. Contrary to the conclusion reached by Abrahams, et al [19] that all states of a two-dimensional system are localized, one has to assume that in a strong magnetic field at least one state of each Landau level is extended in order to observe a quantized Hall resistance. Some calculations indicate that the extended states are connected with edge states [17].

In principle, an explanation of the Hall plateaus without including localized states in the tails of the Landau levels is possible if a reservoir of states is present outside the two-dimensional system [20, 21]. Such a reservoir for electrons, which should be in equilibrium with the 2DEG, fixes the Fermi energy within the energy gap between the Landau levels if the magnetic field or the number of electrons is changed. However, this mechanism seems to be more unlikely than localization in the the tails of the Landau levels due to disorder. The following discussion assumes therefore a model with extended and localized states within one Landau level and a density of states as sketched in Fig. 6.

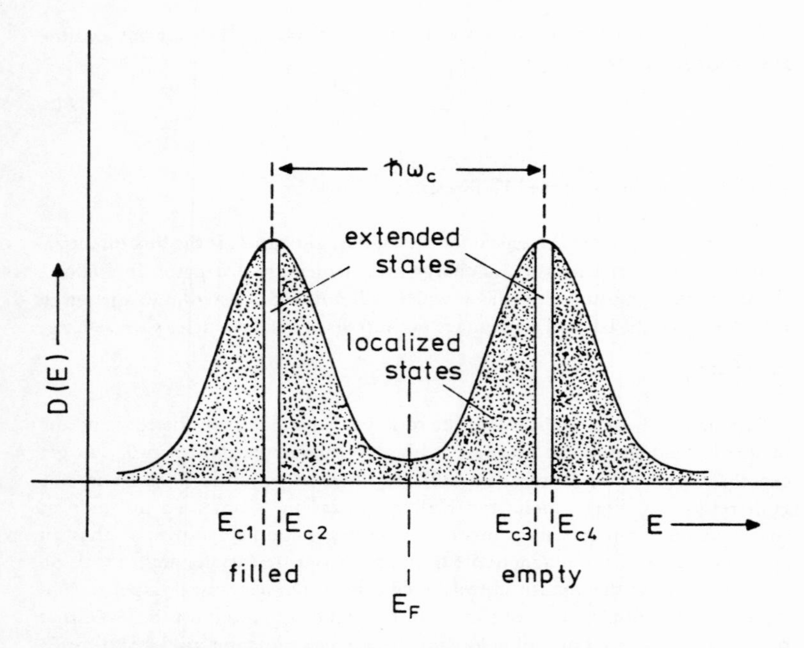

Fig. 6. Model for the broadened density of states of a 2 DEG in a strong magnetic field. Mobility edges close to the center of the Landau levels separate extended states from localized states.

4. *Experimental Data*

Magneto-quantumtransport measurements on two-dimensional systems are known and published for more than 20 years. The first data were obtained with silicon MOSFETs and at the beginning mainly results for the conductivity σ_{xx} as a function of the carrier density (gate voltage) were analyzed. A typical curve is shown in Fig. 7. The conductivity oscillates as a function of the filling of the Landau levels and becomes zero at certain gate voltages V_g. In strong magnetic fields σ_{xx} vanishes not only at a fixed value V_g but in a range ΔV_g and Kawaji was the first one who pointed out that some kind of immobile electrons must be introduced [22] since the conductivity σ_{xx} remains zero even if the carrier density is changed. However, no reliable theory was available for a discussion of localized electrons whereas the peak value of σ_{xx} was well explained by calculations based on the self consistent Born approximation and short range scatterers which predict $\sigma_{xx} \sim (n + 1/2)$ independent of the magnetic field.

The theory for the Hall conductivity is much more complicated and in the lowest approximation one expects that the Hall conductivity σ_{xy} deviates from the classical curve $\sigma_{xy}^{\circ} = -\dfrac{n_s e}{B}$ (where n_s is the total number of electrons in the two-dimensional system per unit area) by an amount $\Delta\sigma_{xy}$ which depends mainly on the third power of the density of states at the Fermi energy [23].

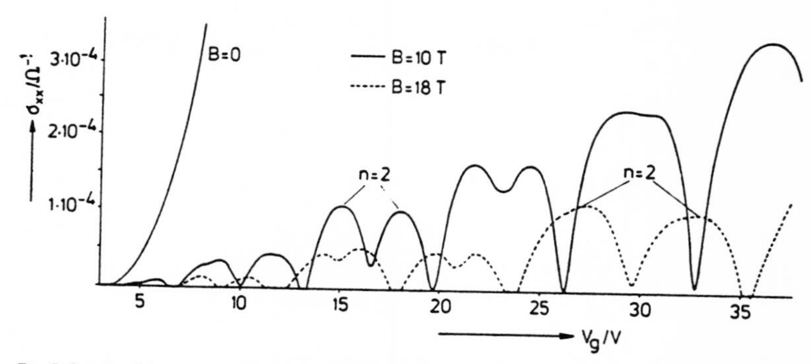

Fig. 7. Conductivity σ_{xx} of a silicon MOSFET at different magnetic fields B as a function of the gate voltage V_g.

However, no agreement between theory and experiment was obtained. Today, it is believed, that $\Delta\sigma_{xy}$ is mainly influenced by localized states which can explain the fact that not only a positive but also a negative sign for $\Delta\sigma_{xy}$ is observed. Up to 1980 all experimental Hall effect data were analyzed on the basis of an incorrect model so that the quantized Hall resistance, which is already visible in the data published in 1978 [24] remained unexplained.

Whereas the conductivity σ_{xx} can be measured directly by using a Corbino disk geometry for the sample, the Hall conductivity is not directly accessible in an experiment but can be calculated from the longitudinal resistivity ϱ_{xx} and the Hall resistivity ϱ_{xy} measured on samples with Hall geometry (see Fig. 3).

$$\sigma_{xy} = -\frac{\rho_{xy}}{\rho_{xx}^2 + \rho_{xy}^2} \qquad \sigma_{xx} = \frac{\rho_{xx}}{\rho_{xx}^2 + \rho_{xy}^2} \tag{16}$$

Fig. 8 shows measurements for ϱ_{xx} and ϱ_{xy} of a silicon MOSFET as a function of the gate voltage at a fixed magnetic field. The corresponding σ_{xx}- and σ_{xy} -data are calculated on the basis of Eq. (16).

The classical curve $\sigma_{xy}{}^\circ = -\frac{n_s e}{B}$ in Fig. 8 is drawn on the basis of the incorrect model, that the experimental data should lie always below the classical curve (= fixed sign for $\Delta\,\sigma_{xy}$) so that the plateau value σ_{xy} = const. (observable in the gate voltage region where σ_{xx} becomes zero) should change with the width of the plateau. Wider plateaus should give smaller values for $|\sigma_{xy}|$. The main discovery in 1980 was [25] that the value of the Hall resistance in the plateau region is not influenced by the plateau width as shown in Fig. 9. Even the aspect ratio L/W (L = length, W = width of the sample), which influences normally the accuracy in Hall effect measurements, becomes unimportant as shown in Fig. 10. Usually, the measured Hall resistance $R_H{}^{exp}$ is always smaller than the theoretical value $R_H{}^{theor} = \rho_{xy}$ [26, 27]

$$R_H{}^{exp} = G \cdot R_H{}^{theor} \quad G < 1 \tag{17}$$

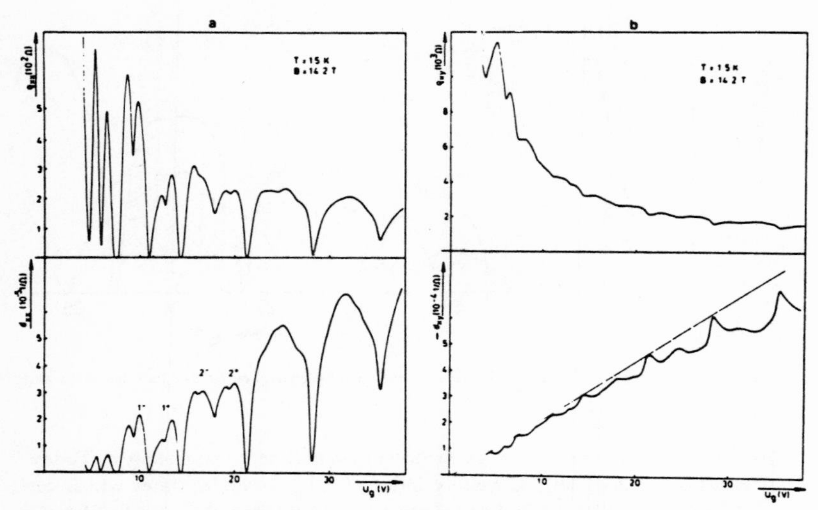

Fig. 8. Measured ϱ_{xx}- and ϱ_{xy}-data of a silicon MOSFET as a function of the gate voltage at B = 14,2 T together with the calculated σ_{xx}- and σ_{xy}-curves.

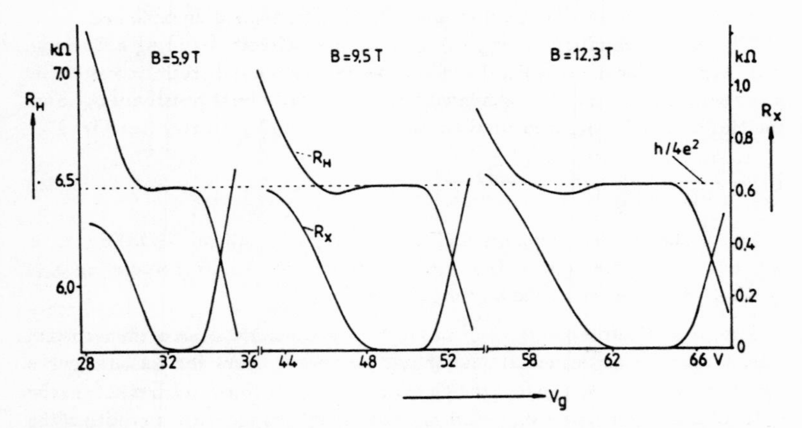

Fig. 9. Measurements of the Hall resistance R_H and the resistivity R_x as a function of the gate voltage at different magnetic field values. The plateau values $R_H = h/4e^2$ are independent of the width of the plateaus.

However, as shown in Fig. 11, the correction 1-G becomes zero (independent of the aspect ratio) if $\sigma_{xx} \rightarrow 0$ or the Hall angle θ approaches 90° ($\tan\theta = \frac{\sigma_{xy}}{\sigma_{xx}}$). This means, that any shape of the sample can be used in QHE-experiments as long as the Hall angle is 90° (or $\sigma_{xx} = 0$). However, outside the plateau region ($\sigma_{xx} \sim \rho_{xx} \neq 0$) the measured Hall resistance $R_H^{exp} = \frac{U_H}{I}$ is indeed always

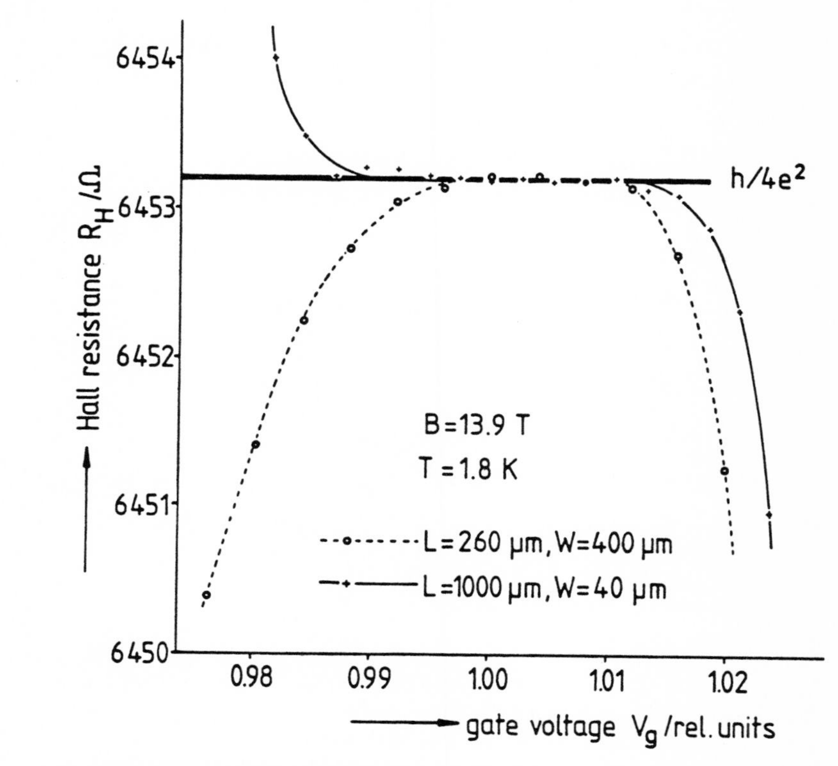

Fig. 10. Hall resistance R_H for two different samples with different aspect ratios L/W as a function of the gate voltage (B = 13,9T).

smaller than the theoretical ρ_{xy}-value [28]. This leads to the experimental result that an additional minimum in R_H^{exp} becomes visible outside the plateau region as shown in Fig. 9 which disappears if the correction due to the finite length of the sample is included (see Fig. 12). The first high precision measurements in 1980 of the plateau value in R_H (V_g) showed already that these resistance values are quantized in integer parts of h/e^2 = 25812,8 Ω within the experimental uncertainty of 3 ppm.

The Hall plateaus are much more pronounced in measurements on GaAs-$Al_xGa_{1-x}As$ heterostructures since the small effective mass m* of the electrons in GaAs (m*(Si)/m*(GaAs)>3) leads to a relatively large energy splitting between Landau levels (Eq. 4) and the high quality of the GaAs-$Al_xGa_{1-x}As$ interface (nearly no surface roughness) leads to a high mobility μ of the electrons so that the condition μB>1 for Landau quantizations is fulfilled already at relatively low magnetic fields. Fig. 13 shows that well developed Hall plateaus are visible for this material already at a magnetic field strength of 4 Tesla. Since a finite carrier density is usually present in heterostructures even

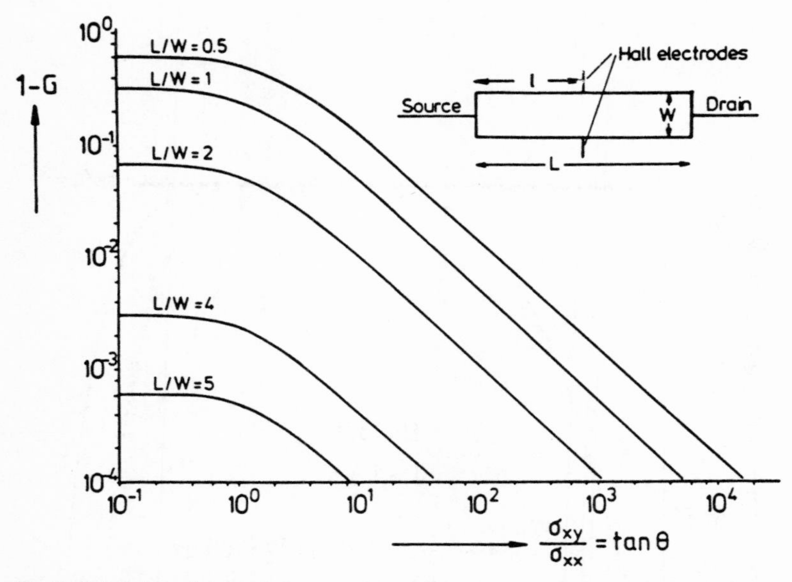

Fig. 11. Calculations of the correction term G in Hall resistance measurements due to the finite length to width ratio L/W of the device (l/L= 0,5).

at a gate voltage $V_g = 0V$, most of the published transport data are based on measurements without applied gate voltage as a function of the magnetic field. A typical result is shown in Fig. 14. The Hall resistance $R_H = \varrho_{xy}$ increases steplike with plateaus in the magnetic field region where the longitudinal resistance ϱ_{xx} vanishes. The width of the ϱ_{xx}-peaks in the limit of zero temperature can be used for a determination of the amount of extended states and the analysis [29] show, that only few percent of the states of a Landau level are not localized. The fraction of extended states within one Landau level decreases with increasing magnetic field (Fig. 15) but the number of extended states within each level remains approximately constant since the degeneracy of each Landau level increases proportional to the magnetic field.

At finite temperatures ϱ_{xx} is never exactly zero and the same is true for the slope of the ϱ_{xy}-curve in the plateau region. But in reality, the slope $d\varrho_{xy}/dB$ at $T<2K$ and magnetic fields above 8 Tesla is so small that the ϱ_{xy}-value stays constant within the experimental uncertainty of $6 \cdot 10^{-8}$ even if the magnetic field is changed by 5%. Simultanously the resistivity ϱ_{xx} is usually smaller than $1m\Omega$. However, at higher temperatures or lower magnetic fields a finite resistivity ϱ_{xx} and a finite slope $d\varrho_{xy}/dn_s$ (or $d\varrho_{xy}/dB$) can be measured. The data are well described within the model of extended states at the energy position of the undisturbed Landau level E_n and a finite density of localized states between the Landau levels (mobility gap). Like in amorphous systems, the temperature dependence of the conductivity σ_{xx} (or resistivity ϱ_{xx}) is

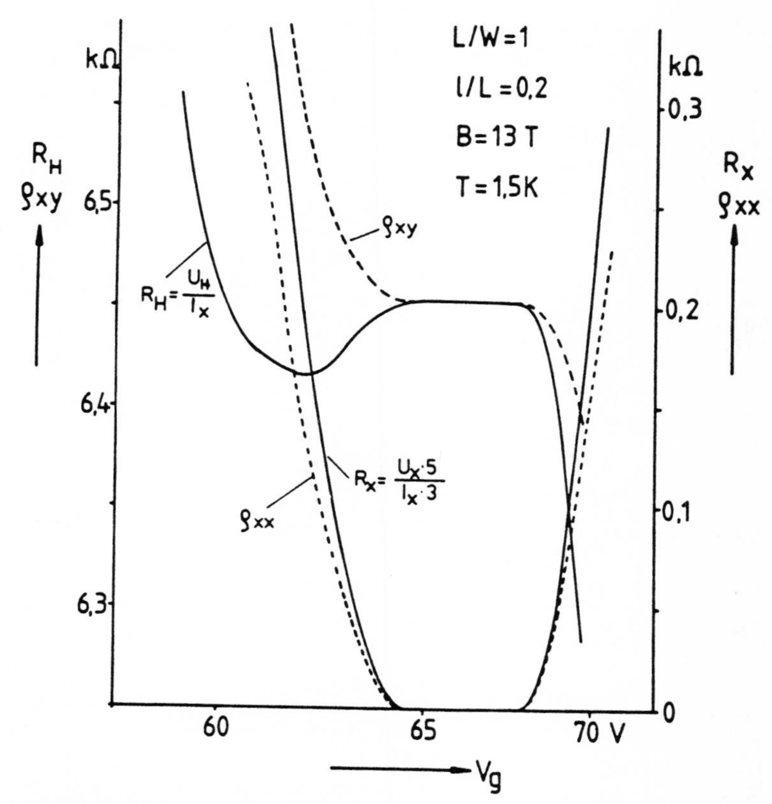

Fig. 12. Comparison between the measured quantities R_H and R_x and the corresponding resistivity components ϱ_{xy} and ϱ_{xx}, respectively.

thermally activated with an activation energy E_a corresponding to the energy difference between the Fermi energy E_F and the mobility edge. The largest activation energy with a value $E_a = 1/2\hbar\omega_c$ (if the spin splitting is negligibly small and the mobility edge is located at the center E_n of a Landau level) is expected if the Fermi energy is located exactly at the midpoint between two Landau levels.

Experimentally, an activated resistivity

$$\varrho_{xx} \sim \exp[-(E_a/kT)] \tag{18}$$

is observed in a wide temperature range for different two-dimensional systems (deviations from this behaviour, which appear mainly at temperatures below 1K, will be discussed separately) and a result is shown in Fig. 16. The activation energies (deduced from these data) are plotted in Fig. 17 for both, silicon MOSFETs and GaAs-$Al_xGa_{1-x}As$ heterostructures as a function of the

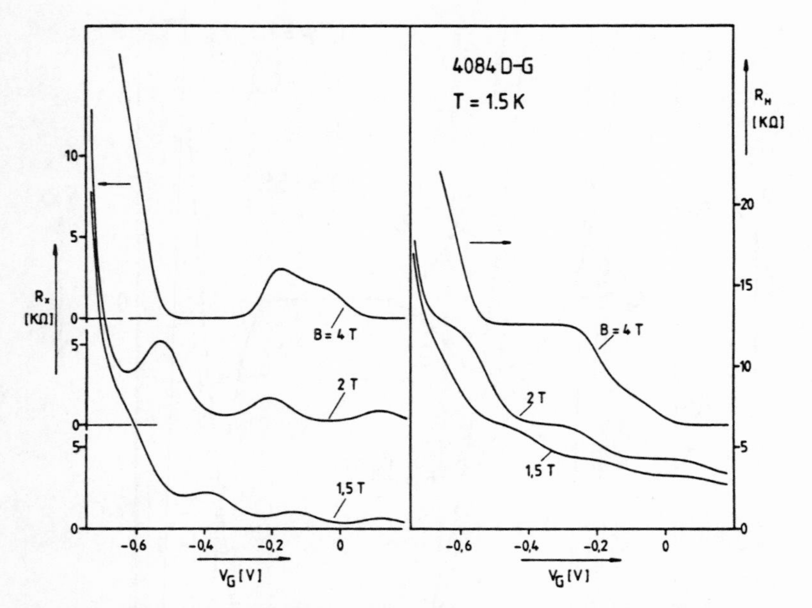

Fig. 13. Measured curves for the Hall resistance R_H and the longitudinal resistance R_x of a GaAs-$Al_xGa_{1-x}As$ heterostructure as a function of the gate voltage at different magnetic fields.

magnetic field and the data agree fairly well with the expected curve $E_a = 1/2\hbar\omega_c$. Up to now, it is not clear whether the small systematic shift of the measured activation energies to higher values originates from a temperature dependent prefactor in Eq. (18) or is a result of the enhancement of the energy gap due to many body effects.

The assumption, that the mobility edge is located close to the center of a Landau level E_n is supported by the fact that for the samples used in the experiments only few percent of the states of a Landau level are extended [29]. From a systematic analysis of the activation energy as a function of the filling factor of a Landau level it is possible to determine the density of states D(E) [30]. The surprising result is, that the density of states (DOS) is finite and approximately constant within 60 % of the mobility gap as shown in Fig. (18). This background DOS depends on the electron mobility as summarized in Fig. (19).

An accurate determination of the DOS close to the center of the Landau level is not possible by this method since the Fermi energy becomes temperatur dependent if the DOS changes drastically within the energy range of 3kT. However, from an analysis of the capacitance C as a function of the Fermi energy the peak value of the DOS and its shape close to E_n can be deduced [31, 32].

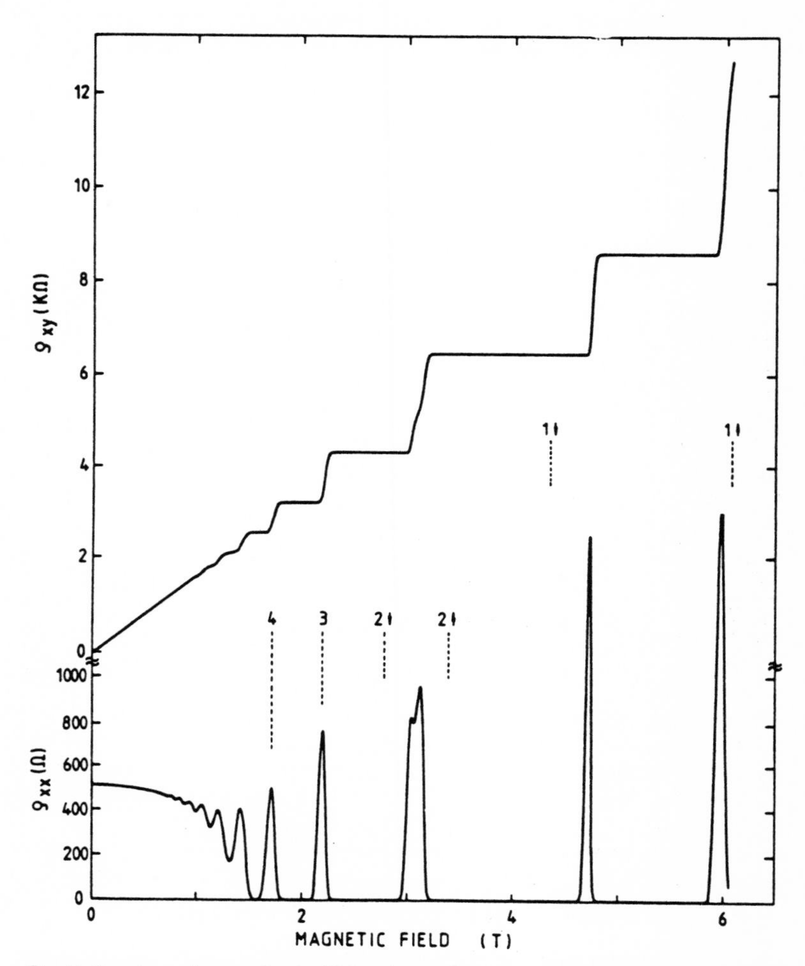

Fig. 14. Experimental curves for the Hall resistance $R_H = \varrho_{xy}$ and the resistivity $\varrho_{xx} \sim R_x$ of a heterostructure as a function of the magnetic field at a fixed carrier density corresponding to a gate voltage $V_g = 0V$. The temperature is about 8mK.

This analysis is based on the equation

$$\frac{1}{C} = \frac{1}{e^2 \cdot D\,(E_F)} + \text{const.} \tag{19}$$

The combination of the different methods for the determination of the DOS leads to a result as shown in Fig. (20). Similar results are obtained from other experiments, too [33, 34] but no theoretical explanation is available.

If one assumes that only the occupation of extended states influences the Hall effect, than the slope $d\varrho_{xy}/dn_s$ in the plateau region should be dominated

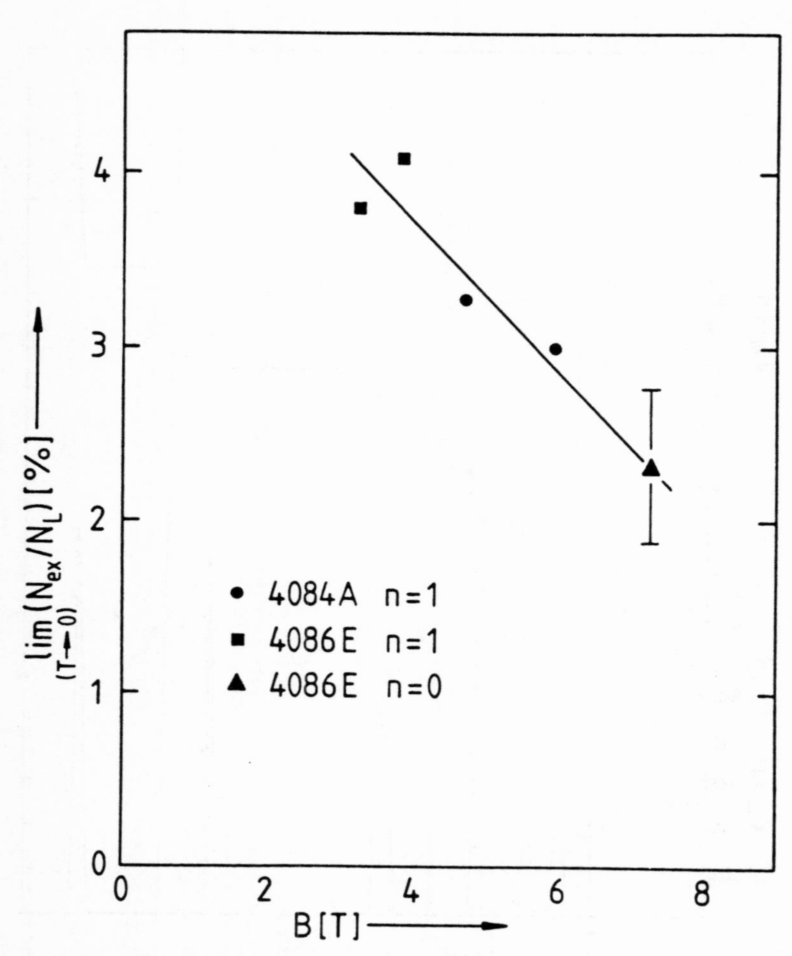

Fig. 15. Fraction of extended states relative to the number of states of one Landau level as a function of the magnetic field.

by the same activation energy as found for $\varrho_{xx}(T)$. Experimentally [35], a one to one relation between the minimal resistivity $\varrho_{xx}{}^{min}$ at integer filling factors and the slope of the Hall plateau has been found (Fig. 21) so that the flatness of the plateau increases with decreasing resistivity, which means lower temperature or higher magnetic fields.

The temperatur dependence of the resistivity for Fermi energies within the mobility gap deviates from an activated behaviour at low temperatures, typically at $T<1K$. Such deviations are found in measurements on disordered

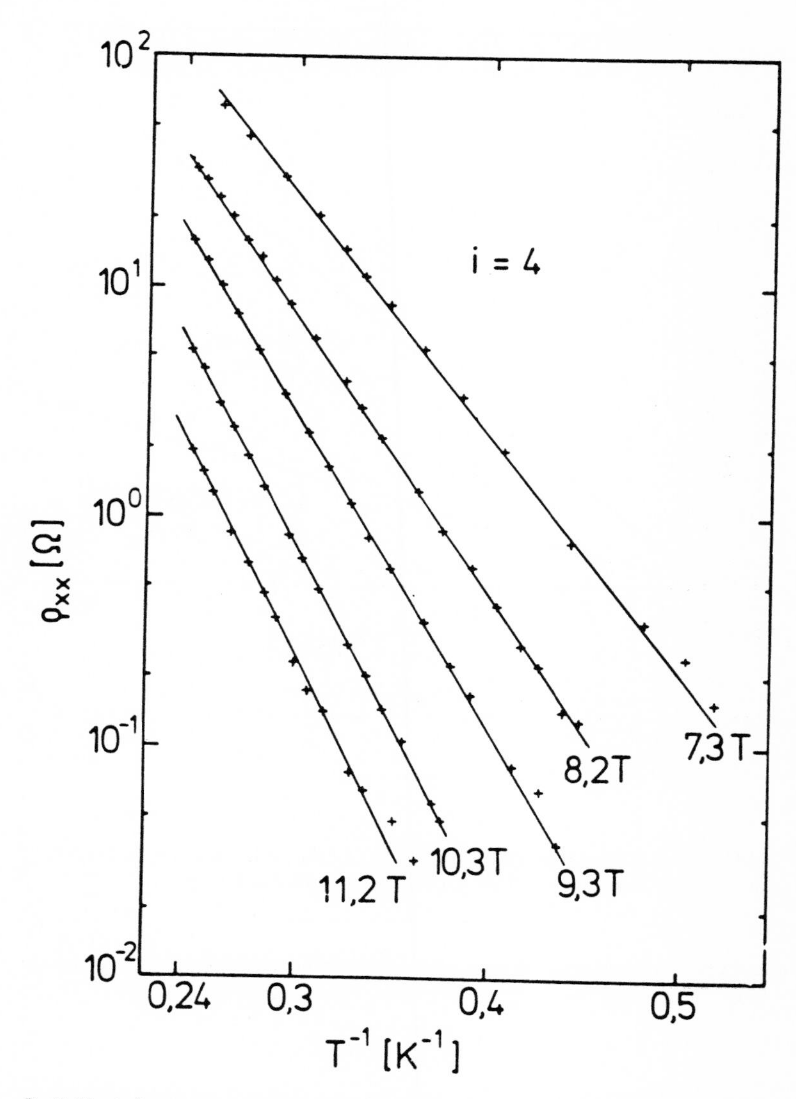

Fig 16. Thermally activated resistivity ρ_{xx} at a filling factor i = 4 for a silicon MOSFET at different magnetic field values.

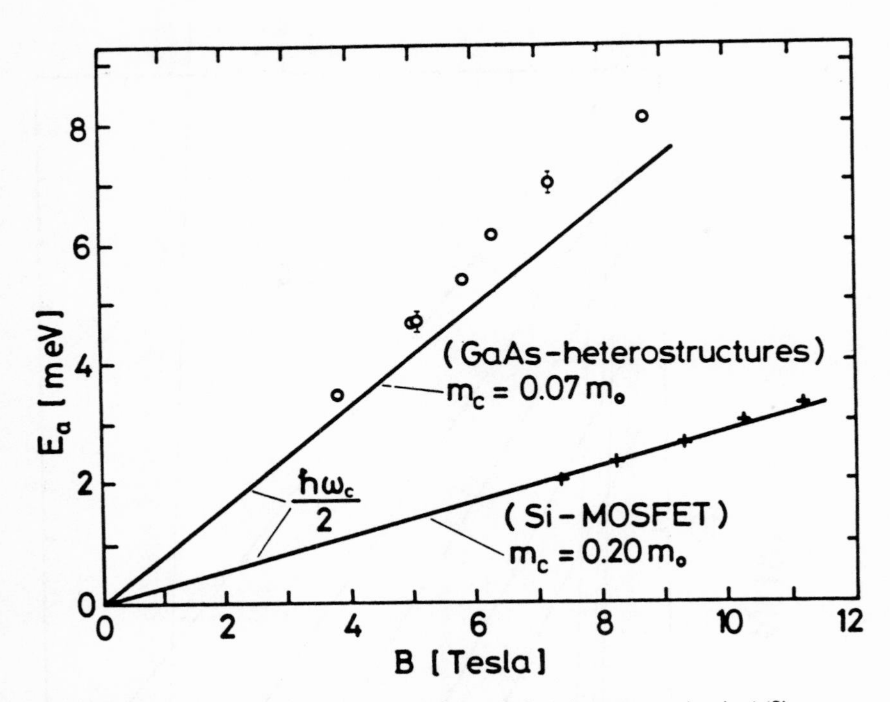

Fig. 17. Measured activation energies at filling factors i = 2 (GaAs heterostructure) or i = 4 (Si-MOSFET) as a function of the magnetic field. The data are compared with the energy $0{,}5\hbar\omega_c$.

systems, too, and are interpreted as variable range hopping. For a two-dimensional system with exponentially localized states a behaviour

$$\varrho_{xx} \sim \exp\left[-(T_0/T)^{1/3}\right] \tag{20}$$

is expected. For a Gaussian localization the following dependence is predicted [36, 37]

$$\rho_{xx} \sim \frac{1}{T} \exp\left\{-(T_0/T)^{1/2}\right\} \tag{21}$$

The analysis of the experimental data demonstrates (Fig. 22) that the measurements are best described on the basis of Eq. (21). The same behaviour has been found in measurements on another two-dimensional system, on InP-InGaAs heterostructures [38].

The contribution of the variable range hopping (VRH) process to the Hall effect is negligibly small [39] so that experimentally the temperature dependence of $d\varrho_{xy}/dn_s$ remains thermally activated even if the resistivity ϱ_{xx} is dominated by VRH.

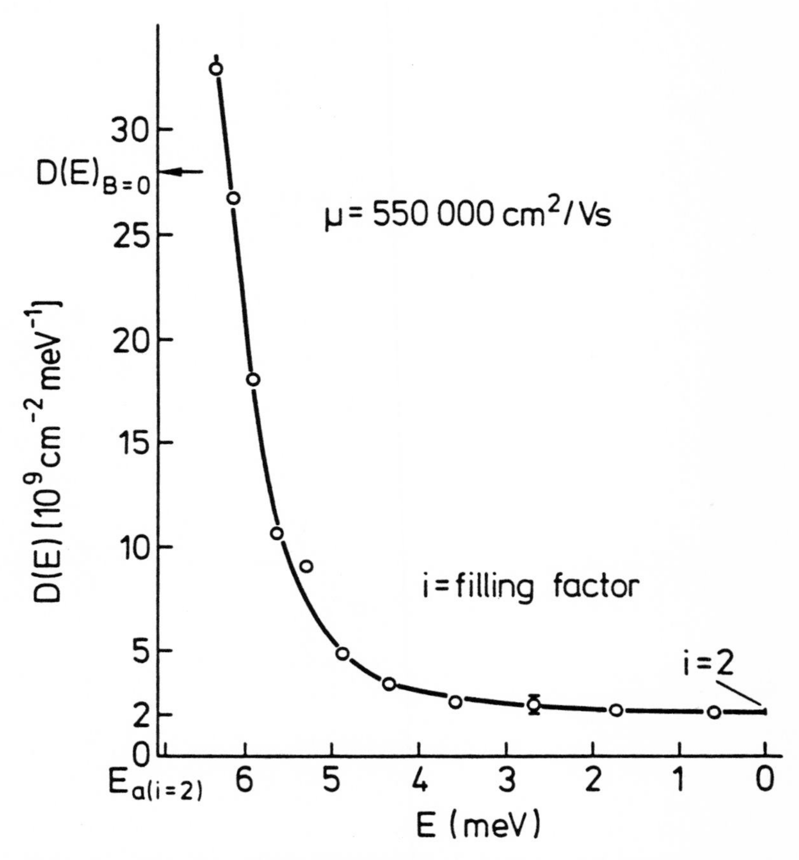

Fig. 18. Measured density of states (deduced from an analysis of the activated resisitivity) as a function of the energy relative to the center between two Landau levels (GaAs-heterostructure).

The QHE breaks down if the Hall field becomes larger than about $E_H = 60V/cm$ at magnetic fields of 5 Tesla.

This corresponds to a classical drift velocity $v_D = \frac{E_H}{B} \approx 1200m/s$. At the critical Hall field E_H (or current density j) the resistivity increases abruptly by orders of magnitude and the Hall plateau disappears. This phenomenon has been observed by different authors for different materials [40–47]. A typical result is shown in Fig. 23. At a current density of $j_c = 0,5$ A/m the resistivity ϱ_{xx} at the center of the plateau (filling factor $i = 2$) increases drastically. This instability, which develops within a time scale of less than 100 ns seems to originate from a runaway in the electron temperature but also other mechanism like electric field dependent delocalization, Zener tunneling or emission of

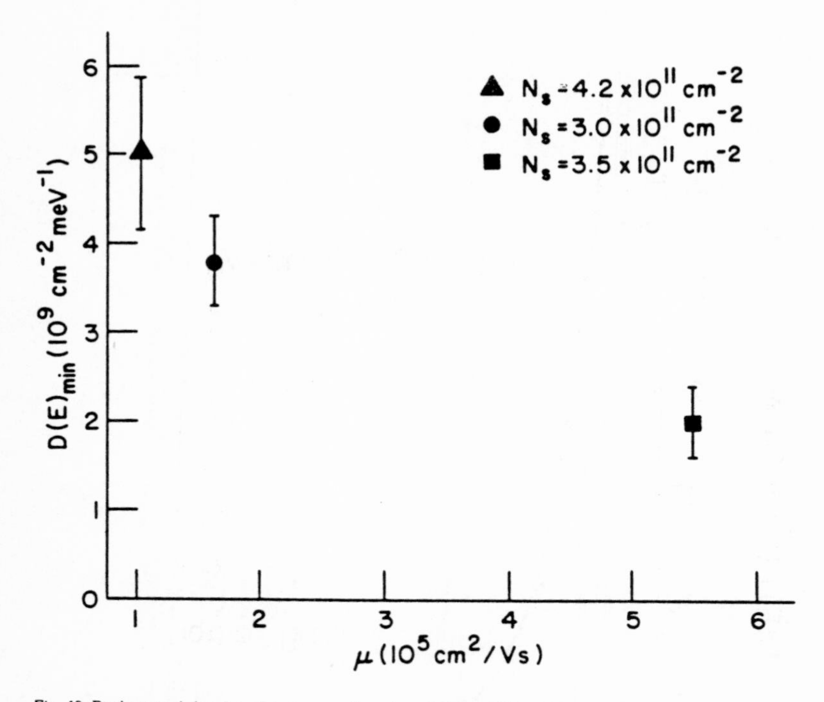

Fig. 19. Background density of states as a function of the mobility of the device.

acoustic phonons, if the drift velocity exceeds the sound velocity, can be used for an explanation [48–50].

Fig. 23 shows that ϱ_{xx} increases already at current densities well below the critical value j_c which may be explained by a broadening of the extended state region and therefore a reduction in the mobility gap ΔE. If the resistivity ϱ_{xx} is thermally activated and the mobility gap changes linearly with the Hall field (which is proportional to the current density j) than a variation

$$\ln \varrho_{xx} \sim j$$

is expected. Such a dependence is seen in Fig. 24 but a quantitative analysis is difficult since the current distribution within the sample is usually inhomogenious and the Hall field, calculated from the Hall voltage and the width of the sample, represents only a mean value. Even for an ideal two-dimensional system an inhomogenious Hall potential distribution across the width of the sample is expected [51–53] with an enhancement of the current density close to the boundaries of the sample.

The experimental situation is still more complicated as shown in Fig. 25. The potential distribution depends strongly on the magnetic field. Within the plateau region the current path moves with increasing magnetic field across the

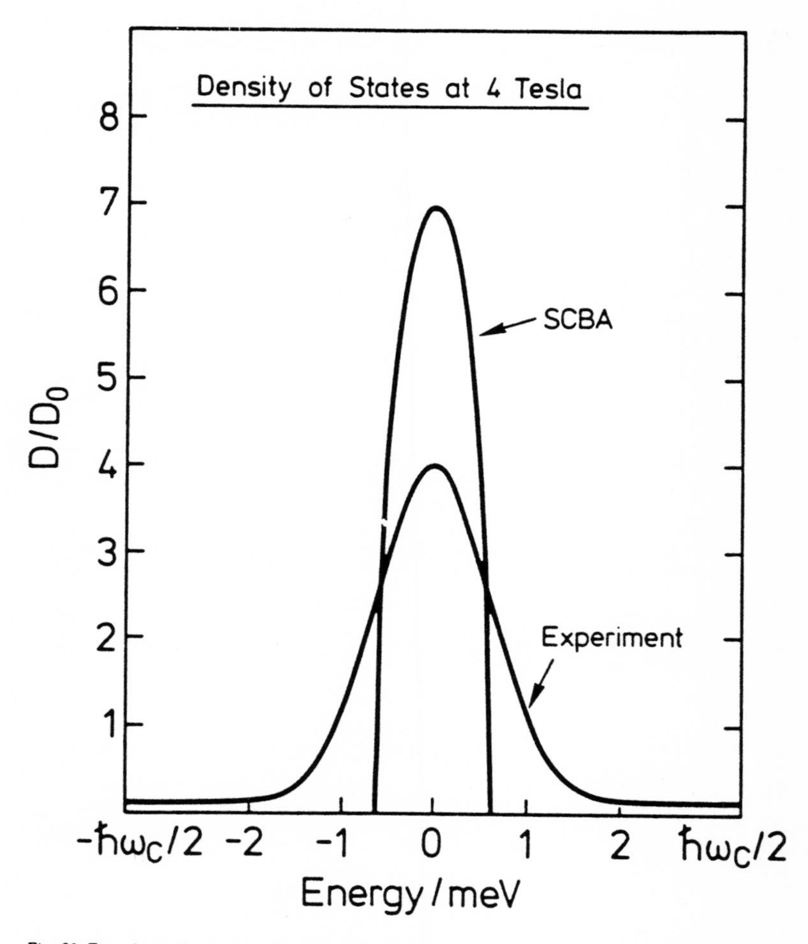

Fig. 20. Experimentally deduced density of states of a GaAs heterostructure at B = 4T compared with the calculated result based on the self-consistent Born approximation (SCBA).

width of the sample from one edge to the other one. A gradient in the carrier density within the two-dimensional system seems to be the most plausible explanation but in addition an inhomogeneity produced by the current itself may play a role. Up to now, not enough microscopic details about the two-dimensional system are known so that at present a microscopic theory, which describes the QHE under real experimental conditions, is not available. However, all experiments and theories indicate that in the limit of vanishing resistivity ϱ_{xx} the value of the quantized Hall resistance depends exclusively on fundamental constants. This leads to a direct application of the QHE in metrology.

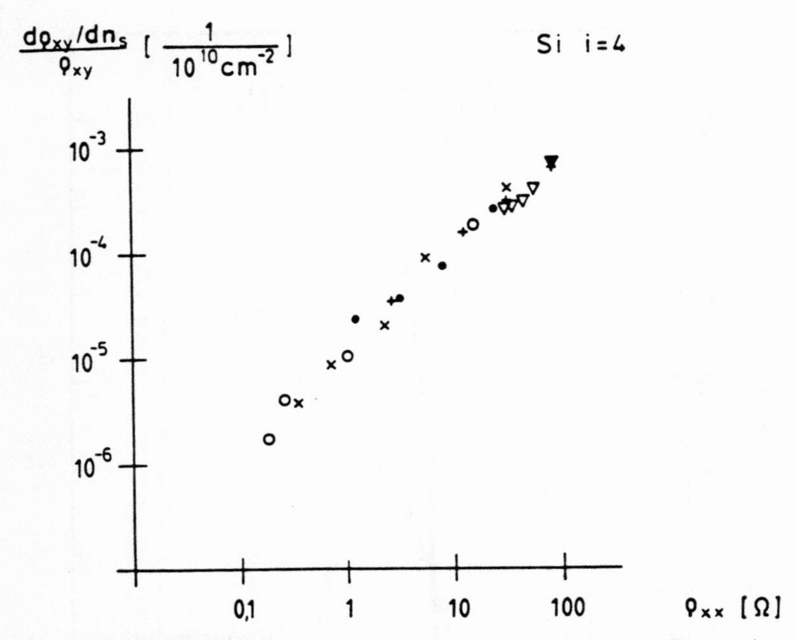

Fig. 21. Relation between the slope of the Hall plateaus $d\varrho_{xy}/dn_s$ and the corresponding ϱ_{xx}-value at integer filling factors.

5. *Application of the Quantum Hall Effect in Metrology*

The applications of the Quantum Hall Effect are very similar to the applications of the Josephson-Effect which can be used for the determination of the fundamental constant h/e or for the realization of a voltage standard. In analogy, the QHE can be used for a determination of h/e^2 or as a resistance standard. [54].

Since the inverse fine structure constant α^{-1} is more or less identical with h/e^2 (the proportional constant is a fixed number which includes the velocity of light), high precision measurements of the quantized Hall resistance are important for all areas in physics which are connected with the finestructure constant.

Experimentally, the precision measurement of α is reduced to the problem of measuring an electrical resistance with high accuracy and the different methods and results are summarized in the Proceedings of the 1984 Conference on Precision Electromagnetic Measurements (CPEM 84) [55]. The mean value of measurements at laboratories in three different countries is

$$\alpha^{-1} = 137{,}035988 \pm 0.00002$$

The internationally recommended value (1973) is

$$\alpha^{-1} = 137{,}03604 \pm 0.00011$$

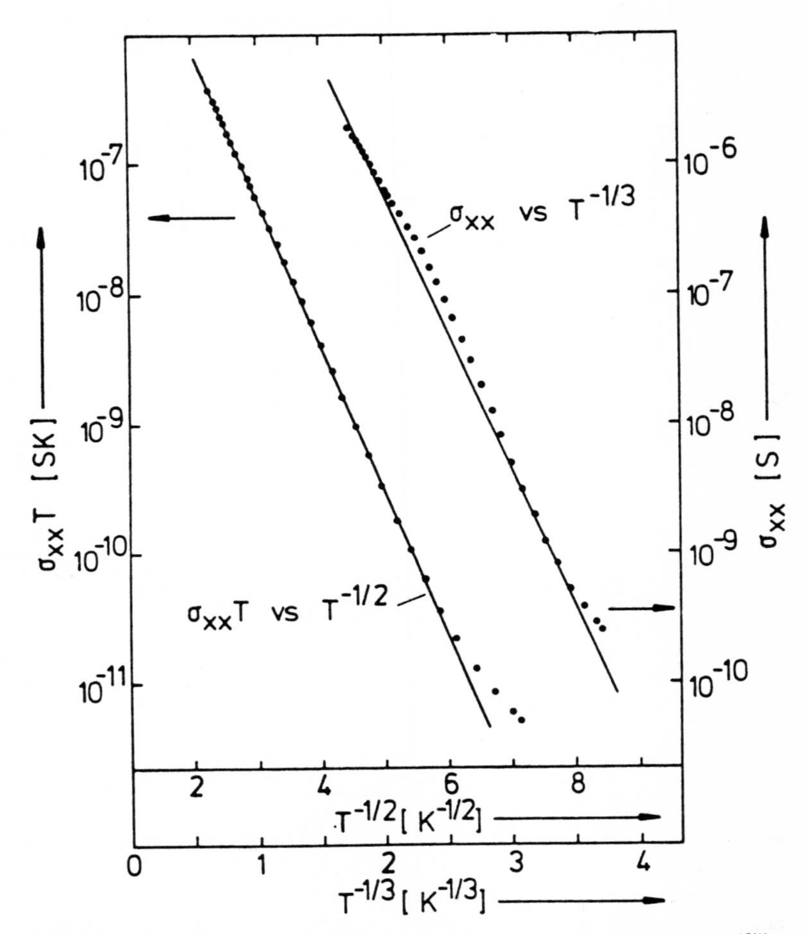

Fig. 22. Analysis of the temperature dependent conductivity of a GaAs heterostructure (filling factor i = 3) at T < 0,2K.

and the preliminary value for the finestructure constant based on a new least square adjustment of fundamental constants (1985) is

$$\alpha^{-1} = 137{,}035991 \pm 0.000008$$

Different groups have demonstrated that the experimental result is within the experimental uncertainty of less than $3.7 \cdot 10^{-8}$ independent of the material (Si, GaAs, $In_{0.53}Ga_{0.47}As$) and of the growing technique of the devices (MBE or MOCVD) [56]. The main problem in high precision measurements of α is—at present—the calibration and stability of the reference resistor. Fig. 26 shows the drift of the maintained 1Ω-resistor at different national laboratories. The

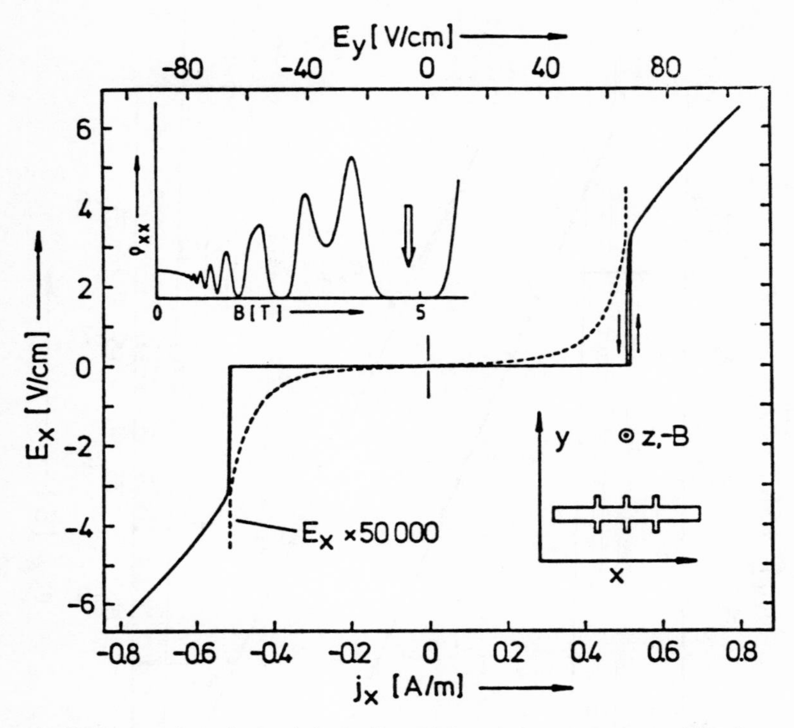

Fig. 23. Current-voltage characteristic of a GaAs-$Al_xGa_{1-x}As$ heterostructure at a filling factor i = 2 (T = 1,4K). The device geometry and the $\varrho_{xx}(B)$-curve are shown in the inserts.

very first application of the QHE is the determination of the drift coefficient of the standard resistors since the quantized Hall resistance is more stable and more reproducible than any wire resistor. A nice demonstration of such an application is shown in Fig. 27. In this experiment the quantized Hall resistance R_H has been measured at the "Physikalisch Technische Bundesanstalt" relative to a reference resistor R_R as a function of time. The ratio R_H/R_R changes approximately linearly with time but the result is independent of the QHE-sample. This demonstrates that the reference resistor changes his value with time. The one standard deviation of the experimental data from the mean value is only $2.4 \cdot 10^{-8}$ so that the QHE can be used already today as a relative standard to maintain a laboratory unit of resistance based on wire-wound resistors. There exists an agreement that the QHE should be used as an absolute resistance standard if three independent laboratories measure the same value for the quantized Hall resistance (in SI-units) with an uncertainty of less than $2 \cdot 10^{-7}$. It is excepted that these measurements will be finished until the end of 1986.

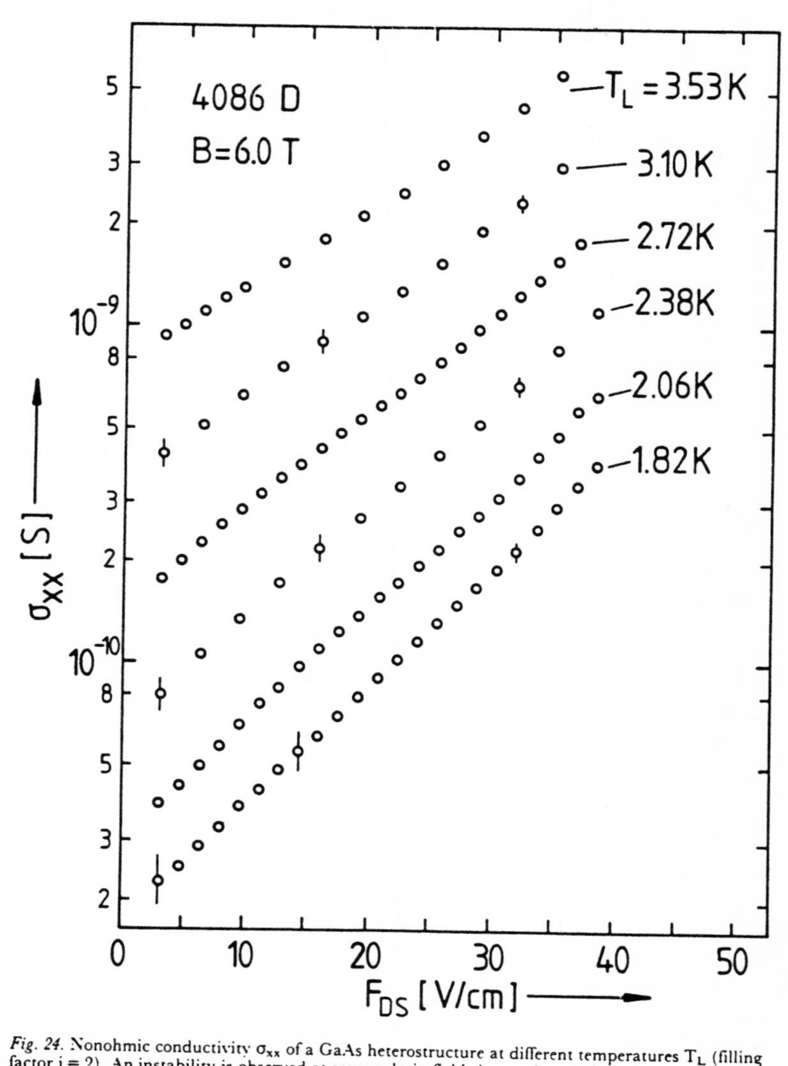

Fig. 24. Nonohmic conductivity σ_{xx} of a GaAs heterostructure at different temperatures T_L (filling factor i = 2). An instability is observed at source-drain fields larger than 40 V/cm.

Acknowledgements

The publicity of the Nobel Prize has made clear that the research work connected with the Quantum Hall Effect was so successful because a tremendous large number of institutions and individuals supported this activity. I would like to thank all of them and I will mention by name only those scientists who supported my research work at the time of the discovery of the QHE in

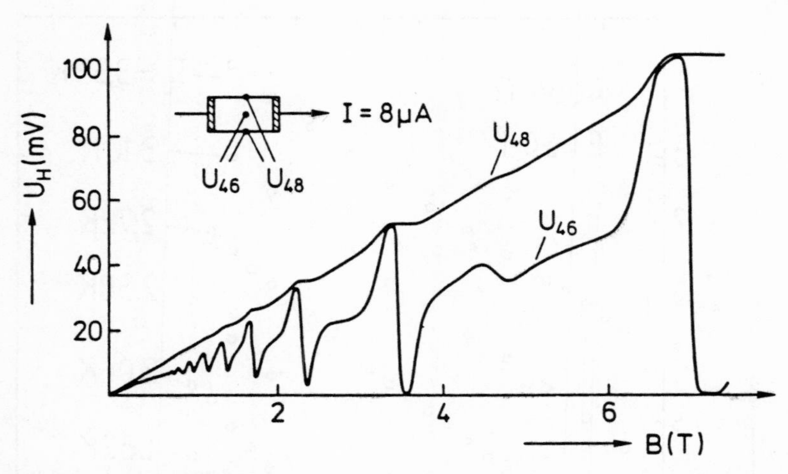

Fig. 25. Measured Hall potential distribution of a GaAs heterostructure as a function of the magnetic field.

1980. Primarily, I would like to thank G. Dorda (Siemens Forschungslaboratorien) and M. Pepper (Cavendish Laboratory, Cambridge) for providing me with high quality MOS-devices. The continuous support of my research work by G. Landwehr and the fruitful discussions with my coworker, Th. Englert, were essential for the discovery of the Quantum Hall Effect and are greatfully acknowledged.

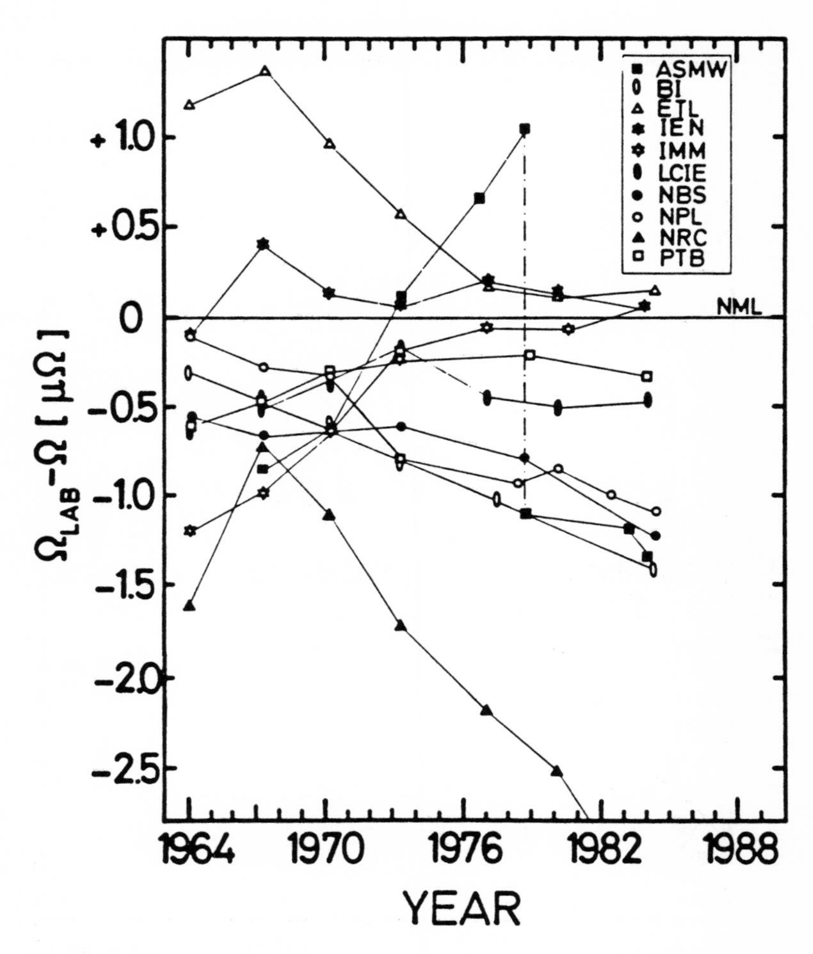

Fig. 26. Time dependence of the 1 Ω standard resistors maintained at the different national laboratories.

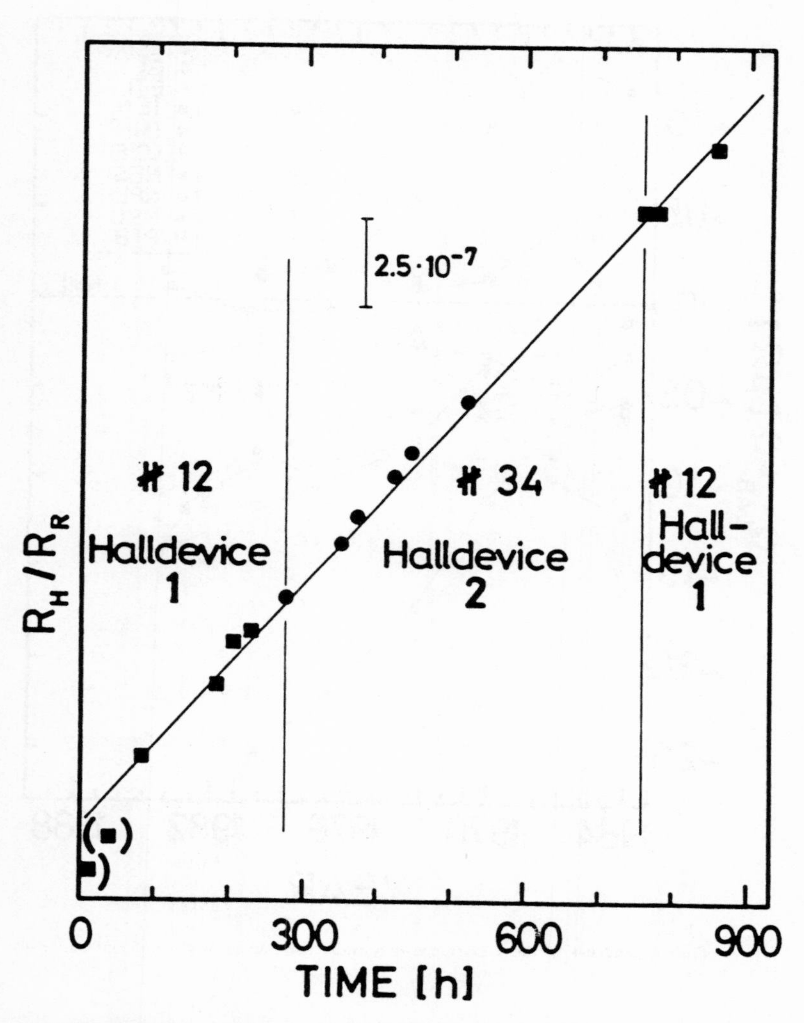

Fig. 27. Ratio R_H/R_R between the quantized Hall resistance R_H and a wire resistor R_R as a function of time. The result is time dependent but independent of the Hall device used in the experiment.

REFERENCES

[1] A.B. Fowler F.F. Fang, W.E. Howard and P.J. Stiles Phys. Rev. Letter *16,* 901 (1966)
[2] For a review see: Proceedings of the Int. Conf. on Electronic Properties of Two-Dimensional Systems, Surf. Sci. *58,* (1976), *73* (1978), *98* (1980, *113* (1982), *142* (1984)
[3] F. Stern and W.E. Howard, Phys. Rev. *163,* 816 (1967)
[4] T. Ando, J. Phys. Soc., Jpn. *51,* 3893 (1982)
[5] J.F. Koch, Festkörperprobleme (Advances in Solid State Physics), H.J. Queisser, Ed. (Pergamon-Vieweg, Braunschweig. 1975) Vol. XV, p. 79
[6] T. Mimura, Surf. Science *113,* 454 (1982)
[7] R.B. Laughlin, Surface Science *113,* 22 (1982)
[8] R. Kubo, S.J. Miyake and N. Hashitsume, in Solid State Physics, Vol. 17, 269 (1965), F. Seitz and D. Turnball, Eds., (Academic Press, New York, 1965)
[9] T. Ando, J. Phys. Soc. Jpn. *37,* 1233 (1974)
[10] R. B. Laughlin in Springer Series in Solid State Sciences *53,* p. 272, G. Bauer, F. Kuchar and H. Heinrich, Eds. (Springer Verlag, 1984)
[11] R.E. Prange, Phys. Rev. B *23,* 4802 (1981)
[12] H. Aoki and T. Ando, Solid State Commun. *38,* 1079 (1981)
[13] J. T. Chalker, J. Phys. C *16,* 4297 (1983)
[14] W. Brenig, Z. Phys. *50B,* 305 (1983)
[15] A. Mac Kinnon, L. Schweitzer and B. Kramer, Surf. Sci. *142,* 189 (1984)
[16] T. Ando, J. Phys. Soc. Jpn. *52,* 1740 (1983)
[17] L. Schweitzer, B. Kramer and A. Mac Kinnon, J. Phys. C *17,* 4111 (1984)
[18] H. Aoki and T. Ando, Phys. Rev. Letters *54,* 831 (1985)
[19] E. Abrahams, P.W. Anderson, D.C. Licciardello and T.V. Ramakrishnan, Phys. Rev. Letters *42,* 673 (1979)
[20] G.A. Baraff and D.C. Tsui, Phys. Rev. B *24,* 2274 (1981)
[21] T. Toyoda, V. Gudmundsson and Y. Takahashi, Phys. Letters *102A,* 130 (1984)
[22] S. Kawaji and J. Wakabayashi, Surf. Schi. *58,* 238 (1976)
[23] S. Kawaji, T. Igarashi and J. Wakabayashi, Progr. in Theoretical Physics *57,* 176 (1975)
[24] T. Englert and K. v. Klitzing, Surf. Sci. *73,* 70 (1978)
[25] K. v. Klitzing, G. Dorda and M. Pepper, Phys. Rev. Letters *45,* 494 (1980)
[26] K. v. Klitzing, H. Obloh, G. Ebert, J. Knecht and K. Ploog, Prec. Measurement and Fundamental Constants II, B.N. Taylor and W.D. Phillips, Eds., Natl. Burl. Stand. (U.S.), Spec. Publ. *617,* (1984) p. 526
[27] R.W. Rendell and S.M. Girvin, Prec. Measurement and Fundamental Constants II, B.N. Taylor and W.D. Phillips. Eds. Natl. Bur. Stand. (U.S.), Spec. Publ. *617,* (1984) p. 557
[28] K. v. Klitzing, Festkörperprobleme (Advances in Solid State Physics), XXI, 1 (1981), J. Treusch, Ed., (Vieweg, Braunschweig)
[29] G. Ebert, K. v. Klitzing, C. Probst and K. Ploog, Solid State Commun. *44,* 95 (1982)
[30] E. Stahl, D. Weiss, G. Weimann, K. v. Klitzing and K. Ploog, J. Phys. C *18,* L 783 (1985)
[31] T.P. Smith, B.B. Goldberg, P.J. Stiles and M. Heiblum, Phys. Rev. B *32,* 2696 (1985)
[32] V. Mosser, D. Weiss, K. v. Klitzing, K. Ploog and G. Weimann to be published in Solid State Commun.
[33] E. Gornik, R. Lassnig, G. Strasser, H.L. Störmer, A.C. Gossard and W. Wiegmann, Phys. Rev. Lett. *54,* 1820 (1985)
[34] J.P. Eisenstein, H.L. Störmer, V. Navayanamurti, A.Y. Cho and A.C. Gossard, Yamada Conf. XIII on Electronic Properties of Two-Dimensional Systems, p. 292 (1985)

[35] B. Tausendfreund and K. v. Klitzing, Surf. Science *142,* 220 (1984)
[36] M. Pepper, Philos, Mag. *37,* 83 (1978)
[37] Y. Ono, J. Phys. Soc. Jpn. *51,* 237 (1982)
[38] Y. Guldner, J.P. Hirtz, A. Briggs, J.P. Vieren, M. Voos and M. Razeghi, Surf. Science *142,* 179 (1984)
[39] K.I. Wysokinski and W. Brenig, Z. Phys. B – Condensed Matter *54,* 11 (1983)
[40] G. Ebert and K. v. Klitzing, J. Phys. C 5441 (1983)
[41] M.E. Cage, R.F. Dziuba, B.F. Field, E.R. Williams, S.M. Girvin, A.C. Gossard, D.C. Tsui and R.J. Wagner, Phys. Rev. Letters *51,* 1374 (1983)
[42] F. Kuchar, G. Bauer, G. Weimann and H. Burkhard, Surf. Science *142*, 196 (1984)
[43] H.L. Störmer, A.M. Chang, D.C. Tsui and J.C.M. Hwang, Proc. 17th ICPS, San Francisco 1984
[44] H. Sakaki, K. Hirakawa, J. Yoshino, S.P. Svensson, Y. Sekiguchi, T. Hotta and S. Nishii, Surf. Science *142,* 306 (1984)
[45] K. v. Klitzing, G. Ebert, N. Kleinmichel, H. Obloh, G. Dorda, G. Weimann, Proc. 17th ICPS, San Francisco 1984
[46] F. Kuchar, R. Meisels, G. Weimann and H. Burkhard, Proc. 17th ICPS, San Francisco 1984
[47] V.M. Pudalov and S.G. Semenchinsky, Solid State Commun. *51,* 19 (1984)
[48] P. Streda and K. v. Klitzing, J. Phys. C *17,* L 483 (1984)
[49] O. Heinonen, P.L. Taylor and S.M. Girvin, Phys. Rev. B *30,* 3016 (1984)
[50] S.A. Trugman, Phys. Rev. B *27,* 7539 (1983)
[51] A.H. Mac Donold, T.M. Rice and W.F. Brinkman, Phys. Rev. B *28,* 3648 (1983)
[52] O. Heinonen and P.L. Taylor, Phys. Rev. B *32*, 633 (1985)
[53] J. Riess, J. Phys. C *17,* L 849 (1984)
[54] K. v. Klitzing and G. Ebert, Metrologia *21,* 11 (1985)
[55] High precision measurements of the quantized Hall resistance are summarized in: IEEE Trans. Instrum. Meas. IM *34*. pp. 301–327
[56] F. Delahaye, D. Dominguez, F. Alexandre, J.P. Andre, J.P. Hirtz and M. Razeghi, Metrologia (to be published)

THE PHYSICS OF QUANTUM WELLS

R.J. Nicholas and D.C. Rogers

Clarendon Laboratory, Parks Road, Oxford, OX1 3PU, England

1. INTRODUCTION

The creation of semiconductor quantum well structures and superlattices is the result of the development in recent years of epitaxial growth techniques such as Molecular Beam Epitaxy (M.B.E.) and Metal-Organic Vapour Phase Epitaxy (M.O.V.P.E.). By growing epitaxial structures consisting of layers of different compound or alloy semiconductors, it is possible to form a one-dimensional potential for carriers which is tailored to some desired function. The simplest form of heterostructure is a single heterojunction, in which an abrupt junction is made between layers of two different semiconductors of the same crystal structure but different energy gaps. This results in step potentials in the conduction and valence bands of height ΔE_c and ΔE_v respectively. Two types of heterojunction may be formed; in a type I system (Fig. 1a) both band edges in the narrower gap material are between the band edges of the wide gap material, whereas in a type II system (Fig. 1b), the bands are staggered. The type of heterojunction formed and the relative sizes of ΔE_c and ΔE_v are specific to the materials used.

The growth of a thin layer (typically 20 - 200 Å) of narrower gap material between thick layers of wide gap material results in a structure known as a quantum well. In a type I system (Fig. 2a), a square finite potential well is formed in both bands, quantizing the motion of electrons and holes perpendicular to the layer plane. Each energy level thus formed becomes the origin of an electric sub-band for motion in the plane of the layers. Structures of this type are of great practical use in the fabrication of solid state laser diodes, in which the quantum well is grown between p-type and n-type layers of wide gap material. Carrier confinement in the active region of such a laser, due to the potential well, gives a substantial increase in lasing efficiency [1], and the zero point energy of the well can be used to tune the effective energy gap to a desired value, with substantial increases being possible in very narrow wells[2]. Further enhancement of the lasing performance may be produced by introducing optical confinement on the scale of the wavelength of the light, making use of the different refractive index of the two materials used (the Separate Confinement Heterostructure). Another feature of interest for device applications is the formation of excitons at room temperature in these structures [3]; this arises from an enhancement of the exciton binding energy due to the reduction in dimensionality of the exciton wavefunction in a quantum well, and has potential applications in electro-optic modulators[4]. For type II heterojunctions there will be a potential barrier for one carrier type (Fig. 2b). If at least one further alternating layer is grown, then a quantum well will be formed in the second material for the other carrier type. This leads to the formation of the spatially indirect 'type II superlattice', in which the lowest conduction band state and highest valence band state are spatially separated from each other.

If the layer growth sequence is repeated several times then a multi-quantum well structure is grown, and once the barrier thickness separating the individual wells becomes thin enough to allow a significant amount of tunnelling then the structure may be said to be a superlattice (Fig. 3). This then looks like the one-dimensional Kroenig-Penney potential along the growth axis, and so by appropriate choices of layer thickness it should be possible to grow a structure with a specific band structure[5]. The various properties of these structures are now a major subject of interest in semiconductor research; however, until recently even such a basic quantity as the

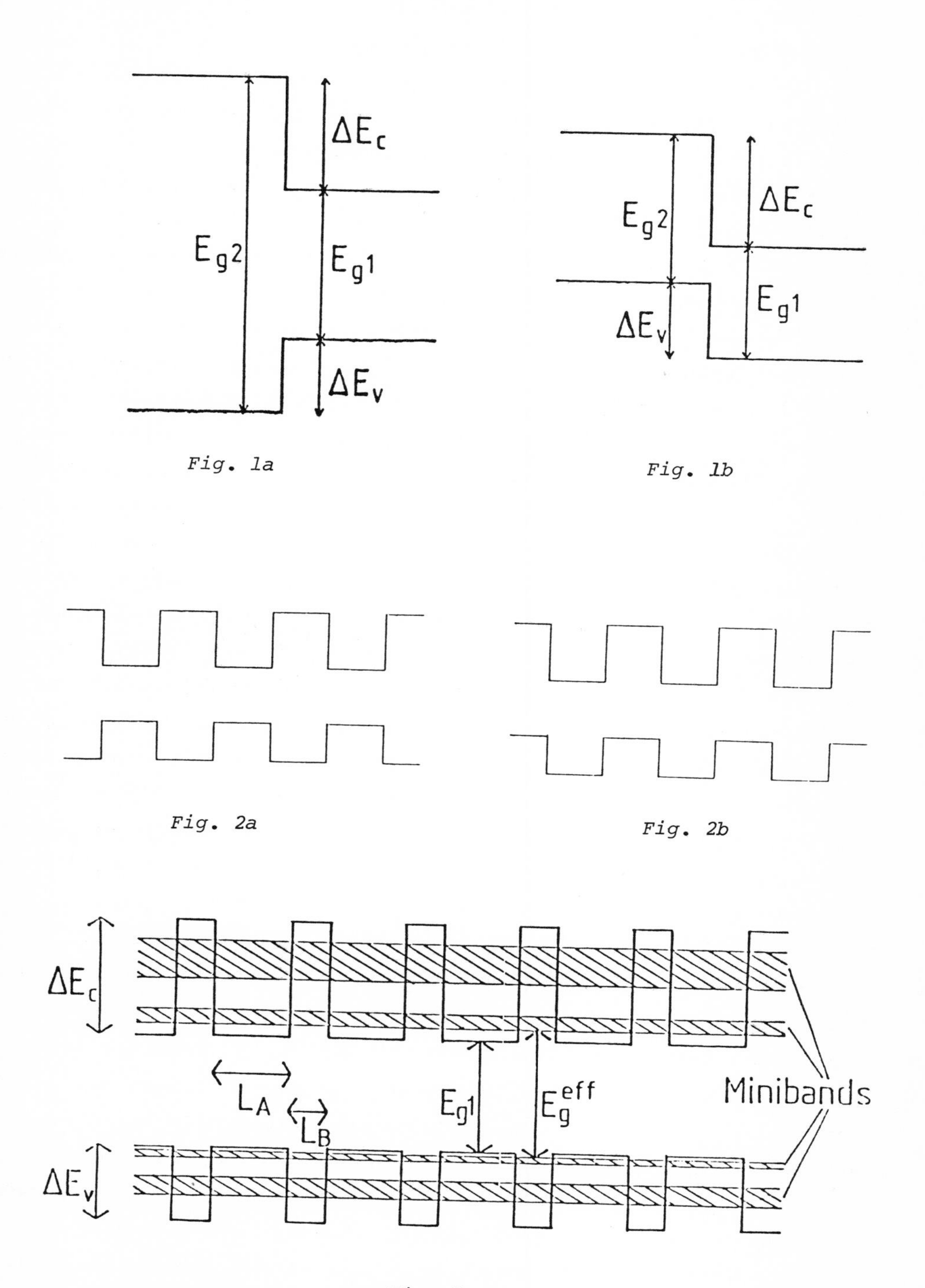

Fig. 1a

Fig. 1b

Fig. 2a

Fig. 2b

Fig. 3.

conduction band discontinuity in the most widely studied system (GaAs-GaAlAs) was still the subject of controversy[6].

The semiconductors most commonly used for the fabrication of heterostructures are III-V compound semiconductors with the Zinc Blende structure, although there has recently been a growing interest in the study of Si/SiGe alloy strained layer structures. The III-V materials have two main advantages. Firstly, the majority of compounds have direct energy gaps at the Brillouin Zone centre, which tends to simplify the electronic properties due to the absence of the valley degeneracies present in Si and Ge, leads to lower effective masses and hence higher mobilities, and also permits the fabrication of solid state lasers. The second advantage is the great range of different materials available, since all the different compounds used may also be alloyed into one another. This leads to the possibility of growing different materials with the same lattice constant but different energy gap, both of which vary smoothly over the range of alloy compositions. For example, the alloys $Ga_{0.47}In_{0.53}As$ and $Al_{0.48}In_{0.53}As$ have the same lattice constant as InP, but the low temperature band gaps are respectively 0.81 eV, 1.62 eV and 1.42 eV, all being direct gaps at the zone centre. These three can therefore be used to grow heterostructures lattice matched on InP substrates, which are readily available commercially. Heterostructures using $Ga_{0.47}In_{0.53}As$ as the narrow gap material are of great practical interest for optoelectronic applications, since the effective band gap, and hence emission wavelength, of a GaInAs quantum well laser can be tuned at room temperature to either of the regions around 1.55μm or 1.3μm which give maximum transmission and minimum dispersion in silica optical fibres. Also important in GaInAs is the relatively high (compared to GaAs) Γ-L valley splitting in the conduction band of 0.55 eV, which allows a much higher electron drift velocity than in GaAs, with a corresponding improvement of F.E.T. performance.

The most technologically developed heterostructure system is GaAs-$Ga_{1-x}Al_xAs$. This has the advantage of being acceptably well lattice matched for any value of x, the alloy composition. GaAs has a low temperature band gap of 1.52 eV, which is direct at the zone centre. $Ga_{1-x}Al_xAs$ has a zone centre direct gap for $x<0.45$ which increases linearly with x; at higher values of x it is indirect, with the lowest conduction band valley near the X point. The majority of work on this system has involved structures with $x<0.45$ in the barriers, so that both materials have direct gaps. Quantum wells of GaAs-GaAlAs have been extensively studied since Dingle et al [7] first reported direct evidence of quantised carrier states, and studies have utilised techniques such as absorption, photoluminescence [8], photoluminescence excitation [9,10] and magneto-optical absorption [11]. The control of the growth is such that the precision of the layer thicknesses are of order a single monolayer, and these lecture notes will contain data from structures containing 60 apparently identical layers with a thickness of 22 Å.

Other systems studied to date include the InAs-GaSb system, which is a type II system with a completely staggered band offset, the conduction band edge of InAs being 150 meV below the valence band of InAs [12,13]. In GaSb-AlSb wells the small lattice mismatch is enough to introduce a strain which will lower the energy gap of the quantum well layers by some 50 meV [14].

Deliberately strained layer superlattices (SLS) have been the object of considerable recent interest, due to the modifications of the properties of the layers which may occur once the requirement for lattice matching in superlattices is lifted. The strain which can be introduced will modify the band gaps, and can lift the degeneracy of any degenerate band edges, such as the valence bands in most zinc blende structure semiconductors, and the conduction bands in indirect materials such as Si or GaP. Typical systems studied to date include $In_xGa_{1-x}As$-GaAs [15], $Al_{1-x}In_xAs$-GaAs [16], and binary systems such as GaAs-InAs [17,18] in which it may be possible to eliminate the presence of alloy disorder, while still retaining some of the tunability of band gap available in alloys such as $Ga_xIn_{1-x}As$.

2. ENERGY LEVELS

The confinement of a particle in a square potential well is one of the simplest problems in quantum mechanics, and its solution is well known. In a semiconductor, there are several complicating factors, mostly arising from the fact that there is a background crystal potential which determines the energy dispersion relation of the particle. The simplest theoretical approach to modelling these systems (the envelope function approximation) is to assume that the crystalline potential determines the band energies and masses in each layer, and that these give rise to a super-potential which may be treated separately. The eigenstates of this super-potential are a set of envelope functions ($f_{1,2}$ which, when multiplied by the appropriate Bloch functions, give the eigenstates of the entire system.

The envelope function model has been studied in detail by Bastard[19], using three band k.p perturbation theory[20]. This gives dispersion relations for the electrons, light holes and split-off band holes defined by the relation,

$$(E-V(x))(E+E_g^i-V(x))(E+E_g^i+\Delta^i-V(x)$$
$$= p^{i2}(\hbar^2 k^{i2}/m_e)(E+E_g^i+2/3\Delta^i-V(x)) \quad (1)$$

where the superscript i refers to the parameters and appropriate k value for the material considered, and V(x) is the position in energy of the conduction band edge of the material. This may be used to define a material dependent, k-dependent effective mass for each carrier type of the form $E = \hbar^2 k^2/2m^*(E)$. For the heavy holes a parabolic approximation is used, with,

$$E-V(x)+E_g^i = \hbar^2 k^2/m_{hh}^{*\,i} \quad (2)$$

In bulk material the description of the valence band in terms of heavy and light holes is due to the different m_J states present, however the absence of a defined axis prevents the assignment of defined m_J states to one or the other hole mass. Once the growth direction has defined an axis, the heavy and light holes are decoupled, and it is possible to assign the $m_J = \pm 1/2$ states to the light holes and $m_J = \pm 3/2$ to the heavy holes[8].

The boundary conditions at the interfaces of the structure are that the wavefunction should be continuous, and that there should be continuity of particle flux across the interface, giving

$$f_1 = f_2, \quad \frac{1}{m_1^*}\frac{df_1}{dz} = \frac{1}{m_2^*}\frac{df_2}{dz} \tag{3}$$

where m_i^* is the material and k-dependent effective mass calculated from equation (1) and f_i is the envelope function in each material. This assumes that the interface is exactly abrupt.

For a regular superlattice of well thickness $L_A (V(x) = 0)$ and barrier thickness $L_B (V(x) = \Delta E_c)$, the Bastard model gives the dispersion equation,

$$\cos qd = \cos k_A L_A \cosh ik_B L_B - \tfrac{1}{2}(-\varepsilon + 1/\varepsilon) \sin k_A L_A \sinh ik_B L_B \tag{4}$$

where $\varepsilon = (k_B m^*_A(k_A))/(k_A m^*_B(k_B))$ and q is the wavevector component for motion along the superlattice direction. For a single quantum well of width L_A between infinitely thick barriers, this gives the result that the energy levels are solutions of the equation,

$$\cos k_A L_A - \tfrac{1}{2}(-\varepsilon + 1/\varepsilon) \sin k_A L_A = 0 \tag{5}$$

This equation can be solved numerically for a given set of band parameters $(E_g^{A,B}, \Delta^{A,B} P^{A,B})$, conduction band discontinuity and well width L_A.

In practice the interfaces between layers may not be perfectly abrupt. In particular there has been evidence to show that interfaces grown by the M.O.V.P.E. technique can be graded over distances of order 40 Å [21]. Dingle [7] used the potential

$$V(z) = \frac{\hbar^2 a(a-1)}{8m^* L_z^2 \sin^2 \pi((z/L_z) - 1/2)} \tag{6}$$

to model the effects of interface grading, which gives eigenvalues

$$E_n = (\hbar^2 \pi^2 / 2mL_z^2)(a+N)^2, \; N = 0, 1, 2 \ldots \tag{7}$$

where a is a parameter equal to unity for a square well and giving an increasing interface grading at higher values.

One of the basic assumptions of envelope function theories is that the well width or superlattice periodicity is large compared with the atomic periodicity of the materials used. In the most extreme cases studied recently this condition hardly holds, for example sections 3 and 4 will show measurements for a well width of 22 Å of GaAs, which consists of only four full lattice constants. For narrow wells a more appropriate method is to use pseudopotential calculations. The application of this technique to GaAs based quantum wells[22,23], has shown significant differences from the conclusions of envelope function theory for well thicknesses below 30 Å. In particular a major drawback of the envelope function approach is the neglect of higher conduction band minima (X and L) states .

A third approach to the calculation of sub-band energies is the tight binding method, where the superlattice wavefunction is built up from a linear combination of atomic orbitals. This technique has also been applied to GaAs-GaAlAs heterostructures[24], and one particularly significant result relates to the valence band sub-bands. As the width of the quantum well is varied, repulsion between the heavy hole N = 2 and light hole N = 1 states results in large discrepancies with envelope function theory. Heavy hole states are found to be in reasonable agreement with envelope function results, but light hole states are significantly different.

3.
MEASURING THE ENERGY LEVELS

The most common methods of determining the energy levels of quantum wells involve the use of either optical absorption, photoluminescence or photoluminescence excitation spectroscopy (PLE), in which transitions are observed from quantised hole levels to electron states. Absorption measurements are not always completely straightforward, due to the fact that the majority of structures grown on GaAs possess a band gap, and quantised energy levels, which lie higher in energy than that of the substrate GaAs material, thus requiring the thinning of specimens to thicknesses of order 1μm. One convenient way to overcome this is to study photoresponse, as has been reported by Rogers and Nicholas[25] for GaAs-GaAlAs heterostructures. These authors studied the photoresponse of sixty independent wells of GaAs, of varying thickness, grown onto a GaAs buffer layer. At low temperatures the majority of the photoresponse was due to photoconductivity in the GaAs buffer layer, which acted as a detector placed directly behind the quantum wells. Any absorption by the wells then produced a minimum in the response. Typical response spectra are shown in Fig. 4 for wells from 22 Å to 110 Å wide, at 55K. The absorptions seen in the quantum well response are interpreted as excitonic transitions between the sub-bands formed in the valence and conduction bands. The spectrum was found to shift rigidly with temperature such that the various features remained at fixed energies above the GaAs band gap. Transition energies can therefore be expressed in the form $E_T' = E_T - E_g$, which is independent of temperature. Each sample shows two series of strong peaks, one involving heavy-hole sub-bands and the other light-hole sub-bands, with the selection rule $\Delta N = 0$, where N is the quantum number of the square well states. For a perfect square potential well with infinite barrier height, these would be the only transitions allowed, however the inclusion of non-parabolicity and wavefunction penetration into the barrier means that weaker transitions are observed corresponding to $\Delta N = 2$. The strongest of these is the HH3 (heavy hole, N = 3) to E1, which appears roughly midway between the $\Delta N = 0$ transitions for N = 1 and 2. $\Delta N = 1$ transitions are forbidden by symmetry within the envelope function approach, nevertheless transitions such as LH1 - E2 can also be observed as weak features in some spectra. This is probably due to the mixing of the light hole N = 1 and heavy hole N = 2 states by off diagonal terms in the Luttinger-Kohn Hamiltonian. Such mixing is very evident when the dispersion relations are calculated for in plane motion[24].

Since the quantum wells used for study in Fig. 4 were undoped, the optical transitions are all excitonic, and the sub-band separations must be deduced by adding the appropriate excitonic Rydberg to each transition energy. This has been calculated by a number of authors [26,27], while recent experiments[28,29], suggest values in agreement with theory, of order 100 meV for a 100 Å well and 12 meV for 50 Å. Therefore these values have been used in deductions of the energy levels for the quantum wells.

In order to deduce the energy levels of the electrons and holes independently from each other, it is necessary to make use of the observation of the $\Delta N \neq 0$ transitions. This is usually sufficient to calculate all the sub-band energies relative to the heavy-hole $N = 1$ sub-band edge. For well widths greater than 50 Å there are still at least three bound heavy hole states in GaAs, and the confinement energy for $N = 1$ can be estimated to within $\pm$ 1 meV from the approximate N dependence given by the solutions to equation (5). This then gives a complete deduction of the energy levels for a given potential well. The results from a number of such wells are given as an example in Table I, for wells with a barrier of $Ga_{0.65}Al_{0.35}As$. The main errors in this type of estimate probably arise from the assumption of the exciton binding energies, which although quite well known for the $N = 1$, $\Delta N = 0$ transitions, have not been calculated for any of the higher or $\Delta N \neq 0$ transitions. This probably leads to errors of order $\pm$ 5 meV.

Once the energy levels in the quantum wells are known, they may be compared with theoretical predictions in order to deduce some of the parameters of the wells. One of the main parameters of interest is the magnitude of the conduction band offset ΔE , which was long thought to be 85% of the band gap difference, following the original deductions of Dingle et al[7]. Another variable parameter is the exact thickness of the wells, since this is not known to an absolute precision of better than a few per cent. Fig. 5 shows a plot of the best fit to the results from a sample of approximately 110 Å width, using a fixed band offset of 77% E. The observed transition energies are marked as solid lines. A unique fit can be obtained for the two $N = 1$, $\Delta N = 0$ transitions using the original assumption of Dingle, that $\Delta E_c = (0.85)\Delta E_g$, however this gives a very bad fit to higher transitions. It was found to be impossible to obtain a good fit for any of the structures studied using this model. The two prime causes for this disagreement with theory are almost certainly the mixing of the light and heavy hole bands, as mentioned above, and uncertainties in the values of the input parameters to the model. In particular the values of the masses of the light and

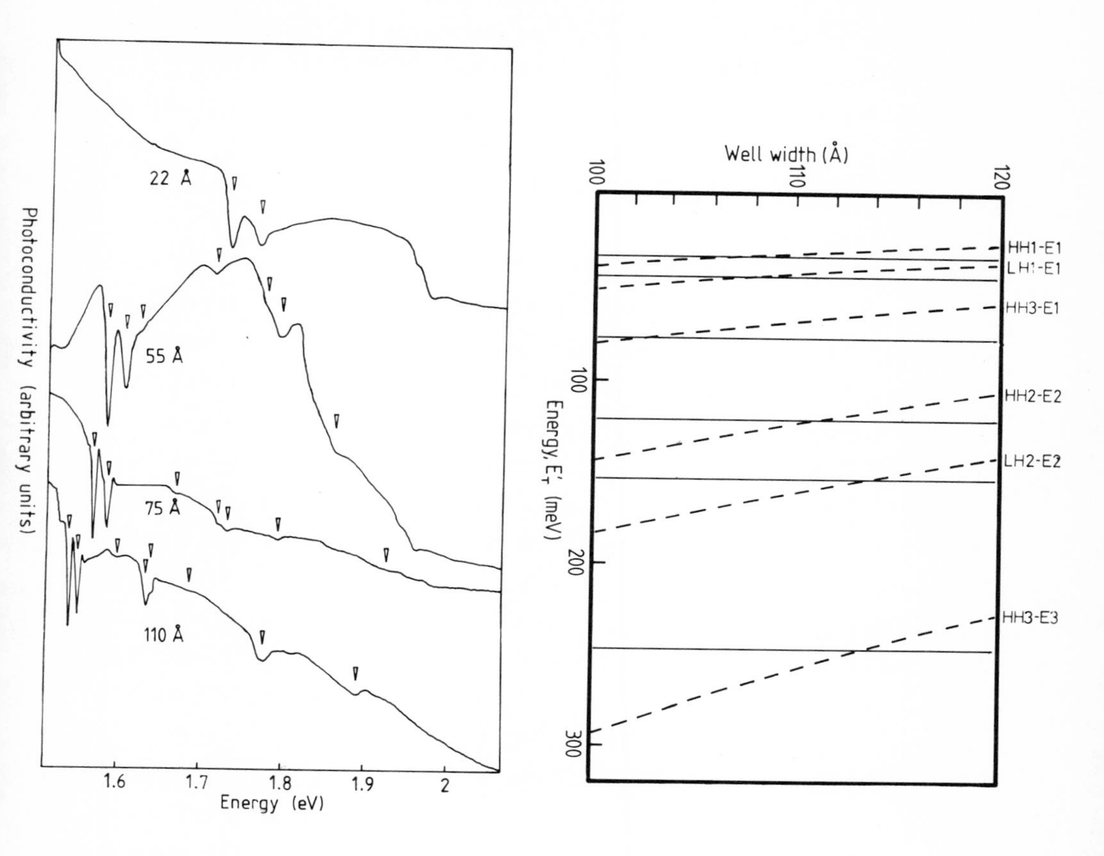

Fig. 4: Zero field spectra at 55K for four different multi-quantum well samples at 55K.

Fig. 5: An envelope function fit to the transitions in a 110A well (Dashed lines). The solid lines are experiment.

TABLE I

Well thickness		22Å	55Å	60Å	75Å	110Å
δL		1Å	3Å	3Å	5Å	5Å
Sub-band						
HH	N = 1	40	18	17	13	7
	2		67	65	43	26
	3		154	142	111	59
LH	N = 1	76	42	38	30	17
	2		139	126	106	60
	3					109
SO	N = 1				32	17
E	N = 1	198	82	77	55	28
	2		257	244	188	102
	3					209

heavy holes in bulk GaAs have been determined by cyclotron resonance [(30)], but are still uncertain to some degree. A study of photoluminescence from simulated parabolic quantum wells of GaAs-GaAlAs led Miller et al to suggest values for ΔE_c as low as $0.5\Delta E_g$, together with some suggested reductions in the previously accepted values for the heavy and light hole masses.

It is however possible to use the observed transition energies to extract information on the magnitude of the band offsets which is independent of modelling, provided that we assume that all the observed transitions involve bound states of the quantum wells. It has been reported by Bastard et al[(31)] that virtual states exist in the continuum below the GaAlAs valence band edge, and Zucker et al[(32)] report delocalised excitons arising from virtual states; however, the calculations of Jaros and Wong[(33)] show the low-lying virtual state to be largely localised in the barriers, which would lead to a very small optical matrix element between bound and virtual states. Furthermore, Zucker et al report that the virtual state transition is barely visible in absorption measurements. It is therefore most unlikely that the strong features attributed in the spectra to N = 0 transitions are due to virtual energy levels. Having established this, it is then possible to conclude that the discontinuity in each band must be great enough to bind all the levels observed. Using the results of Table I then sets quite stringent limits on the band offsets of

$$0.57\Delta E_g < \Delta E_c < 0.66\Delta E_g,$$

which are independent of any theoretical model.

The result shown above is consistent with other recently reported values for ΔE_c in the range $0.60 - 0.65\Delta E_g$, from measurements such as the analysis of carrier density in a 2-D hole gas[(34)], or from thermally activated transport[(35)]. Using the value 0.65, and adjusting the matrix elements in the envelope function theory to give the heavy and light hole transport masses for <100> of $0.087m_e$, and $0.40m_e$ (as deduced from (30)), the transition energies have been calculated from equation (5). These are plotted in Fig. 6, together with the experimental values. It is clear that the main difficulty lies in the calculation of the LH1 and HH3 levels relative to HH1, probably for the

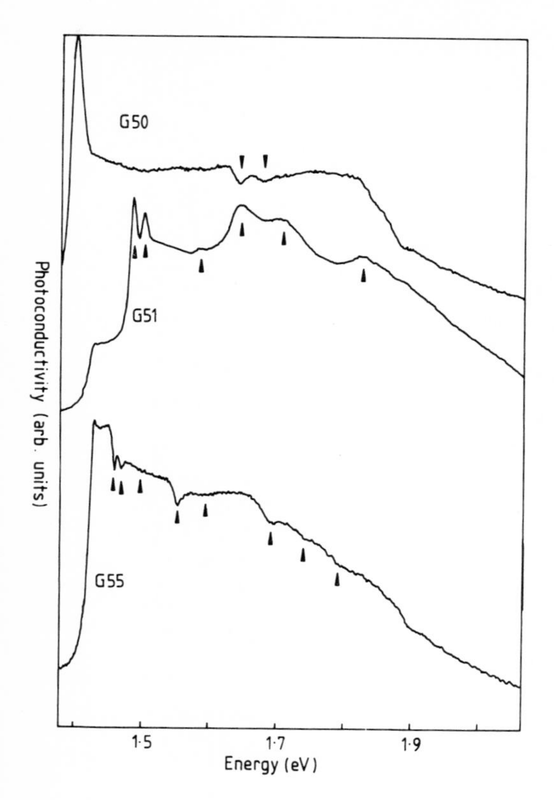

Fig. 6: Room temperature photoresponse from 110Å, 75Å and 22Å multiple quantum well samples. The two wider wells show transitions at around 1.8 eV, which are thought to be due to excitation from the spin-orbit split off band. The transitions in sample G51 correspond to maxima as a result of its higher quantum well photoconductivity.

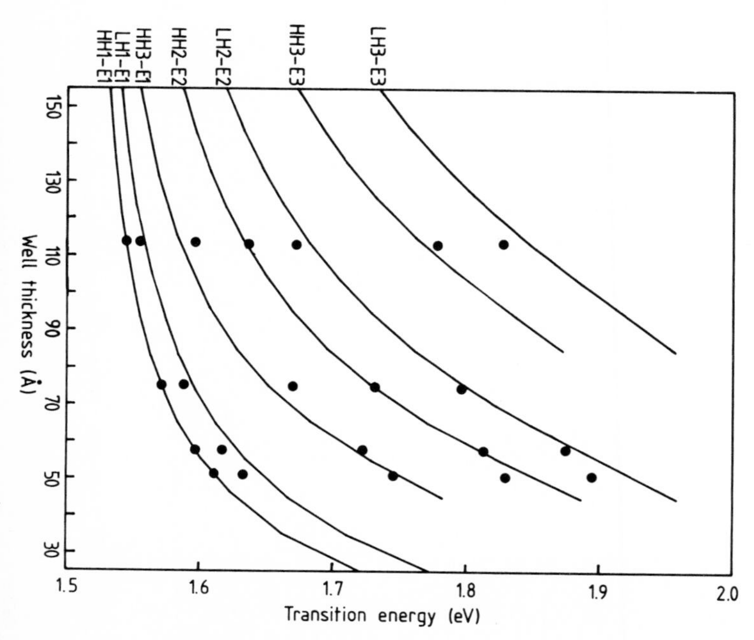

Fig. 7: A plot of the transition energies as a function of well thickness for the envelope function model, with a conduction band offset of 65% E_g.

reasons discussed above.

In practice it is possible to extend these measurements right up to room temperature, and spectra are shown for three samples in Fig. 7. In this case the spectrum for the 75 Å quantum wells shows the interesting behaviour that the photoconductive response from the quantum wells themselves has become larger than that from the buffer layer, and so the transitions then appear as peaks in the response. A further interesting feature is the appearance of transitions at around 1.8 eV for the 75 and 110 Å wells. This is thought to be due to a $\Delta N = 0$ transition from the $N = 1$ level of the spin-orbit split off band to the E1 level.

One further useful piece of information which may be derived from these spectra comes from the linewidths. Fluctuations in the width of the wells across the slice, or variations in the well thickness from layer to layer in a multiple well sample, will broaden the observed transitions. Since the confinement energies are approximately proportional to $1/L$, we have $\delta E/E \approx 2\delta L/L$. This allows us to estimate the mean width of any fluctuations, and this is also shown in Table I. The quite remarkable result is that for some of the structures studied, which were grown by M.B.E. at the Philips Research Laboratories, Redhill, the fluctuations are of order 1 Å, which is considerably less than one unit of inter-atomic spacing.

Studies of optical transitions have also been reported on a number of other materials systems, such as $Ga_{0.47}In_{0.53}As$-InP[36-38]. A photoconductivity spectrum is shown in Fig. 8 for a 180 Å single quantum well of $Ga_{0.47}In_{0.53}As$-InP, grown by M.O.V.P.E. [39]. This shows the presence of four bound states, which is consistent with a band offset for this system of $\Delta E_c = (0.4)\Delta E_g$, as deduced from internal photoemission measurements in biassed Schottky diodes[37,40], and in agreement with earlier C(V) and I(V) measurements[40]. In this case the envelope function calculations of section 2 give good agreement with the measured transitions (arising from heavy hole levels), using the known parameters of GaInAs[41]. The transition widths for this structure give a fluctuation in effective thickness of ± 12 Å, which is probably due to the additional uncertainties caused by both alloy composition fluctuations in the well, and the slightly lower homogeneity of the M.O.V.P.E. growth technique.

Fig. 9 shows a photoconductivity spectrum from eight quantum wells of $Ga_{0.47}In_{0.53}As$, but with confining layers of $Al_{0.48}In_{0.52}As$, grown by M.B.E.[42]. The well thickness in this case is 190 Å, and five heavy hole to electron transitions are seen. The conduction band offset is rather larger for this system, with $\Delta E_c = (0.6)\Delta E_g$[43]. One important difference in this case is due to the fact that the quantum wells have been doped n-type with a carrier concentration of $1.6 \times 10^{11} cm^{-2}$ per well, and consequently have a Fermi energy of 8 meV. The lowest optical transitions then correspond to free electron excitation, and are Moss-Burstein shifted up in energy from the subband edge. This is the reason for the weakness of the N = 1 transition. The higher transitions are also likely to be free carrier like, due to the strong screening introduced by the electrons.

4.
MAGNETO-OPTICS

The study of inter-band optical absorption in high magnetic fields is well known to be a powerful means of band structure determination. The technique has been applied to many bulk semiconductors (for example GaAs[43]) to determine basic parameters such as electron non-parabolicity and hole effective mass, and also to give detailed information on the exciton states. When the magnetic field is applied parallel to the growth axis of the quantum wells, the motion in the plane of the wells is quantised into Landau levels, and optical transitions occur between the quantised hole and electron levels. This can cause some difficulties in interpretation, since the transition energies observed are the sum of two contributions, and in addition there will be an excitonic contribution in undoped wells. Nevertheless a careful analysis can yield information on both the electron and hole levels, particularly when combined with information derived from inter-band spectroscopy of a single carrier type, such as cyclotron resonance.

Typical spectra are shown in Fig. 10, for the sixty 75 Å GaAs wells studied in section 3, for various magnetic fields applied parallel to the growth axis. The spectra are again plots of photoconductivity from an underlying GaAs buffer layer, which is modulated by the absorption in

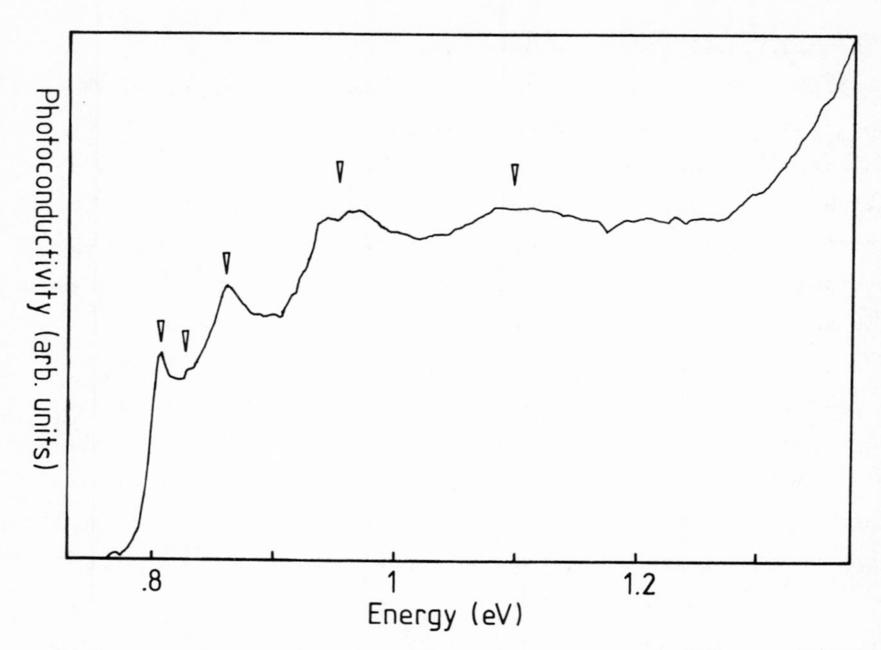

Fig. 8: Photoconductive response from a single quantum well of GaInAs-InP 180 Å wide, at 77 K.

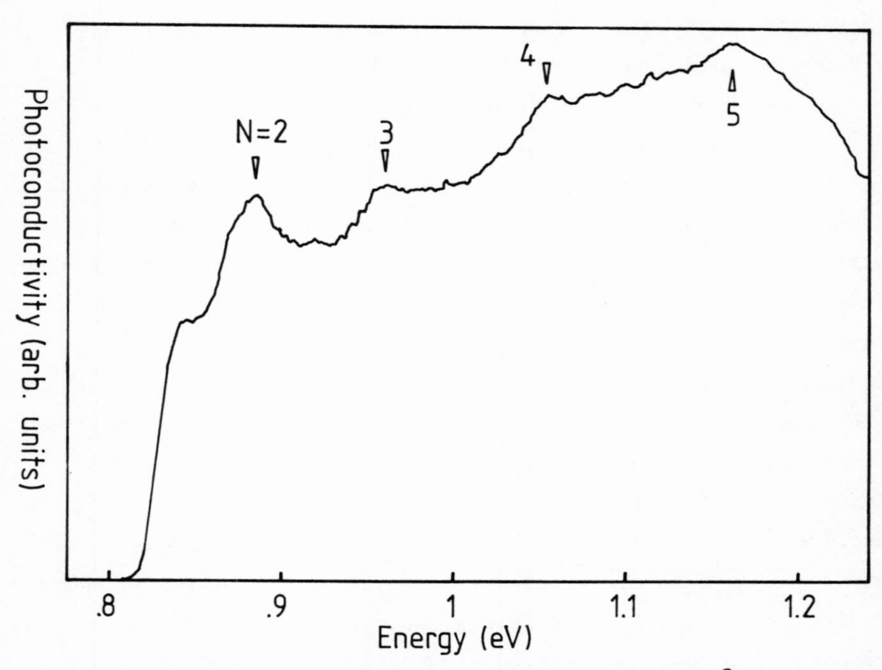

Fig 9: Photoconductive response from eighty 190 Å width quantum wells of GaInAs-AlInAs, at 4.2 K.

the quantum wells. A series of transitions associated with the Landau levels evolves from the HH1-E1 exciton transition, and all the transitions previously observed at zero field are shifted up in energy proportionally to the square of the magnetic field. The transition energies are plotted against magnetic field in Fig. 11, which shows the evolution of the characteristic 'fan' of Landau levels originating from the HH1 - E1 transition. Also visible are a few weaker features, which probably arise from transitions between the light hole and electron Landau levels. At higher energies there appears to be structure from the Landau levels of the heavy hole and electron N = 2 subbands, but the complexity of the system makes quantitative conclusions very difficult. The situation becomes more complex in the 110 Å wells, for which the spectra and transition energy plots are shown in Figs. 12 and 13. The absorptions no longer form such a simple regular series, and the 'fan' diagram has become considerably more complex as the heavy and light hole transitions come closer together and the N = 1 and N = 2 transitions cross at a lower energy. In contrast studies of 22 Å wells are much simplified by the fact that only one bound state exists in the well for each of the three major carrier types, as is clear from the plot of transition energies versus field shown in Fig. 14. An interesting feature of this is that the transition associated with the electron and heavy hole n = 1 Landau levels is seen to persist to very low fields.

In order to calculate the electron and hole levels, it is first necessary to calculate the exciton energies. Rogers et al[29] have analysed the spectra using the work of Akimoto and Hasegawa[44] for highly anisotropic systems, in which the Zeeman shifted states of the hydrogenic exciton at low fields transform in the high field limit to excitonic states bound between the electron and hole Landau levels. In the limit of high magnetic field the binding energy was found to be

$$E_B \approx 3\left[\frac{\hbar eB}{2(2n+1)\mu^*_{ex}\,R^*}\right]^{\frac{1}{2}} R^* D_1, \tag{6}$$

where μ^*_{ex} is the exciton reduced mass and R^* is the excitonic effective Rydberg. D_1 is a parameter related to the dimensionality of the exciton, taking values from 0.25 (3D) to 1 (2D); the zero field exciton

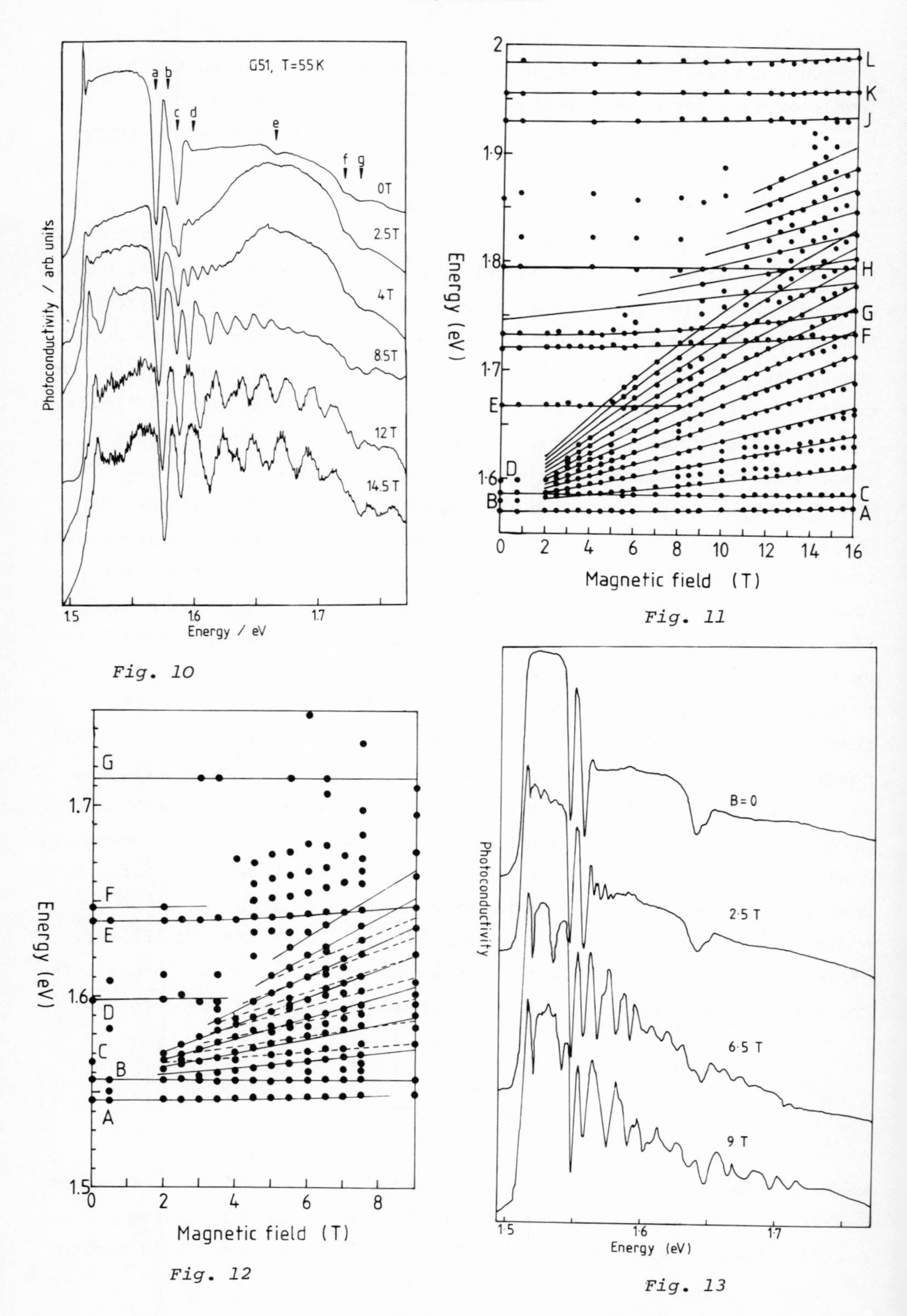

Fig. 10

Fig. 11

Fig. 12

Fig. 13

binding energy is given by $E_B(0) = 4R^*D_1$.

Next we need to consider the energy levels of the electrons. Using three band $\underline{k}.\underline{p}$ theory[20], Palik et al[45] have derived the expression

$$E_e(B,n,K_2) = \left[\frac{(2n+1)\hbar eB}{2m_e^*} + \frac{\hbar^2 k_z^2}{2m_e^*}\right]\left[1 + \frac{K_2}{E_g}\left\{\frac{(2n+1)\hbar eB}{2m_e^*} + \frac{\hbar^2 k_z^2}{2m_e^*}\right\}\right] \tag{7}$$

where the terms in k_z will be the result of the quantum well confinement, and the non-parabolicity parameter K_2 describes the fourth-order contribution to the band curvature. For GaAs three band $\underline{k}.\underline{p}$ theory gives $K_2 = -0.83$, and this expression was found to give a good description of the conduction band[43]. In some recent reports[45] there have been suggestions that the non-parabolicity of bulk GaAs may be higher than predicted by equation (7), in which case the parameter K_2 may be used as a variable parameter to model this increase. Recent 5 band $\underline{k}.\underline{p}$ theories have suggested using $K_2 \sim 1.0$[46]. Equation (7) is however only a good description at energies relatively close to the band edge, since at higher energies it predicts an inversion of the Landau level ordering, with a maximum energy E of $-E_g/4K_2$. An alternative description is that derived by Bowers and Yafet[47] also from three band $\underline{k}.\underline{p}$ theory, in the approximation $E << 2/3$. This gives

$$E^e_{n,\pm} = \frac{E_g}{2}\left\{-1 + \left[1 + \frac{4}{E_g}\left[\frac{\hbar eB}{m^*}(n+1/2) + \frac{\hbar^2 k_z^2}{2m^*} \mp \frac{1}{2}\mu_B |g^*| B\right]\right]\right\} \tag{8}$$

At low energy this expression corresponds to (7), with $K_2 = -1$, but with no variable parameter which can be adjusted to change the degree of non-parabolicity. It has the advantage of having no singularity at high energies, and has therefore been used in the fitting of data taken in GaInAs quantum wells where the transition energies reach almost 50% of the band gap.

The final contribution to the energies comes from the holes. The valence band is complicated by interactions between the heavy and light holes, and the sub-band dispersion relations do not have a simple analytical form even at $B = 0$[24]. Expressions for the valence band masses can, however, be derived in the limit of decoupled heavy and light hole bands, which is equivalent to the limit of a high applied uniaxial strain. This reduces the crystal symmetry and splits the band-edge degeneracy, in the same way as the quantum confinement which has introduced a preferred direction and split the heavy and light holes. In the high stress limit, Hensel and Suzuki[47] have shown that the hole masses are given by

$$\begin{aligned} m_e/m_\perp &= \gamma_1-\gamma_2, \quad m_e/m_{\parallel} = \gamma_1+2\gamma_2 \quad M_J = \pm 1/2 \\ m_e/m_\perp &= \gamma_1+\gamma_2, \quad m_e/m_{\parallel} = \gamma_1-2\gamma_2 \quad M_J = \pm 3/2 \end{aligned} \tag{9}$$

for motion parallel and perpendicular to a <100> strain axis, where γ_1 and γ_2 are the well known Luttinger-Kohn parameters. In a quantum well the confinement energies of the $M_J = \pm 3/2$ states correspond to a mass in the direction of confinement of $(\gamma_1-2\gamma_2)^{-1}m_e$, and are therefore commonly referred to as heavy hole sub-bands; however, from (9) we see that in the limit of complete decoupling, the perpendicular mass of the $M_J = 3/2$ state is light hole like, given by $(\gamma_1+\gamma_2)^{-1}m_e$. This suggests that the effective mass for motion in the layer plane in the N = 1 heavy hole sub-band in a GaAs quantum well will be reduced from its 3-D value of $0.465m_e$ to a value of $0.11m_e$ by quantum confinement, in the limit of complete decoupling. In a real system, the M_J states will not be completely decoupled, so the effective mass in this subband may need to be used as a variable parameter. This analogy is not adequate to describe the higher subbands, all of which will be complicated by mutual interactions.

The next step is to fit the experimentally determined transition energies using the theory described above. The adjustable parameters used are the excitonic dimensionality parameter, D_1, the non-parabolicity factor K_2, and the heavy hole mass m^*_{hh} in the Landau level expression

$$E_{hh}(B,n,m^*_{hh}) = E_o + (2n+1)\hbar eB/m^*_{hh}, \tag{10}$$

where E_o is the heavy hole sub-band confinement energy. The transition energy is given by

$$E_T = E_g + E_e(B,n,K_2) + E_{hh}(B,n,m^*_{hh}) - E_B(B,n,D_1), \tag{11}$$

where the expression for E_B is corrected for electron non-parabolicity, and the parameters K_2, m^*_{hh} and D_1 are varied to obtain the best possible fit. These are shown as solid lines in Figs. 11, 13 and 14, using equation (7) for the electron levels.

As a further check on the values of the exciton binding energy, careful extrapolations of the results at low magnetic fields should be made. These are shown in Fig. 15 for the 75 Å wells. At low fields the first Landau level transition becomes the $2p^+$ exciton state[48], and this extrapolates to an energy a little below that at which an absorption onset can be seen in zero field (Fig. 10). Extrapolations of the second ($3d^{2+}$) and third ($4f^{3+}$) Landau level transitions approach this onset, which is thought to be the continuum, representing the onset of free carrier absorption. It should be stressed that low field data should be used for this process (<5T), since the non-linear field dependence of the transitions can give rise to significantly higher intercept values when only high field data is used. The difference between the continuum edge and the exciton transition then gives the excitonic binding energy, which may be compared with that deduced from equation (6). The values of the exciton binding energy deduced by both methods are plotted as a function of well width in Fig. 16. The fitted values from the high field data are slightly larger than the more reliable low-field values from the extrapolations made for the 75 Å and 110 Å wells; this is thought to be because the measurements were really taken in the intermediate field regime, rather than the absolute high field limit ($\hbar\omega_c \gg R^*$).

There has been some disagreement between the different methods of determining the exciton binding energy in quantum wells. Theoretical calculations give binding energies for the N = 1 heavy hole exciton rising from the bulk value in a very wide well, to a maximum of around 10 - 13 meV at a well thickness of about 30 Å [26,28], and falling for narrower wells. These results are a little lower than those described above, which are in good agreement with those of Miller et al. [49], who interpret the absorption edge above the exciton as a 2s state, unresolved from the continuum without magnetic field; since a point half way up the absorption onset is chosen, this is roughly equivalent to our identification of the upper edge as the series limit, and the two

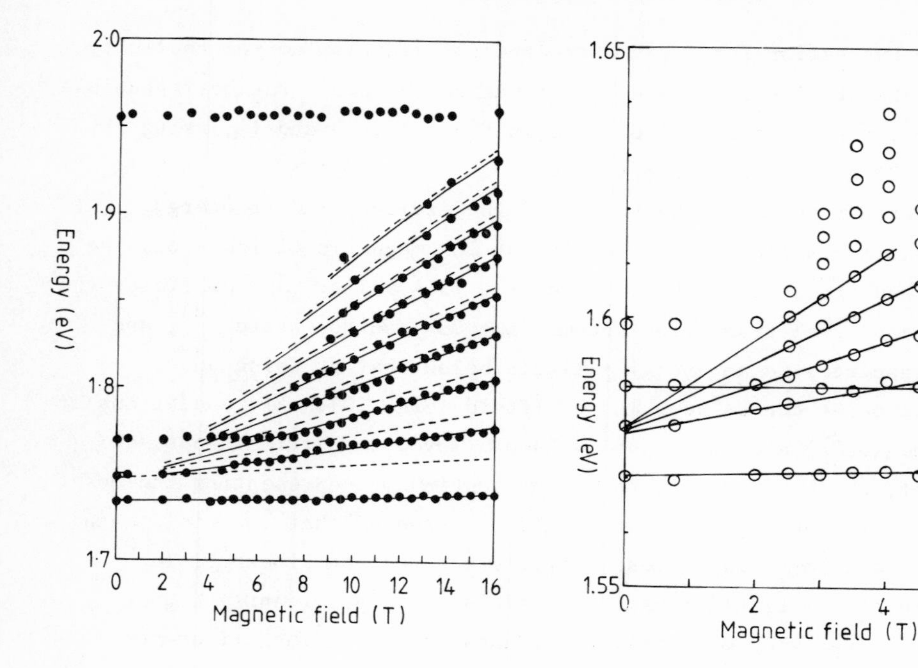

Fig. 14: The transition energies as a function of field for 22Å quantum wells. The solid lines show fits including exciton binding, while the dashed lines show the energies for free carrier transitions.

Fig. 15: An expanded plot of the transition energies for the 75Å GaAs-GaAlAs quantum wells.

TABLE II

Magneto-optics fitting parameters and exciton binding energies

Well thickness	22Å	55Å	60Å	75Å	110Å
K_2	-1.2±0.1	-1.2±0.1	-1.2±0.1	-1.2±0.1	-1.2±0.1
m^*_{hh}	-.22±.03	-.47±.06	-.55±.05	-.59±.08	-.63±.06
m^*_{lh}				+ .25	+ .5
D_1 (see text)	.67±.16	.79±.21	.69±.05	.60±.05	.48±.03

sets of measurements are in good agreement[29]. In contrast magneto-optical measurements by Maan et al. [13] and Miura et al [11] give results considerably higher than theory. These measurements have not taken into account Coulomb effects in the higher Landau levels, which will affect the slope of the Landau level transition as a function of field, and also use results taken at rather too large values of field.

Also varied in the fits are the electron non-parabolicity and the heavy hole mass. The results of this are shown in Table II. A K_2 of -1.2 is used for all the wells studied, which is higher than the value for the bulk. This is thought to be due to some breakdown of 3 band k.p theory[46], and the effect of quantum well confinement. The heavy hole effective mass shows a steady decrease with decreasing well width, which is fully consistent with a progressive decoupling of the light and heavy hole subbands. However, the heavy hole mass in wide wells is significantly greater than the bulk value of $0.45m_e$.

The diamagnetic shift of the 1s exciton states may also be used to make some deductions of the behaviour of the excitons and hole levels [29], and the reduced masses. These measurements suggest that the light hole may even become electron like in some cases.

Magneto-absorption measurements have also been performed on 190 Å width GaInAs-AlInAs quantum wells[42], as shown in Fig. 17. In this case the wells are doped, so that all the transitions were free-carrier like, and no excitonic contributions needed to be considered. At fields above 5T the transition to the n = 0 electron Landau level becomes significant once the ultra quantum limit is reached, and there is a significant number of available final states for the transition from the n = 0 heavy hole level. The positions of the observed transmission minima are plotted as a function of field in Fig. 18. Extrapolation of the n = 0 transition to zero field gives an intercept approximately 8 meV below the onset of the zero field absorption, in agreement with a Moss-Bursten shift by the known Fermi energy.

The field dependence of the Landau levels was fitted using the Bowers and Yafet expression[8], due to the unphysical maximum energy predicted by Eq. (7). The value of the bulk conduction band edge effective mass in lattice matched GaInAs is known to be $0.0405m_e$ from cyclotron resonance measurements made below the optic phonon energy [41]. This value is known to give a good fit to similar inter-band magneto-optical data on bulk GaInAs [50,51]. The only unknown parameter in the fit is the heavy hole mass, which only makes a relatively small contribution to the total transition energy. The solid lines shown in Fig. 18 give the best fit, using a heavy hole mass $m^*_{hh} = 0.75\ m_e$, which is substantially larger than for the bulk material. This is again likely to be a consequence of the repulsion between the energy levels of the light and heavy holes.

Interesting modifications occur to the valence bands of III-V semiconductors when they are subjected to a uniaxial strain, as described above. One way in which this may occur is in a strained layer superlattice. The lattice mismatch leads to a strain which may be considered to be a combination of hydrostatic and uniaxial components. There is a change in the overall magnitude of the band gap, due to the hydrostatic stress, and a splitting of the J = 3/2 valence band states by the uniaxial component. When one layer is considerably thinner than the other, and/or it is constrained by the lattice constant of the substrate, then it will take up all of the strain present in the structure. This is the case for a combination such as $Ga_{1-x}In_xAs$ with InP grown on an InP substrate, when the alloy is not lattice matched. Fig. 19 shows the absorption spectra of ten 150 Å quantum wells of In-rich GaInAs. At zero magnetic field the only strong feature present is an absorption onset at approximately 705 meV, corresponding to the effective band gap of the superlattice layers, which is well below the band gap (810 meV) of lattice matched GaInAs. Also visible is a very weak onset at 790 meV. At high magnetic fields a series of much stronger transmission minima evolves, and the positions of these minima are plotted as a function of applied magnetic field in Fig. 20. The wells are lightly n-type, and so these are interpreted as free carrier transitions between the highest hole subband and the lowest electron subband in the same manner as the unstrained GaInAs well studied above. Extrapolation of the lowest transition back to zero field then gives the effective band gap of the structure as 705 meV. The sum of the confinement energies is approximately 20 meV for a well of this thickness, which gives the band edge position of the GaInAs as 685 meV.

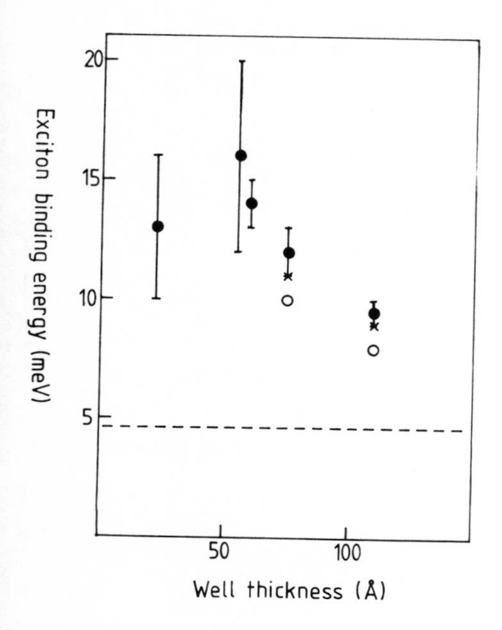

Fig. 16: Heavy hole exciton binding energy as a function of well width. Closed circles are the results of high field fits, and open circles come from low field fits.

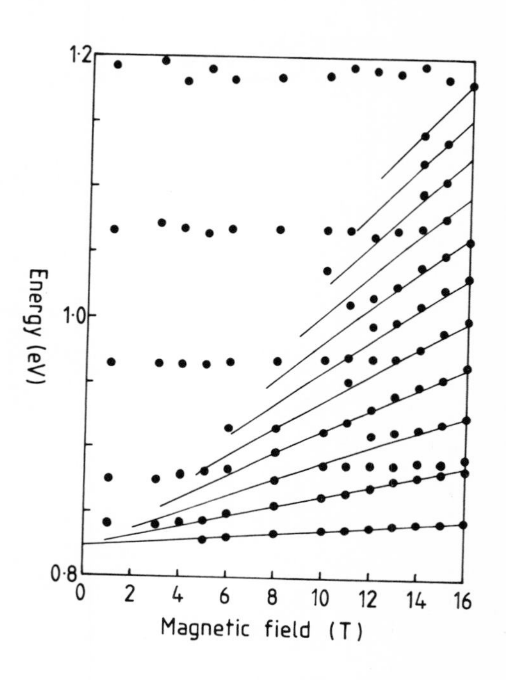

Fig. 18: The positions of the transmission minima as a function of field for 190Å width quantum wells of $Ga_{.47}In_{.53}As$-$Al_{.48}In_{.52}As$.

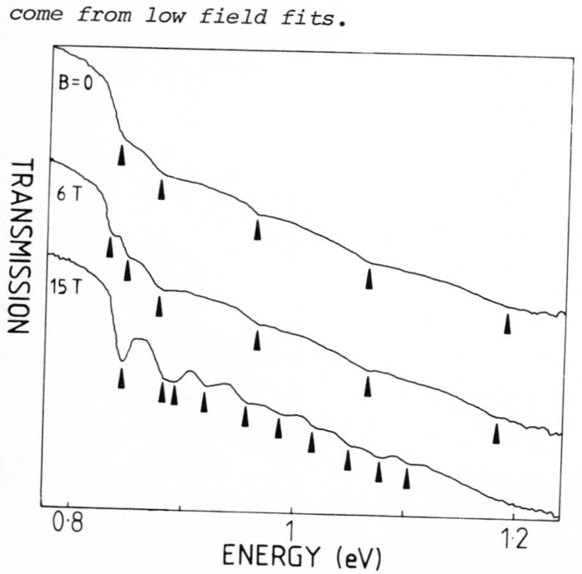

Fig. 17: Magnetotransmission of eight 190Å width quantum wells of GaInAs-AlInAs at 4.2 K. The different transitions seen at B = 0 arise from the different energy levels of the quantum well.

This value may then be used to deduce an accurate value of the alloy composition and strain present in the structure.

The energy gaps in a strained layer superlattice with thick barriers have been calculated by Asi and Oe[52] to be,

$$E_g^{3/2} = E_g^o - \left[2a\left(\frac{C_{11}-C_{12}}{C_{11}}\right) + b\left(\frac{C_{11}+2C_{12}}{C_{11}}\right)\right]\varepsilon \qquad (12)$$

$$E_g^{1/2} = E_g^o - \left[2a\left(\frac{C_{11}-C_{12}}{C_{11}}\right) - b\left(\frac{C_{11}+2C_{12}}{C_{11}}\right)\right]\varepsilon \qquad (13)$$

for the heavy and light hole states respectively, where a and b are the hydrostatic and shear deformation potentials, C_{11} and C_{12} are elastic stiffness constants and ε is the strain in the layer. A calculation using values for these parameters and the lattice constant, interpolated linearly between the appropriate values for GaAs and InAs, gives the exact alloy composition to be $x = 0.32$, giving a lattice mismatch of 1.1%. This gives energy gaps of 685 meV and 760 meV respectively for the heavy and light hole bands. The weak onset seen in the zero field absorption at around 790 meV is therefore consistent with transitions corresponding to the effective band gap between the lowest electron and highest light hole subbands.

Another effect of the strain appears to be a change in the higher Landau level transitions. If these are extrapolated to zero field, they do not have a common intercept as found for the unstrained well. The energy differences involved, up to 17 meV, are too great to be due to exciton effects, which should not in any case be present, as discussed earlier. Neither can this behaviour be attributed to the field dependence of the electron Landau levels, which have been studied independently in cyclotron resonance experiments[53]. Instead the hole states must have been modified by the strain.

The hole levels may be deduced from the inter-band transitions by subtracting the electron energies, as given by equation (8), which works well for both the bulk GaInAs[50,51] and the unstrained quantum wells[42]. The band edge electron mass is estimated from the alloy composition and strain to be $0.035m_e$, which is in good agreement with the subband edge mass of 0.038 found in cyclotron resonance[53], provided that the effects of non-parabolicity and polaron coupling are taken into account. This procedure gives the hole Landau levels shown in Fig. 21.

The solid lines in Fig. 21 are an empirical fit to the levels given by

$$E(n,B) = (n+1/2)\frac{\hbar eB}{m^*_{hh}} + E(n,0), \qquad (14)$$

where E(n,0) is an energy offset relative to the band edge. The bulk heavy hole mass of $0.47m_e$[50] gives an acceptable fit to all the levels in the range $5T<B<16T$, with energy offsets of 8, 13 and 17 meV for the n = 1, 2 and 3 Landau levels. At low fields this description should break down as the levels converge to a common intercept, with considerably greater slopes, corresponding to a lower effective mass. This is thought to be due to the modifications to the valence band caused by the stress, as discussed above, since the effective mass for motion in the plane should be much lower at low energies.

The form of the levels shown in Fig. 21 is evidence of the highly non-parabolic nature of the lowest hole subband, as has been predicted in strained GaInAs quantum wells[54]. The origin of this non-parabolicity is the repulsion between sub-bands at finite k-values[55], so that it should also occur in unstrained quantum wells, where the degeneracy is lifted by the confinement energies. Iwasa et al.[56] have studied hole cyclotron resonance in GaAs-GaAlAs quantum wells, and find for a 50 Å well that the cyclotron effective mass increases rapidly as the resonance field is first increased, and then begins to saturate close to the heavy hole value. This is in qualitative agreement with the results reported here. Strained layer superlattice studies are only as yet in their infancy, and offer the prospect of a number of very interesting developments in the future.

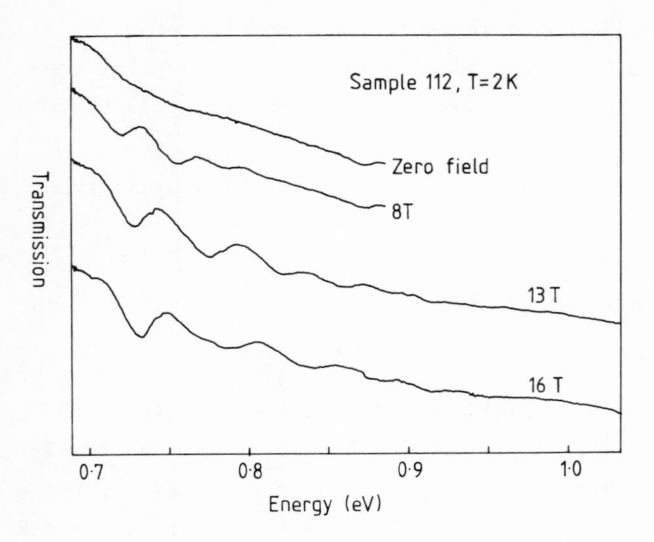

Fig. 19: Magnetotransmission of ten strained quantum wells of 150Å of GaInAs-InP at 2.0 K.

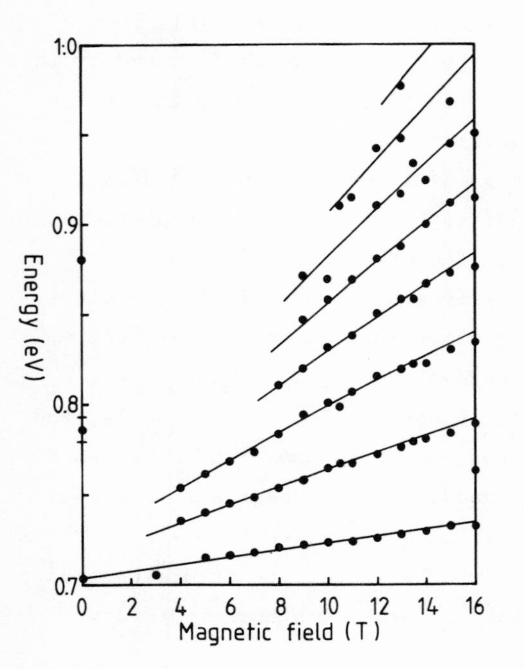

Fig. 20: Positions of the transmission minima as a function of field for the 150 Å GaInAs-InP strained layer superlattice.

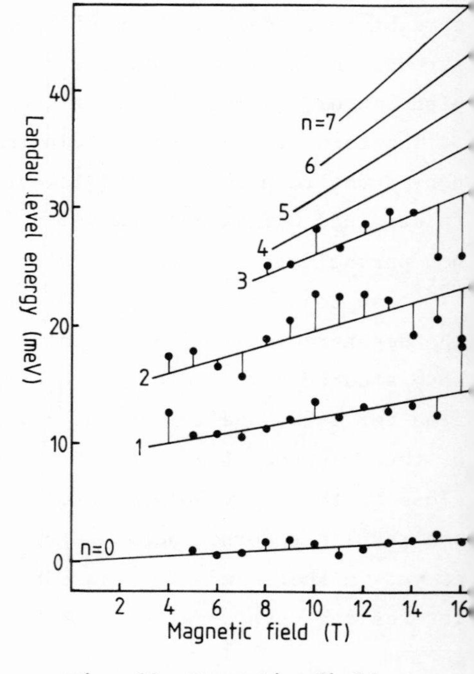

Fig. 21: Magnetic field dependence of the hole Landau levels deduced from the data of Fig. 20. The solid lines are given by Eq. 14.

ACKNOWLEDGEMENTS

We would like to thank the large number of people who have contributed to the work on quantum wells which has been discussed above, and in particular S. Ben Amor, M.A. Brummell, L.C. Brunel, A.Y. Cho, G. Duggan, C.T. Foxon, M.A. Hopkins, S. Huant, T.M. Kerr, J.C. Portal, H. Ralph, M. Razeghi, D. Sivco, J. Singleton and K. Woodbridge.

REFERENCES

1. N. Holonyak, R.M. Kolbas, R.D. Dupuis and P.D. Dapkus, I.E.E.E. J. Quantum Elec. 16 170 (1980).
2. D.F. Welsch, G.W. Wicks and L.F. Eastman, Appl. Phys. Lett. 43 762 (1983).
3. D.A.B. Miller, D.S. Chemla, D.J. Eilenberger, P.W. Smith, A.C. Gossard and W.T. Tsang, Appl. Phys. Lett. 41 679 (1982).
4. D.A.B. Miller, Surf. Sci. to be published (1986).
5. L. Esaki and R. Tsu, I.B.M. J. Res. Develop. 14 61 (1970).
6. H. Kroemer, Surf. Sci. to be published, (1986).
7. R. Dingle, W. Wiegmann and C.H. Henry, Phys. Rev. Lett. 33 827 (1974).
8. C. Weisbuch, R.C. Miller, R. Dingle, A.C. Gossard and W. Wiegmann, Solid State Commun 37 219 (1981).
9. R.C. Miller, D.A. Kleinman, W.A. Nordland and A.C. Gossard, Phys. Rev. B22 863 (1980).
10. P. Dawson, G. Duggan, H.I. Ralph, K. Woodbridge and G.W. t'Hooft, J. Sup. Microstruct. 1 231 (1985).
11. N. Miura, Y. Iawas, S. Taruca and H. Okamoto, Proc. 17th. Int. Conf. Physics of Semiconductors, (Springer-New York) (1985).
12. Y. Guldner, J.P. Vieren, P. Voisin, M. Voos and L.L. Chang, Phys. Rev. Lett. 45 1719 (1980).
13. J.C. Maan, Y. Guldner, J.P. Vieren, P. Voisin, M. Voos, L.L. Chang and L. Esaki, Solid State Commun. 39 683 (1981).
14. P. Voisin, Springer series in Solid State Sciences 53 192 (1984).
15. H. Kato, M. Nakayama, S. Chika, N. Ichiguchi, K. Kubota and N. Sano, Surf. Sci. to be published (1986).
16. M. Sauvage, C. Delalande, P. Voisin, P. Etienne and P. Delescluse, Surf. Sci. to be published (1986).
17. Y. Matsui, H. Hiyashi, K. Kibuchi and K. Yoshida, Surf. Sci. to be published (1986).
18. P. Voisin, M. Voos, M.C. Tamargo, R.E. Nahory and A.Y. Cho, Surf. Sci. to be published (1986).
19. G. Bastard, Acta Electronica 25 147 (1983); Phys. Rev. B12 7584 (1982).

20. E.D. Kane, J. Phys. Chem. Solids 1 249 (1957).
21. R. Bisaro, G. Laurencin, A. Friederich and M. Razeghi, Appl. Phys. Lett. 40 978 (1982).
22. A.C. Marsh and J.C. Inkson, J. Phys. C: Solid State Phys. 19 43 (1986).
23. K.B. Wong, M. Jaros, M.A. Gell and D. Ninno, J. Phys. C: Solid State Phys. 19 53 (1986).
24. Y.C. Chang and J.N. Schulman, Phys. Rev. B31 2069 (1985).
25. D.C. Rogers and R.J. Nicholas, J. Phys. C: Solid State Phys. 18 L891 (1985).
26. R.L. Greene and K.K. Bajaj, Solid State Commun. 45 831 (1983).
27. C. Priester, G. Allan and M. Lannoo, Phys. Rev. B30 7302 (1984).
28. P. Dawson, K.J. Moore, G. Duggan, H.I. Ralph and C.T.B. Foxon, to be published (1986).
29. D.C. Rogers, J. Singleton, R.J. Nicholas, C.T. Foxon and K. Woodbridge, Phys. Rev. to be published (1986).
30. M.S. Skolnick, A.K. Jain, R.A. Stradling, L. Leotin, J.C. Ousset and S. Askenazy, J. Phys. C: Solid State Phys. 9 2809 (1976).
31. G. Bastard, U.O. Ziemelis, C. Delalande, M. Voos, A.C. Gossard and W. Wiegmann, Solid State Commun 49 671 (1984).
32. J.E. Zucker, A. Pinczuk, D.S. Chemla, A.C. Gossard and W. Wiegmann, Phys. Rev. B29 7065 (1984).
33. M. Jaros and K.B. Wong, J. Phys. C: Solid State Phys. 17 L765 (1984).
34. W.I. Wang, E.E. Mendez and F. Stern, Appl. Phys. Lett. 45 639 (1984).
35. J. Batey, S.L. Wright and D.J. DiMaria, J. Appl. Phys. 57 484 (1985).
36. M. Razeghi, J. Nagle and C. Weisbuch, GaAs and Related Compounds 1984 (I.O.P. Conf. Ser. 74) p379 (1985).
37. M.S. Skolnick, P.R. Tapster, S.J. Bass, A.D. Pitt, N. Apsley and S.P. Aldred, Semicond. Sci. Technol. 1 29 (1986).
38. M.S. Skolnick, P.R. Tapster, S.J. Bass, N. Apsley, A.D. Pitt, N.G. Chew, A.G. Cullis, S.P. Aldred and C.A. Warwick, Appl. Phys. Lett. to be published (1986).
39. M. Razeghi, J.P. Hirtz, U.O. Ziemelis, C. Delalande, B. Etienne and M. Voos, Appl. Phys. Lett. 43 585 (1983).
40. S.R. Forrest and O.K. Kim, J. Appl. Phys. 52 5838 (1981).

41. R.J. Nicholas, C.K. Sarkar, L.C. Brunel, S. Huant, J.C. Portal, M. Razeghi, J. Chevrier, J. Massies and H.M. Cox, J. Phys. C: Solid State Phys. 18 L427 (1985).

42. D.C. Rogers, R.J. Nicholas, S. Ben Amor, J.C. Portal, A.Y. Cho and D. Sivco, Solid State Commun. to be published (1986).

43. Q.H.F. Vrehen, J. Phys. Chem. Solids 29 129 (1968).

44. O. Akimoto and H. Hasegawa, J. Phys. Soc. Japan 22 181 (1967).

45. M.A. Hopkins and R.J. Nicholas, to be published (1986).

46. M. Braun and U. Rossler, J. Phys. C: Solid State Phys. 18 3365 (1985).

47. J.C. Hensel and K. Suzuki, Phys. Rev. B9 4219 (1974).

48. P.J. Lin-Chung and B.W. Henvis, Phys. Rev. B12 630 (1975).

49. R.C. Miller, D.A. Kleinman, W.T. Tsang and A.C. Gossard, Phys. Rev. B24 1134 (1981).

50. K. Alavi, R.L. Aggarwal and S.H. Groves, Phys. Rev. B21 1311 (1980).

51. D.C. Rogers, D.Phil thesis, Oxford University (1986).

52. H. Asai and K. Oe, J. Appl. Phys. 54 2052 (1983).

53. L.C. Brunel, S. Huant, R.J. Nicholas, M.A. Hopkins, M.A. Brummell, K. Karrai, J.C. Portal, M. Razeghi, K.Y. Cheng and A.Y. Cho, Surf. Sci. 170 542 (1986).

54. Y.C. Chang, Appl. Phys. Lett. 46 710 (1985).

55. G.D. Sanders and Y.C. Chang, Phys. Rev. B31 6892 (1985).

56. Y. Iwasa, N. Miura, S. Tarucha, H. Okamoto and T. Ando, Surf. Sci. to be published (1986).

THE THEORY OF THE INTEGER QUANTUM HALL EFFECT

J. Hajdu

Institut für Theoretische Physik,
Universität zu Köln, D-5000 Köln 41,
Federal Republic of Germany

1.
FACTS

In 1980 von Klitzing, Dorda and Pepper[1] reported that the Hall voltage measured on an n-channel Silicon MOSFET as a function of the gate voltage U_G (which is proportional to the carrier concentration) at low temperatures (~ 1K) and high magnetic fields (~ 20T) shows plateaux at which the corresponding Hall resistance takes the quantized values

$$R_H = \frac{1}{i}\frac{h}{e^2} \qquad i = 1,2,\ldots \tag{1.1}$$

The centres of the plateaux lie at values of the carrier (electron) concentration $n = N/Ar$ in the 2d conducting layer which correspond to integer values of the filling factor

$$\eta = n.2\pi\ell^2 \tag{1.2}$$

$\ell^2 = \hbar/eB$, and B denotes the magnetic field. Furthermore, in the plateau regimes of R_H the voltage drop U along the applied current vanishes ($U \leqslant 10^{-14}$V). Thereby the corresponding parallel resistance

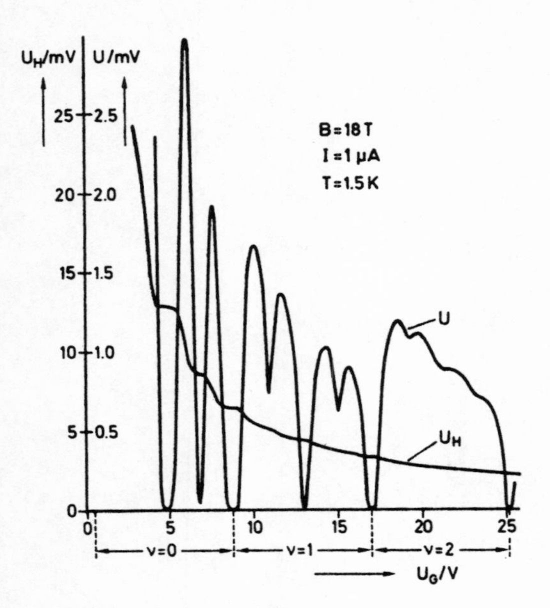

Fig. 1. The quantum Hall effect [1].

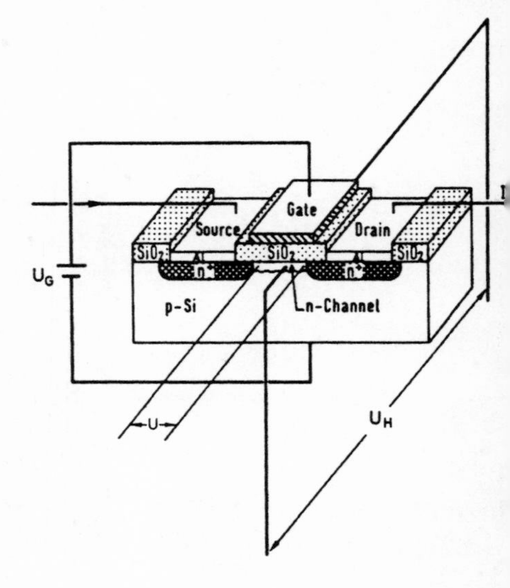

Fig. 2. Hall measurement on a MOSFET. Imposing the current I and measuring the Hall voltage U implies two circuit loops.

$$R = 0 \tag{1.3}$$

This integer quantum Hall effect (IQHE) is a universal phenomenon. The accuracy of the quantization (1.1) is of the order 10^{-8} (the reproducibility is much higher. (Fig. 1)

2.

A FEW REMARKS

(1) In the case of a rectangular 2d system ($Ar = L_x L_y$) the resistances are related to the resistivities by $R = \rho\,(L_y/L_x)$; $R_H = \rho_H$.

(2) In an isotropic 2d system the resistivities and conductivities are related by

$$\rho = \frac{\sigma}{\sigma^2+\sigma_H^{\ 2}} \qquad \rho_H = \frac{\sigma_H}{\sigma^2+\sigma_H^{\ 2}} \tag{2.1}$$

(3) Using the conductivities the IQHE is characterised by the plateau values

$$\sigma_H = i\,\frac{e^2}{h}\ , \quad i = 0,1,2,\ldots \tag{2.2}$$

$$\sigma = 0 \tag{2.3}$$

(4) In two dimensions the atomic unit of the conductivity is e^2/h. Dimensionless conductivities in these units are defined by

$$\sigma = \xi\frac{e^2}{h}\quad , \quad \sigma_H = \xi_H\,\frac{e^2}{h} \tag{2.4}$$

The task for experiment and theory is to determine the functions

$$\xi = \xi(\eta)\quad , \quad \xi_H = \xi_H(\eta) \tag{2.5}$$

(5) According to both classical and quantum theory the conductivities of an infinite free electron system are $\sigma^o = 0$ and

$$\sigma_H^o = \frac{en}{B} = \eta\,\frac{e^2}{h} \tag{2.6}$$

and consequently

$$\xi_H^o = \eta \tag{2.7}$$

(straight line, no QHE). For $\eta = i$ the observed Hall conductivity (2.2) coincides with the values calculated for free electrons (2.7).

(6) Consider a system which is infinite in the x direction and assume periodic boundary conditions for the wave function in the y direction,

$$\psi(x,y+L_y) = \psi(x,y) \tag{2.8}$$

Then the free electron energy eigenfunctions $\psi_\alpha(x,y)$ are products of a plane wave in the y direction with wave number $k = (2\pi/L_y)$ integer and an oscillator wave function in the x direction centered around $X = \ell^2 k$. The corresponding energy eigenvalues (Landau levels)

$$\varepsilon_\alpha = \hbar\omega_c(\nu+\tfrac{1}{2}) = \varepsilon_\nu \qquad (2.9)$$

$\alpha = (\nu,k)$, $\nu = 0,1,2,\ldots$, $\omega_c = eB/m$ are degenerate. The number of states per cm² with fixed ν is $1/2\pi\ell^2 = eB/h$. This explains why η is called the filling factor.

(7) For free electrons the Fermi energy ε_F is, as a function of the carrier density n (or η), a step function. Consequently $\sigma_H^o(\varepsilon_F)$ is also a step function. This is, however, not the observed QHE - the idealization of which is a step function behaviour of $\sigma_H(\eta)$. The QHE results if

(a) $\sigma_H(\varepsilon_F)$ is a step function but
(b) $\varepsilon_F(\eta)$ is smooth

A random potential (brought about by impurities) produces both features: (b) by broadening of the Landau levels (2.9) to bands and (a) by (strong) localization at low temperatures.

3.

THE DEVELOPMENT OF THE THEORY

Most of the different theoretical approaches to the IQHE can be directly traced back to pioneer works which were published within only one year after the paper by von Klitzing et al[1] had appeared (August 11, 1980). These approaches and works are:

Scattering approach, Prange[2]
Phenomenological localization approach, Aoki and Ando[3], Tsui and Allen[4]
Gauge argument, Laughlin[5]
Percolation approach, Tsui and Allen[4], Ono[6]
Numerical analysis, Ando[7]

The new approaches invented in the later development of the theory and the corresponding landmarks are:

Topological approach, Thouless, Kohmoto, Nightingale and den Nijs[8]
Středa formula[9]
Field theory, Levine, Libby and Pruisken[10]

In the following sections I will try to summarize these approaches and make a few comments on their significance and limitations.

4.

SCATTERING APPROACH

Prange[2] has pointed out that the addition of a single point scatterer to a free electron gas does not give rise to any change of the contribution of a fully occupied Landau band to the Hall conductivity: the loss of current due to the bound state is exactly compensated by an increase of the current carried by the scattering states. Later scattering theory has been applied to prove this compensation in the case of a finite scattering regime[11] and in the limit of vanishing concentration of point scatterers[12]* and arbitrary short range potentials[15]. Chalker[11] has pointed out that in the limit of weak electric and strong magnetic fields only forward scattering is possible. Brenig[12] and Joynt and Prange[15] applied the Levinson theorem which relates the scattering phase shift to the number of bound states. In a recent work Brenig and Wysokiński[16] estimated the possible corrections to the exact compensation to be of the order $(\ell^2/L_xL_y)^2 \simeq 10^{-16}$.

The main objection against the scattering approach is that it fails to explain the vanishing parallel conductivity in the plateau regimes. In fact scattering theory is not the appropriate tool to treat transport phenomena if (strong) localization takes place. Furthermore, the models of finite scattering range and low density short range scatterers do not represent the physical situation in a real 2d system.

Moreover in the scattering approach the density operator of the system is assumed to be diagonal in the energy representation,

$$\langle\alpha|\zeta|\alpha'\rangle = f_\alpha\, \delta_{\alpha\alpha'} \tag{4.1}$$

where f_α describes a spatially homogeneous non-equilibrium distribution.

* *This result was established earlier by using perturbation theory*[13]*.*
For an alternative non-perturbative proof of the compensation cf.[14]*.*

The scattering approach does not make any effort to explain how such a distribution comes about.

5.

PHENOMENOLOGICAL LOCALIZATION APPROACH

Aoki and Ando[3] and Tsui and Allen[4] seem to be the first to perceive that the QHE is due to localization of electrons in a random potential. The existence of two mobility edges in each Landau band (separating a small symmetric range of mobile states at the centre of the band from localized states at other energies) was predicted by Tsukada[17] for the limit of a high magnetic field by referring to classical percolation theory. Aoki and Ando[3] argued that localized states do not contribute to the current and therefore, at zero temperature, the Hall conductivity must remain constant as long as the Fermi energy varies in a range of localized states (where $\sigma = 0$). Applying a high field version[18] of the Kubo formula and simulating localization by assuming bound states these authors proved that indeed $\sigma_H = 0$ if ε_F lies in the lowest localization regime. Unfortunately their proof for the quantized values of σ_H corresponding to $i = 1,2,...$ is incomplete and must be supplemented by further assumptions (a sort of electron-hole symmetry for each Landau band). Furthermore, the high field Kubo formula is an approximation, the accuracy of which is not sufficiently well known. In view of the high accuracy of the observed quantization the corrections may not be negligible.

Tsui and Allen[4] refer explicitly to classical percolation theory. Their argument is, however, entirely qualitative. In particular the high accuracy of the quantization remains unexplained.

6.

GAUGE ARGUMENT

The gauge argument formulated by Laughlin[5] refers to a system of cylindric (or topologically equivalent) geometry with an additional magnetic field which is parallel to the cylinder axis and vanishes at the surface (2d system). The magnetic field B is perpendicular to the cylinder surface. At the surface the flux φ and the (azimuthal) vector potential $\underline{A} = (o, A_y, o)$ of the additional magnetic field are related by $A_y = \varphi/L_y$. This vector potential can be eliminated by the gauge transformation

$$\begin{aligned} \underline{A} \rightarrow \underline{A}' &= \underline{A} + \text{grad}\,\chi = 0 \quad \text{i.e. } \chi = -A_y y \\ \psi \rightarrow \psi' &= \exp(-ie\chi/\hbar)\psi = \exp(ieA_y y/\hbar)\psi \end{aligned} \tag{6.1}$$

provided

$$eA_y L_y/\hbar = 2\pi\nu, \; \nu = 0, \pm 1, \ldots$$

or

$$\varphi = \nu\varphi_o, \quad \varphi_o = h/e \tag{6.2}$$

since ψ' has to satisfy the same periodic boundary coundition as ψ: $\psi(x, y+L_y) = \psi(x,y)$. Changing the flux by φ_o maps the system back to its original state (gauge invariance). On the other hand, changing the flux in time leads to an induced emf in the cylinder,

$$\frac{d\varphi}{dt} = -E_y L_y \tag{6.3}$$

In the case of vanishing dissipation, $\zeta = \sigma = 0$ (the Fermi energy lies in a range of localized states)

$$-E_y L_y = J_x L_y/\sigma_H = I_x/\sigma_H = \frac{1}{\sigma_H}\frac{dQ}{dt} \tag{6.4}$$

or

$$\sigma_H = \frac{dQ}{dt} \tag{6.5}$$

where Q is the charge transfer in the x direction. Averaging (6.5) over a unit flux quantum φ_o

$$\sigma_H = \frac{e}{h}\Delta Q \tag{6.6}$$

Now gauge invariance for $\Delta\varphi = \varphi_o$ requires $\Delta Q = ie$ since moving i electrons from one edge of the cylinder to the other and reinjecting them at the initial position ("pumping") is the only charge transfer which maps the system back to its original state. (Gauge argument). Thus,

$$\sigma_H = i\,\frac{e^2}{h}$$

[5]. For a generalization of this result to systems with discretely degenerate ground state see[19].

Again, several objections can be raised. First of all the integer i occurring in (6.7) is absolutely arbitrary. To relate it to the value of ε_F or η the gauge argument has to be supplemented by an additional principle (e.g. the principle of least action[20]). Secondly, the change of the physical state of the system when increasing the flux from φ to $\varphi + \varphi_o$ is not defined rigorously. In most investigations[5], [20] the gauge argument is based on the thermal equilibrium current defined by

$$J_y = -\frac{1}{Ar}\frac{\partial\Omega}{\partial A_y} = \mathrm{Tr}\{\rho \hat{J}_y\} \tag{6.8}$$

where η is the thermodynamic potential, ρ the grand canonical density operator and

$$\hat{J}y = -\frac{e}{Ar}\,v_y \tag{6.9}$$

the average current density. For a cylinder of finite length L_x (the wave function is subject to the Dirichlet boundary conditions $\psi(x = \pm L_x/2, y) = 0$) - required by the "pumping" - the total equilibrium Hall current vanishes identically because the edge currents[21], [22], exactly compensate the current in the bulk[23], [24]. Furthermore, the identification of the averaged conductivity over a flux quantum with the measured conductivity requires some justification. Last but not least the "pumping" process appears to be rather obscure.

In the topological approach (sec. 8) which was motivated by the desire to give a rigorous proof for the gauge argument the above-mentioned weaknesses are indeed eliminated.

7.

PERCOLATION APPROACH

Tsukada[17] has pointed out that in the limit $B\to\infty$ the "centre coordinates" X, Y, defined by

$$x = X + v_y/\omega_c \;\; , \;\; y = Y - v_x/\omega_c \tag{7.1}$$

(where $\underline{v}$ is the velocity), can be treated as classical variables because their commutator

$$[X,Y] = i\ell^2 \tag{7.2}$$

vanishes like 1/B. The Hall conductivity in the limit $B\to\infty$ was first investigated by Ono[6]. Using the high-field Kubo formula and neglecting transitions between Landau bands he obtained for the case that the chemical potential ζ lies in the $\nu = 0$ Landau band, in leading order in ℓ^2,

$$\sigma_H = \frac{e^2}{h}\, f(U_c) \tag{7.3}$$

where f is the Fermi distribution function and U_c the percolation threshold for the random potential U(x,y). According to Ono only percolating states at the edges of the system (at $|X| = L_x/2$) contribute to σ_H - provided that $\sigma = 0$. The same result was obtained by Apenko and Losovik[25] (cf. also[26]). The assumptions, however, are rather different. In Ono's treatment the electrons are not confined by potential walls at $|X| = L_x/2$. Consequently the edges of the system are not well-defined. Furthermore, Ono assumes that "the system is ergodic" in the sense that the equilibrium correlation function $\langle Y(t=o)X(t\to\infty)\rangle$ factorizes,

$$\langle Y(o)X(\infty)\rangle = \langle Y\rangle\,\langle X(\infty)\rangle \tag{7.4}$$

Apenko and Losovik, on the other hand, consider a confined system which is not ergodic in the sense of (7.4), and deal with the confinement in an entirely classical way. Technically, the treatment of the confinement potential is rather delicate since it is rapidly varying. A possible way to take the confinement into account is to start with the high field Hamiltonian

$$H = \frac{m}{2} v^2 + U(X,Y) \tag{7.5}$$

and impose Dirichlet boundary conditions on the eigenfunctions of the kinetic part. This leads to a set of classical Hamiltonians

$$H = \varepsilon_\nu(X) + U(X,Y) \ , \quad \nu = 0,1,2,\ldots \tag{7.6}$$

The spectrum $\varepsilon_\nu(X)$ is schematically given in Fig. 3 [21]. A set of classical confinement potentials can be defined by

$$\varepsilon_\nu(X) = \varepsilon_\nu + U_c(X;\nu) \tag{7.7}$$

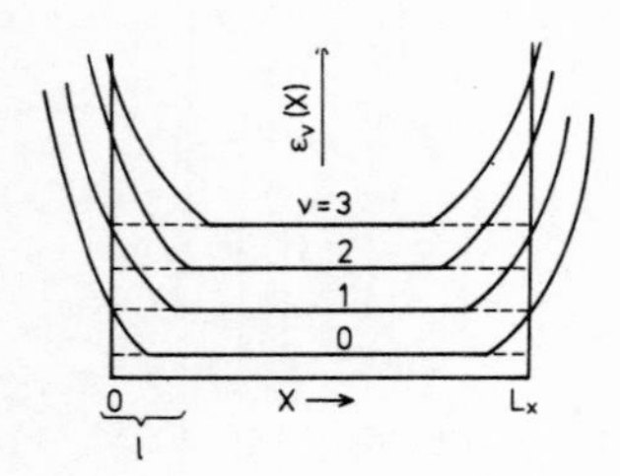

Fig. 3. Schematic energy spectrum of free electrons confined to a finite interval in x direction, in the presence of a strong magnetic field.

For the Hamiltonian (7.6) linear response theory (Kubo formula) yields

$$\sigma_H = \frac{e^2}{h} \sum_\nu \frac{1}{Ar} \iint dXdY \left(-\frac{\partial f}{\partial U}\right)\frac{\partial U}{\partial X}\, \bar{X} \tag{7.8}$$

where $f = f(\varepsilon\nu + U_c + U)$ and $\bar{X}$ is the infinite time average of $X(-t)$ with initial conditions $X(o) = X, Y(o) = Y$ [25]. At $T = 0$

$$\frac{\partial f}{\partial U} = \delta(\varepsilon_\nu + U_c + U - \varepsilon_F) \tag{7.9}$$

and the integration with respect of X and Y can be transformed to integrations along and perpendicular to the equipotential lines

$$\varepsilon_F = \varepsilon_\nu + U_c(X;\nu) + U(X,Y) \tag{7.10}$$

$$\sigma_H = \frac{e^2}{h} \Sigma \frac{1}{Ar} \int d\ell n_x(\ell) X(\ell) \tag{7.11}$$

where $n_x = (\partial U/\partial X)/|\text{grad } U|$. On closed equipotential lines $\bar{X}$ is independent of X and Y. Consequently their contribution to σ_H vanishes (since $\int n_x d\ell = 0$). At the edges percolating equipotential lines exist at all values $\varepsilon_F > \varepsilon_o$ of the Fermi energy. Their contribution can easily be estimated to be

$$\sigma_H = i \frac{e^2}{h} \tag{7.12}$$

where the integer i is determined by $\varepsilon_{i-1} < \varepsilon_F < \varepsilon_i$ $(\varepsilon_{-1} \equiv 0)$. Since Ar is the effective area of the system defined by the confinement there are no corrections to the above estimation of the order ℓ^2/L_x^2. According to classical percolation theory percolating lines in the bulk exist only at the energy $\varepsilon = U(X,Y) \simeq 0$, i.e. at the centres of the bands, $\varepsilon_F \simeq \varepsilon_\nu$ (the average of U is assumed to be zero). For the same reason $\sigma \neq 0$ is expected at $\varepsilon_F \simeq \varepsilon_\nu$ via hopping processes. (For a recent account cf. [27]). Eq. (7.12) describes the Hall conductivity as function ε_F. Since, however, ε_F is a smooth function of the electron concentration (the occupation of all possible closed equipotential lines being a continuous process) (7.12) explains the observed behaviour of the Hall conductivity.

Some of the percolation considerations [4,15,28-31] are not based on linear response theory. By a classical kinetic treatment Lordansky [28] established $\sigma_H = \sigma_H^o$ for fully occupied Landau bands (compensation). Kazarinov and Luryi[29] (cf. also [20]) combined the percolation picture with the gauge argument. Joynt and Prange[15] investigated the conditions of applicability of the classical description, and Luryi and Kazarinov[30] estimated the effect of charge fluctuations on the quantization. Trugman[31] studied the percolation scaling behaviour for zero and finite electric field and predicted a non-linear broadening of the steps in the QHE as a function of the electric field. The essential advantage of these considerations is to demonstrate that the Hall current is driven only by the drop of the electrochemical potential across the channels of percolating equipotential lines, and, therefore, in a random potential, fewer electrons can carry the same Hall current as free electrons.

It has been pointed out by different authors[29, 25] that the high accuracy of the QHE is due to the fact that open and closed percolating lines are topologically different. (On a cylinder they have winding numbers 1 and 0, respectively). Although this topological argument is rather appealing one should not forget to ask how accurate the conductivity formula (e.g. (7.11)) is on which the argument is based.

8.

NUMERICAL ANALYSIS

In his early numerical work Ando[7] demonstrated that the Hall conductivity of a system of independent electrons interacting with randomly distributed fixed impurities indeed shows quantized plateaux - in agreement with the observations. With this result the Coulomb interaction between the electrons has been ruled out as a possible source of the IQHE. Aoki[32] studied the behaviour of electronic states in disordered systems under gauge transformation and confirmed the requirements of the gauge argument. MacKinnon et al.[33] showed that, in a confined system, the edge states behave essentially in the same way as in the case of free electrons - provided the effective disorder is weak as compared to the magnetic energy $\hbar\omega_c$. Schweitzer et al.[34] found that in such systems, close to each other, two mobility gaps exist in each Landau band (in qualitative agreement with previous estimations [17]), and the delocalized bulk states do not contribute to the Hall conductivity. The combination of these results establishes again the QHE. In contradiction to[34], Ando[35] and Aoki and Ando [36] found that in high magnetic field all states are exponentially localized except those with energies at the centre of the Landau bands. The reason for this disagreement (which is irrelevant for the qualitative understanding of the QHE) might be due to the fact that the authors consider different models (lattice respectively free-space systems) and use slightly different localization criteria.

Cf. Schweitzer et al.[37] and the recent review by Ando[38].

Although an accuracy of 10^{-8} is unattainable for numerical analysis such studies are very useful to check approximations and intuitive arguments. An example for this is a recent work by Ando[39] which indicates that σ and σ_H are related by a universal scaling relation which depends on the band index ν - in contradiction to the two-parameter scaling theory (cf. sec. 11).

9.

TOPOLOGICAL APPROACH

Working out the Kubo formula (for σ_H) for independent electrons in an perfect 2d lattice at T = 0 Thouless et al.[8]* observed that the Hall conductivity is determined by a topological invariant over the first Brillouin zone if the Fermi energy is situated in an energy gap, and takes the values (2.2) - in accordance with the gauge argument. A formulation of this relation in the language of differential geometry was given by Avron, Seiler and Simon[41]. The next step in the development of the topological approach was the generalization to arbitrary systems with an appropriate double-periodic structure by Niu, Thouless and Wu[42] (generalized periodic boundary conditions both in x and in y direction) and by Avron and Seiler[43] (Hamiltonian depends periodically on two parameters corresponding to the two current loop Hall measurement, Fig. 2). In the following I summarize the topological approach as fomulated by Pook[44].

Let us consider for the sake of simplicity a system of independent electrons occupying a rectangular area $Ar = L_xL_y$. Assuming that the random potential satisfies periodic boundary conditions,

$$U(x+pL_x, Y+qL_y) = U(x,y) \tag{9.1}$$

p, q integer we repeat the system in both directions infinitely many times. The resulting large system has the discrete translational symmetry of a 2d rectangular lattice with lattice constants Lx and Ly. In the presence of a perpendicular magnetic field the symmetry

* Cf. also [40].

operations are so-called magnetic translations. For rational values of the filling factor these translations commute and the well-known Bloch theorem of solid state physics applies. Using this theorem the Kubo formula for σ_H can be rewritten as

$$\sigma_H = \frac{e^2}{h} \sum_n \{\frac{i}{2\pi} \int_\Gamma \langle du_n | f | du_n \rangle\} \tag{9.2}$$

Here $i = \sqrt{-1}$,

$$|du_n\rangle = |\partial_1 u_n\rangle \, d\theta_1 + |\partial_2 u_n\rangle \, d\theta_2 \tag{9.3}$$

$u_n(\theta_1, \theta_2)$ are the Bloch factors of the energy eigenstates, f is the Fermi function depending on the Bloch Hamiltonian $H(\theta_1, \theta_2)$; $Hu_n = \varepsilon_n u_n$, n is the band index, θ_i are quantum numbers related to the Bloch wave numbers $k_x = \theta_1/L_x$, $k_y = \theta_2/L_y$, $\Gamma = [-\pi,\pi] \times [-\pi,\pi]$ is the domain corresponding to the first Brillouin zone, and $\partial_i = \partial/\partial\theta_i$. Eq. (9.2) differs from the corresponding expression derived in[43] by the factor f.

Provided that gaps exist in the spectrum $\varepsilon_n(\theta_1,\theta_2)$ and, at T = 0, the Fermi energy lies in one of those, (9.2) takes the form

$$\sigma_H = \frac{e^2}{h} \sum_{\varepsilon_n < \varepsilon_F} \{\frac{i}{2\pi} \int_\Gamma \langle du_n | du_n \rangle\} \tag{9.4}$$

Now it can be shown that $\langle du_n | du_n \rangle$ is a geometrical quantity and its integral over Γ a topological invariant (Chern character; cf.[40]). Thus, the Hall conductivity as a function of ε_F is topologically quantized. This is, of course, not yet the QHE. In order to get plateaux of σ_H as a function of n, localization has to be taken into account. A phenomenological way to do this is to assume that for sufficiently large L_x and L_y localization regimes exist in the band tails such that the Bloch velocity vanishes,

$$\partial_i \varepsilon_n(\theta_1, \theta_2) = 0 \quad , \quad i = 1,2 \tag{9.5}$$

Since states with this property do not contribute to σ_H, the topological quantization holds as long as ε_F varies in a localization regime. (As far as topological quantization is concerned band gaps and mobility gaps are equivalent.) Unfortunately this is still not the observed QHE since the succession of the plateau values as a function of n is, so far, completely undetermined. To get the strictly monotonic succession of the plateaux, an additional argument is needed. For instance, perturbation theory can be used to select the exact quantized value of σ_H which corresponds to a given value of n. Notice that the proof of the topological quantization given above requires that the mobility gaps survive the thermodynamic limit. (With $L_x, L_y \to \infty$ the Brillouin zone shrinks to a single point.)

The topological quantization described above is sometimes considered to be the rigorous version or the proof of the gauge argument. In fact in the topological approach the change of state is described as an adiabatic process (inherent in the Kubo formula), the "pumping" is replaced by a periodic boundary condition (the geometry of the system is equivalent to a torus - and not to a cylinder) and the average over the gauge flux is part of the trace [44]. (Cf. the comments made in sec. 6.)

10.

STŘEDA FORMULA

By rearranging the Kubo formula for independent electrons at T = 0 Středa [9] obtained the following expression

$$\sigma_H = -e\frac{\partial N}{\partial B}\Big)_{\varepsilon_F} + i\frac{e^2}{2}\ \{v_y G^-(\varepsilon_F) v_x \delta(H-\varepsilon_F) - (x \leftrightarrow y,\ - \to +)\} \quad (10.1)$$

Here $N(\varepsilon,B)$ is the number of states with energy below ε, $G^{\pm}(\varepsilon) = (H-\varepsilon \mp i\eta)^{-1}$, $\eta \to +0$ are single electron Green's functions, H being the Hamiltonian of the system. If ε_F lies in an energy gap the second term in (10.1) vanishes. Assuming that there is a gap corresponding to $N = i(eB/h)$ (i.e. separated Landau bands exist) (10.1) yields (2.2). The same result is obtained [45] for independent electrons in a perfect 2d infinite lattice if ε_F lies in one of the gaps of the spectrum which are characterized by the equation $\Omega N = i\Omega\ (eB/h) + k$, where Ω is the unit cell area and i, k integer [46] - in agreement with [8].

In spite of this success careful analysis revealed[47] that Středa's derivation of the formula (10.1) is inconsistent. The reasons are the following: on a cylinder the multiplication with the coordinate y is not defined in Hilbert space ($y\psi$ does not satisfy the periodic boundary condition (2.8)). Furthermore, for an infinite interval the multiplication with the coordinate x is an unbounded operation and, therefore, interchanging the order of certain mathematical operations leads to different results. Moreover, the adiabatic switching on process (inherent in linear response theory) may not be identified with the limit occurring in the formal identity

$$\delta(\varepsilon-H) = \lim_{\eta\to+0} \frac{i}{2\pi} [G^{+}(\varepsilon) - G^{-}(\varepsilon)] \tag{10.2}$$

and may not be interchanged with the trace (without explicit justification). In Středa's derivation of (10.1) the consequences of some incorrect operations are compensated by intuitively ignoring certain contributions.

After this heavy criticism one may expect that the Středa formula is completely incorrect. In fact the opposite is the case : for independent electrons, (10.1) is exactly equivalent to the (T = 0) Kubo formula *if*, in the x direction, the electrons are confined to a finite interval and vy is defined by

$$v_y = \frac{i}{m} (-i\hbar\frac{\partial}{\partial y} + eBx + \theta) \tag{10.3}$$

(The meaning of θ is the same as in the topological approach, sec. 9.) Thus, the relevant question to ask is not how large the corrections are if (10.1) is applied to a confined system instead of to a non-confined one[48]. In the case of free electrons confined by Dirichlet boundary conditions (10.1) yields (2.6) without any corrections in ℓ/L_x. The same is true for electrons in a perfect lattice[49]. What should be asked is, rather, how the exact quantization (2.2) can be established by applying (10.1) to a disordered system and assuming that ε_F lies in a mobility gap. It seems that another equivalent way to write the Kubo formula is more appropriate to investigate this question than the Středa formula (10.1)[24].

11.

FIELD THEORY

Numerical analysis (cf. sec. 8) as well as perturbation theory carried out by Ono[50] and Hikami[51] indicate that in a 2d random system at high magnetic fields and zero temperature all states are localized except those with energies at the centre of the Landau bands (ε_ν). It seems that Levine, Libby, and Pruisken[10] had succeeded to prove this rigorously by applying field theoretical methods to a system of independent electrons in a short range correlated random potential,

$$\langle U(\underline{r})U(\underline{r}')\rangle_p = U_0^2 a^2 \delta(\underline{r}-\underline{r}') \tag{11.1}$$

The essential features of this theory are the following. The critical properties of the system are described by an effective Lagrangian

$$L = \sigma^{(0)} L_1 + \sigma_H^{(0)} L_2 \tag{11.2}$$

where the coupling constants $\sigma^{(0)}$ and $\sigma_H^{(0)}$ are the mean field values of the conductivities

$$\sigma^{(0)} = \frac{e^2 n\tau/m}{1+\omega_c^2\tau^2} \quad , \quad \sigma_H^{(0)} = \omega_c \tau \sigma^{(0)} \tag{11.3}$$

with a certain relaxation time τ (cf. e.g.[52]). L_2 appears because B is an axial vector. For certain boundary conditions on the fields (which are compatible with edge currents[24]) L_2 can be shown to be a topological invariant. Levine et al.[10] showed that for all energies $\varepsilon \neq E_\nu$ $\sigma^{(0)}$ is renormalized to $\sigma = 0$ and $\sigma_H^{(0)}$ is renormalized to $\sigma_H = 0$ modulo ie^2/h, and at $\varepsilon = E_\nu$ $\sigma \neq 0$ and $\sigma_H = (e^2/h)(i+\frac{1}{2})$. The integer i corresponding to a certain given value of η can be determined by comparing the exact quantized values of σ_H when the mean field value is $\sigma_H^{(0)}$. Doing so one finds for any integer filling $\eta = k, k = i$ - in agreement with the observed succession of the plateaux. Again, the quantization of σ_H is due to topological reasons. However, the topological structure is now an internal property of a pure gauge theory; the Hall conductivity is not a Chern character as in the topological approach. (In the theory only gauge potentials and no gauge fields appear. Thus, the topological invariant is not of Yang-Mills type.)

The field theory of Levine et al.(10) is essentially a two-parameter scaling theory. The β functions of the renormalization group equations (cf. the lectures of MacKinnon in this volume)

$$\frac{d\sigma}{d\xi} = \beta(\sigma,\sigma_H) \quad , \quad \frac{d\sigma_H}{d\xi} = \beta_H(\sigma,\sigma_H) \tag{11.4}$$

$\xi = \ell n\, L$, $L=L_x=L_y$ are periodic functions of σ_H with period e^2/h. The flow diagram calculated by Levine and Libby(53) in the so-called dilute instanton approximation is shown in Fig. 4. The fix points on the line $\sigma_H = (i+\frac{1}{2})(e^2/h)$ are unstable. Khmelnitskii(54) anticipated a similar diagram, however, with finite slope at $\sigma = 0$, $\sigma_H = i(e^2/h)$.

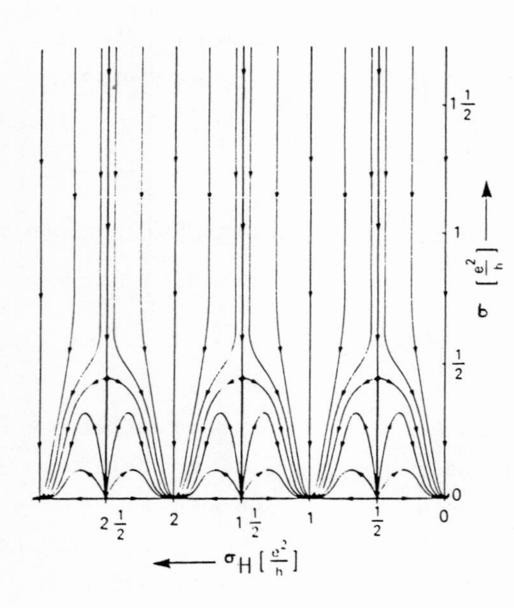

Fig. 4. Flow diagram predicted by the field theory(53).

According to Khmelnitskii(55) and Laughlin(56) the energy E_ν at which $\sigma \neq 0$ is given by

$$E_\nu = \varepsilon_\nu \frac{1+\omega_c^2\tau^2}{\omega_c^2\tau^2} \tag{11.5}$$

Thus, for $B\to\infty$ $E_\nu = \varepsilon_\nu$, and for $B\to 0$ $E_\nu\to\infty$ as required by the B = 0 localization theory (which predicts that in two dimensions at $T = 0$ for all finite energies $\sigma = 0$).

It should be pointed out that the strict periodicity of the flow diagram contradicts the experimental data ($\sigma(\varepsilon_F)$ depends on ν). Furthermore, as mentioned in sec. 8, numerical results by Ando [39] are incompatible with two-parameter scaling. Where is the error, in the field theory or in the numerical work? Hopefully not in both!

12.

CONCLUSIONS

The different theoretical approaches indicate that the IQHE is due to localization in a random potential and the existence of topological invariants. So far the interrelation between these approaches has not yet been clarified. For the time being there are too many theories - none of which is fully satisfactory. In particular the transport behaviour in the transition region between neighbouring plateau regimes has not been explained quantitatively. Also the role played by the Coulomb interaction between the electrons in the IQHE (although investigated in several papers [57]) is by no means fully understood. A still open question is the transition to fractional quantization.

REFERENCES

1. K. von Klitzing, G. Dorda, and M. Pepper, Phys. Rev. Lett. 45, 494 (1980).
2. R.E. Prange, Phys. Rev. B23, 4802 (1981).
3. H. Aoki and T. Ando, Solid State Comm. 38, 1079 (1981).
4. D. Tsui and S.J. Allen, Phys. Ref. B24, 4082 (1981).
5. R.B. Laughlin, Phys. Rev. B23, 5632 (1981).
6. Y. Ono, in: Anderson Localization, edited by Y. Nagaoka and H. Fukuyama, Solid State Sciences 39 (Springer, Berlin, Heidelberg, New York 1982) p. 207.
7. T. Ando, in: ref. (5) p. 68.
8. D.J. Thouless, M. Kohmoto, M.P. Nightingale, and M. den Nijs, Phys. Rev. Lett. 49, 405 (1982).
9. P. Středa, J. Phys. C15, L717 (1982).
10. H. Levine, S.B. Libby, and A.M.M. Pruisken, Phys. Ref. Lett. 51, 1915 (1983); Nucl. Phys. B240, 30, 49, 71 (1984).
11. J.T. Chalker, J. Phys. C16, 4297 (1983); Surf. Sci. 142, 182 (1984).
12. W. Brenig, Z. Phys. B50, 305 (1983).
13. T. Ando, Y. Matsumoto, and Y. Uemura, J. Phys. Soc. Jpn. 39, 279 (1975).
14. J. Kosch, U. Gummich, and J. Hajdu, Z. Phys. B62, 295 (1986).
15. R. Joynt and R.E. Prange, Phys. Ref. B39, 3303 (1984).
16. W. Brenig and K. Wysokinski, Z. Phys. B63, 149 (1986).
17. M. Tsukada, J. Phys. Soc. Jpn. 41, 1466 (1976).
18. R. Kubo, H. Hasegawa, and N. Hashitsume, J. Phys. Soc. Jpn. 14, 56 (1959).
19. R. Tao and Yong-Shi Wu, Phys. Rev. B30, 1097 (1984).
20. S. Luryi, The Quantum Hall Effect. Unpublished Lecture Notes, 1984. Cf. also (29).
21. M. Heuser and J. Hajdu, Z. Phys. 270, 289 (1974).
22. B. Halperin, Phys. Rev. B25, 2185 (1982).
23. J. Hajdu and U. Gummich, Solid State Comm. 52, 985 (1984).
24. M. Janssen, Diplomarbeit, Köln 1986 (unpublished).
25. S.M. Apenko and Yu E. Lozovik, J. Phys. C18, 1197 (1985); Sov. Phys. JETP 62, 328 (1985).
26. J. Hajdu and U. Gummich, Acta Phys. Polon. A69, 859 (1986).

27. B. Shapiro, Phys. Rev. B33, 8447 (1986).
28. S.V. Lordansky, Solid State Comm. 48, 1 (1982).
29. R.F. Kazarinov and S. Luryi, Phys. Rev. B25, 7626 (1982).
30. S. Luryi and R.F. Kazarinov, Phys. Rev. B27, 1386 (1983).
31. S.A. Trugman, Phys. Rev. B27, 7539 (1983).
32. H. Aoki, J. Phys. C16, L205 (1983); C18, L67 (1985).
33. A. MacKinnon, L. Schweitzer, and B. Kramer, Surf. Sci. 142, 189 (1984).
34. L. Schweitzer, B. Kramer, and A. MacKinnon, J. Phys. C17, 4111 (1984).
35. T. Ando, J. Phys. Soc. Jpn. 52, 1740 (1983); 53, 3101, 3216 (1984).
36. H. Aoki and T. Ando, Phys. Rev. Lett. 54, 831 (1985); Techn. Rep. ISSP, Ser. A, No. 1566 (1985).
37. L. Schweitzer, B. Kramer, and A. MacKinnon, Z. Phys. B59, 379 (1985).
38. T. Ando, Progr. Theoret. Phys. Suppl. No. 84, 69 (1985).
39. T. Ando, Techn. Rep. ISSP, Ser. A, No. 1565 (1985).
40. M. Kohmoto, Ann. Phys. 160, 343 (1985).
41. J.E. Avron, R. Seiler, and B. Simon, Phys. Rev. Lett. 51, 51 (1983).
42. Q. Niu, D.J. Thouless, and Y.S. Wu, Phys. Rev. B31, 3372 (1985).
43. J.E. Avron and R. Seiler, Phys. Rev. Lett. 54, 259 (1985).
44. W. Pook, Diplomarbeit, Köln 1986 (unpublished).
45. P. Středa, J. Phys. C15, L1299 (1982).
46. G. Wannier, Phys. Stat. Solidi b88, 757 (1978).
47. O. Vieweger, Diplomarbeit, Köln 1986 (unpublished).
48. L. Smrčka, J. Phys. C17, L63 (1984).
49. R. Rammal, G. Toulouse, M.T. Jaekel, and B. Halperin, Phys. Rev. B27, 5142 (1982).
50. Y. Ono, J. Phys. Soc. Jpn. 51, 2055, 3544 (1982); 53, 2342 (1984); Progr. Theoret. Phys. Supplement No. 84, 138 (1985).
51. S. Hikami, Phys. Rev. B29, 3726 (1984).
52. J. Hajdu and U. Paulus, Lecture Notes in Physics 177, (1983).
53. H. Levine and S.B. Libby, Phys. Lett. 150B, 182 (1985).
54. D.E. Khmelnitskii, JETP Lett. 38, 553 (1983).
55. D.E. Khmelnitskii, Phys. Lett. 106A, 182 (1984).
56. R.B. Laughlin, Phys. Rev. Lett. 52, 2304 (1984).

57. A.H. MacDonald, T.M. Rice, and W.F. Brinkman, Phys. Rev. B28, 3648 (1983);
O. Heinonen and P.L. Taylor, Phys. Rev. B32, 633 (1985);
J. Riess, Phys. Rev. B31, 8265 (1985).

THE MAGNETIC FIELD INDUCED METAL-INSULATOR TRANSITION

M. Pepper

Cavendish Laboratory, Cambridge

1. INTRODUCTION

These notes give some background to the two lectures on the role of the magnetic field in studies of localisation and the metal-insulator transition.

In a topic as wide as this it is only possible to consider a few salient points. Those selected are intended to show how different themes interact. For example, quantum interference, and its dependence on dimensionality, partially determines the location of the mobility edge as well as allowing determination of inelastic scattering times.

The emphasis is on physical principles and follows on from previous lectures in the Summer School on both 3D and 2D systems.

2. THE METAL-INSULATOR TRANSITION IN ORDERED SYSTEMS

This was the first type of transition considered (Mott 1949 and subsequently).[1,2,3] The initial approach was to consider the transition from the point of view of screening. We consider a uniform array of positive centres (each with an electron) and consider the conditions for an electron to be bound to the positive centre.

The screened potential energy of an electron-positive ion is

$$U(r) = -\frac{e^2}{r} e^{-\lambda r}$$

In Thomas-Fermi screening

$$\lambda^2 = \frac{6\pi n_o e^2}{E_F}$$

n_o is the carrier concentration and E_F is the Fermi energy. This can be written $\lambda^2 = 4N_o^{1/3}/a_H$ where a_H is the Bohr orbit. This potential is known to produce a bound state only if $\lambda < a_H$. We therefore have a bound state if $\underline{a} > 4a_H$, where $\underline{a}$ is the distance between centres, $(N_o^{-1/3})$. Mott argued that as the transition is based on screening so every free electron helps to liberate another bound electron and the transition is discontinuous, giving a metal when $\underline{a}$ drops below $\cong 4a_H$.

A different approach to the transition is that based upon the Hubbard[3] bands formed by D and D^- states. As the distance between the centres is reduced so these levels (discrete for an isolated impurity) broaden and form bands. The condition for a metallic system is band overlap and

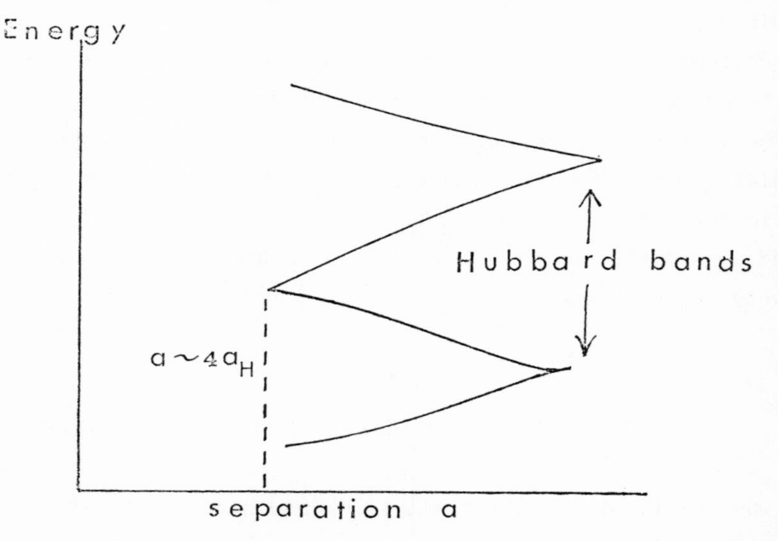

Fig. 1. The formation of Hubbard bands as the separation of the centres decreases. At large values of separation the bands are discrete levels corresponding to H and H^-. In vacuum the energies are 13.6eV and 0.7eV. The transition occurs when the bands overlap at $a \approx 4a_H$.

this occurs at roughly the same value as the screening criterion, namely $a_0 \cong 4a_H$. It should be noted that this transition arises due to the short range interaction, the long range interaction is considered elsewhere.

3.

DISORDER

Professor Mott's lectures have contained a discussion of the criterion for the transition based upon pure disorder (Anderson transition)[4,5] i.e. that occurring in a random array of donors. The criterion for the transition is similar to that in the screening and Hubbard treatments. If the disorder is not sufficient to localise all states in the band then localisation occurs in the tail and a mobility edge separates localised from extended states.

A major problem is the nature of the density of states at E_F at the transition point. This is small if the role of disorder is small (a pseudo-gap) increasing as the role of the disorder increases. However, this neglects the long range Coulomb interaction which can dominate transport properties at the transition[6].

4.

THE WIGNER TRANSITION

Originally proposed on the basis of a jellium model[7], i.e. electrons moving in a smooth background of positive charge, the Wigner transition is relevant to the case of an impurity band with low electronic mass (large Bohr orbit) which is compensated, i.e. the Bohr orbit overlaps large numbers of positive and negative centres where the net charge is small and positive. At very low temperatures electrons will crystallise into a regular array in order to minimise energy.

The stability criterion can be estimated as follows:

If there are n electrons, then localisation in a volume 1/n costs an energy per electron E_0,

$$E_0 \cong \hbar^2 n^{2/3}/m$$

The correlation energy E_C which is gained is $\cong e^2 n^{1/3}$.

The transition occurs when

$$E_c \cong E_o$$

or $n^{1/3} a_H \cong$ constant.

There is a considerable literature on estimates of the constant and many estimates suggest it is ≅0.01 or less[8]. In the presence of a magnetic field the localisation energy E_o is reduced and the transition becomes more favourable. Nevertheless it is not fully clear as to how the characteristics of the transition differ from the Mott or Anderson, although there have been many suggestions that the transition has been observed in the light mass semiconductors Indium Antimonide,[9], and Cadmium Mercury Telluride[10]. In two dimensions in strong magnetic field a series of instabilities occur as the magnetic field is increased. These lead to energy gaps appearing at certain fractional values of occupation of Landau levels and the phenomenon of fractional quantisation[11].

5.

CONDUCTIVITY IN THE LOCALISED REGIME

Prior to the recent theories of quantum interference and interactions (see Professor Mott's lecture for a review), it was thought that a minimum metallic conductivity exists, even in the absence of a magnetic field. When the Fermi energy E_F is below the mobility edge conduction is by two mechanisms in parallel[12], Figure 2.

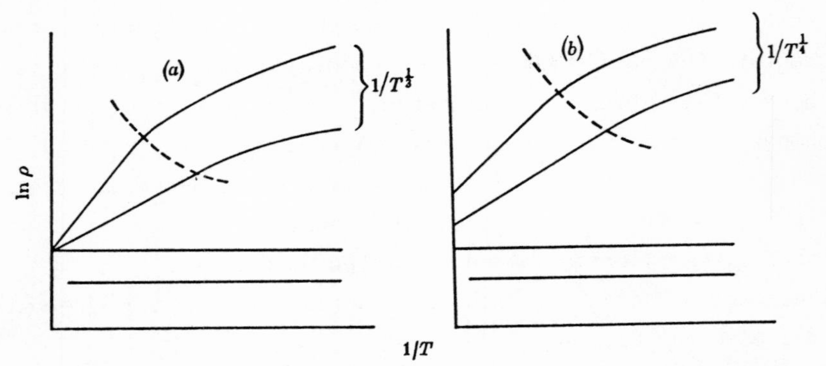

Fig. 2. Behaviour of the resistivity at an Anderson transition in (a) a two and (b) a three dimensional system. The dotted lines show the boundary temperature where conduction is by hopping and by excitation to E_o.

a) Excitation to the mobility edge

$\sigma = \sigma_{min} \exp - W/kT$

$W = |E_c - E_F|$

Corrections due to this formula have been discussed by Professor Mott

b) Hopping Conduction

At low values of doping such that the two "Hubbard bands" have not merged then hopping can only occur in the presence of compensation. The presence of compensating ions is required to pull down E_F into a region of finite density of states.

When the Hubbard bands overlap, hopping occurs and at low temperatures becomes variable range hopping with $\ln \sigma \propto T^{-1/4}$ in 3D and $\ln \sigma \propto T^{-1/3}$ in 2D[13].

6.

MAGNETIC FIELD EFFECTS ON THE TRANSITION

Systems can be pushed through the metal-insulator by the application of uniaxial stress or compression (see lectures by Dr. G. Thomas) as well as by the application of a magnetic field, B. The earliest work on this topic was performed in InSb[14,15], which with its low effective mass ($\cong .016\ m_o$) and large Bohr orbit (600A), shows pronounced shrinkage of wave functions and magnetic freezeout at low values of applied field. A convenient parameter to take is γ ($\gamma = \hbar\omega_c/2R^*$)) where $\hbar\omega_c/2$ is the zero point energy in a magnetic field and R^* is the ionisation energy.

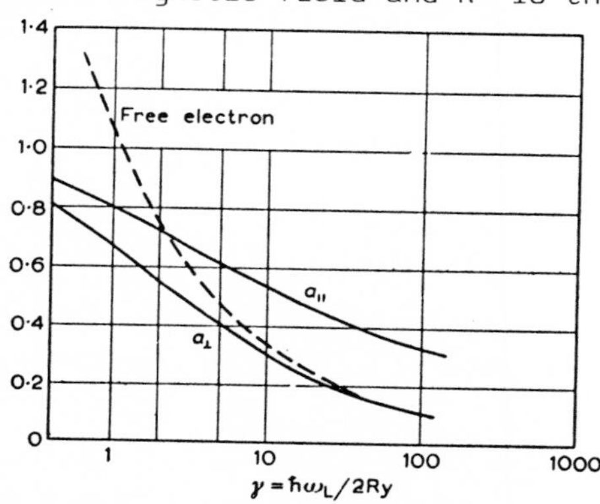

Fig. 3. This figure (from Yafet et al) illustrates the change in Bohr orbit for both transverse and parallel directions ($a_{\perp}$, $a_{\parallel}$). The free electron orbit radius is shown.

In InSb, $\gamma = 1$ for $B \cong 0.2$ Tesla; increasing values of γ reflect the magnetic field becoming dominant. In GaAs $\gamma \cong 1$ is possible but in Silicon γ is always <1. Just on the metallic side of the transition a system can be pushed through the transition for $\gamma \ll 1$; recent work on Silicon[16,17,1 (both Sb and As doped) shows this effect, which arises as the transition is very sensitive to the form of the wave function at distances $\cong 4a_H$. It should be noted that due to Gaussian decay, the wave function is affected by a magnetic field which is sufficiently small that the effective Bohr orbit is not modified significantly. This has been discussed by Efros and Shklovskii in relation to hopping conductivity[19].

The magnetic field induced transition has been investigated in a variety of materials where the dominant effect is wave function shrinkage (see Long and Pepper[57] for a review of earlier work), including both two and three dimensions in GaAs[20,21,22,23,24].

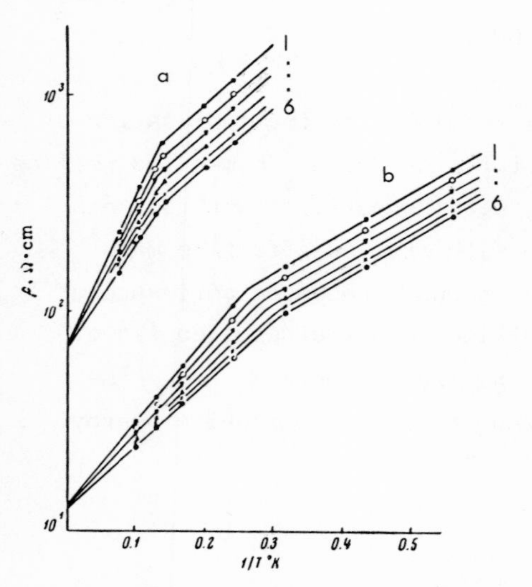

Fig. 4. The results of Gershenzo et al[31] showing the temperature dep endence of the resistivity of doped, compensated p-type InSb as a function of magnetic field. The fields in Tesla are: (1)0, (2)2, (3)2.5, (4)4.0, (5)3.5, (6)4.0. Sample (a) had $N_A=10^{16}$ and K=0.7, sample (b) had $N_A=1.5\times10^{16}$ and K= 0.65. The intercept values of σ_{min} are lower than predicted but appear to increase considerably for a small change in N_A.

The magnetic semiconductors show an interesting difference in behaviour. Here the application of a magnetic field can cause an insulating system to become metallic. The earliest work in this area was on $Gd_{3-x}V_xS_4$ [25], (here V indicates a vacancy). The Gd^{3+} ion has a moment and Gd_2S_3 is an insulating anti-ferromagnet, whereas Gd_3S_4 is metallic and ferromagnetic. The electrons form a degenerate gas of magnetic polarons over the composition range for which the material is anti-ferromagnetic. The polaron enhancement of the mass enhances Anderson localisation in the random field of the vacancies. Application of a magnetic field will line up the spins and the spin polaron cannot form. This can be regarded as a reduction in mass and the Anderson transition can occur.

Materials such as p-type $Hg_{0.85}Mn_{0.15}Te$ show an Anderson transition[26] (from insulating to metallic) as the magnetic field increases and modifies the exchange interaction resulting in an increase in the size of the acceptor wave function and a reduction in localisation.

Recent experiments on magnetic semiconductors have investigated critical behaviour and the critical exponent ν is found to be 1, ($\sigma = \sigma_c(H/H_c-1)^\nu$ in agreement with theories of the transition when the spin splitting is appreciable[27].

There has been considerable discussion on the nature of the transition in CdHgTe. It has been argued that the transition is a Wigner transition[10]; evidence has been presented suggesting that non-linear conductivity at very low fields is due to collective transport produced by the sliding of a pinned Wigner lattice[28]. On the other hand, electrical and optical work on this material has been interpreted as suggesting a conventional magnetic freeze out and an Anderson transition[29].

Theoretical work on the metal-insulator transitions has established that in the presence of a magnetic field, a two parameter scaling function is required.

7.

CONDUCTIVITY MECHANISMS IN THE LOCALISED REGIME

When the Fermi energy is in the localised states conduction proceeds by two mechanisms in parallel.

1. Excitation to the mobility edge,

$$\sigma = \sigma_{min} \exp(- E_2/kT)$$

NB. the activation energy for excitation to the conducting band is termed E_1.

Here σ_{min} is a valid concept at finite temperatures where the phase coherence length (L_{IN}) is less than the elastic mean free path, ℓ.

2. Fixed range (Miller-Abrahams) hopping at the Fermi energy[30],

$\sigma = \sigma_o \exp(-E_3/kT)$, where σ_o is determined by the wave function overlap and electron-phonon coupling. The activation energy E_3 is determined by the doping and compensation. At low temperatures hopping becomes variable range hopping and $\sigma = \sigma_o \exp[-(\text{constant}/T)^{1/4}]$.

Application of a magnetic field clearly distinguishes between these regimes as σ_{min} is not affected (for $L_{IN} < \sigma$) but E_2 increases with B due to a reduction in wave function overlap. On the other hand, in the hopping regime E_3 is unaffected by the magnetic field as it depends on the random field but σ_o is reduced by wave function shrinkage[31,32,33]. The net result is a sharpening of the difference between the regimes of transport. When hopping is variable range, the constant term in the exponent is reduced as this contains the localisation length $(1/\alpha)$ (the constant is $\cong [\pi NkT\alpha]^{-1}$).

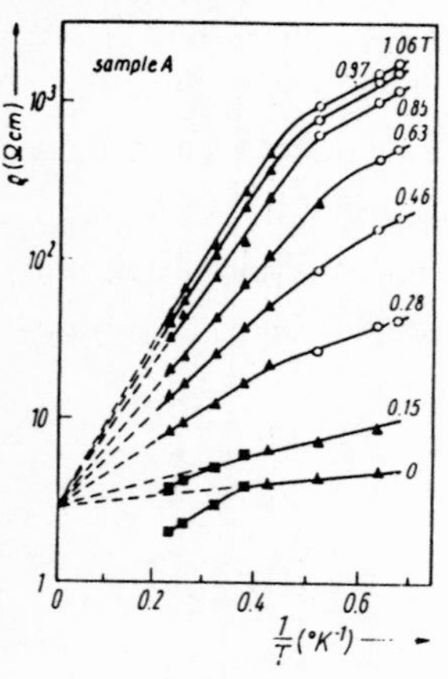

Fig. 5. The results of Ferre et al[32] showing the variation in resistivity of InSb with reciprocal temperature as a function of magnetic field (in T), $|N_D-N_a|=5.7\times10^{13}cm^{-3}$. Other results in the original paper show a transition to metallic to localised behaviour and confirm that ε_1 is unaffected by the field. At higher temperatures the ε_1 process is observed, these points are omitted here.

The application of a magnetic field may help to distinguish between single and many electron hopping[34]. Experiments on n-type CdMnSe have shown that E_3 decreases with increasing B[35]. This result is predicted by the theory of multi-carrier hopping but is difficult to understand on the basis of single carrier transport.

8.

QUANTUM INTERFERENCE AND THE ELECTRON-ELECTRON INTERACTION

Quantum Interference produces two effects on a 3D system:

1. The location of the mobility edge, E_c, is increased above the value

it would take due to disorder alone[36].

2. Metallic conductivity σ, is reduced[37,38] the value for σ being

$$\sigma - \sigma_B g^2 \left[1 - \frac{1}{g^2 (k_F \ell)^2} (1 - \ell/L)\right]$$

L is the inelastic length, ℓ is the elastic length and g is the reduction in the density of states (discussed by Professor Mott).

The shift of the mobility edge has been incorporated into a phase diagram of the transition by Shapiro[39], Figure 6. Initially, the magnetic field reduces quantum interference and the mobility edge is reduced in energy so decreasing N_c the critical concentration for the transition. As the magnetic field increases, N_c then increases as expected due to shrinkage. The initial reduction in N_c has not been found in Si but has been observed in InP[40].

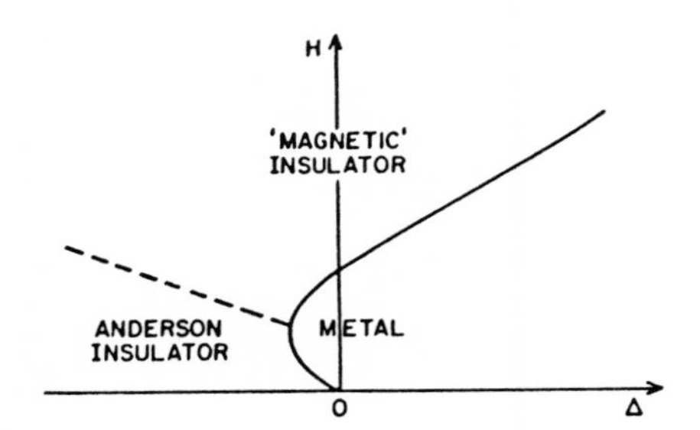

Fig. 6. Shapiro's Phase Diagram

The reduction in metallic conductivity due to interference has been observed and good agreement has been found with the negative magnetoresistance formula of Kawabata[38]

$$\delta\sigma = \left[\frac{\sigma_B}{12\sqrt{3}}\right] \left[\tau_{IN}/\tau\right]^{3/2} \left[\frac{e\tau}{m^*}\right] B^2 \qquad \text{at low fields}$$

and

$$\delta\sigma = \frac{e^2}{2\pi^2 \hbar} \, 0.605 \sqrt{\frac{eB}{\hbar}} \; ; \quad \delta\sigma = 289.3 \, B^{\frac{1}{2}} (\Omega m)^{-1} \, (\text{Tesla})^{-\frac{1}{2}}$$

at high fields

This formula has been verified to 5%[41]. This allows extraction of the inelastic scattering time τ_{IN}[42]. Evidence has been found supporting the existence of a disorder induced $T^{3/2}$ term as well as the Landau-Baber T^2 law[43].

Work on InSb indicates that the T = 0 conductivity $\sigma(0,B)$ for a metallic sample obeys the scaling law

$$\sigma(0,B) = \sigma_c \left[\frac{a_\perp^2 a_\parallel}{a_c}\right]^{\nu} - 1$$

$a_c = 1/4\, N^{-1/3}$ and, $a_\perp$ and $a_\parallel$ are the transverse and parallel values of the Bohr orbit. ν is found to be $\cong 1.2$ which is in agreement with theories suggesting $\nu = 1$ in strong fields[44].

9.

INTERACTION EFFECTS

The interaction corrections first developed by Altshuler and colleagues[6] have been discussed in previous lectures. In 3D

$$\delta\sigma(T) = \frac{0.23e^2}{\pi^2\hbar} \left[4/3 - 2F\right] \left(\frac{k}{\hbar D}\right)^{\frac{1}{2}} T^{\frac{1}{2}}$$

The factor 4/3 comes from the Exchange interaction and the F term is due to the Hartree interaction. A further factor of F arises from particle-particle scattering. Here F is defined as the exact scattering amplitude of a particle and a hole with total spin unity. It decreases from 1 to 0 as the carrier concentration increases.

In the presence of magnetic field the conductivity $\delta(B,T)$ can be written

$$\delta(B,T) = \delta\sigma'(T) + \delta\sigma''(B,T)$$

$\delta\sigma'(T)$ is the same as the non magnetic case except the factor (4/3-2F) is replaced by 2-F and the term $\delta\sigma'$, which arises from the lifting of the spin degeneracy is given by

$$\delta\sigma = (B,T) - \delta\sigma''(0,T) = -\frac{e^2}{\hbar}\frac{F}{4\pi^2} g_3 \sqrt{\frac{T}{2D}}$$

g_3 is a coupling constant which has limiting values $(g\mu B/kT-13)$ in high fields and $0.053(g\mu B/kT)^2$ in low fields. The criterion of a weak (strong field being $(gNB/hT)<(>)kT$. Here g is g value and B the Bohr magneton.

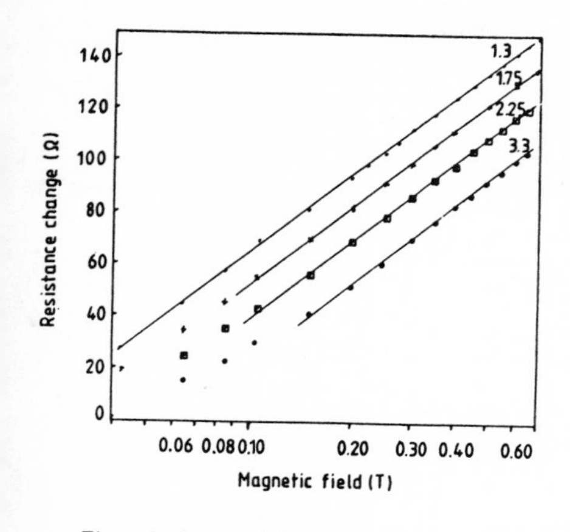

Fig. 7. Resistance versus magnetic field for different temperatures which are indicated. These results were obtained with GaAs and agree with the Kawabata formula 70~5%.

The interaction corrections have been extensively investigated in both 2D and 3D and it has been found that the characteristics of the magnetic field induced transition are determined by the interaction effects. An exception appears to be InP[45,46], Figures 8 and 9: here work has shown that a minimum metallic conductivity is found at temperatures down to ≅ .03K, i.e. $\sigma = \sigma_{min}$ exp-W/kT because m ≅ 0. However, at lower temperatures the interaction effects should be altered and give metallic conductivity below σ_{min}. InP doped near the transition shows $T^{1/2}$ behaviour

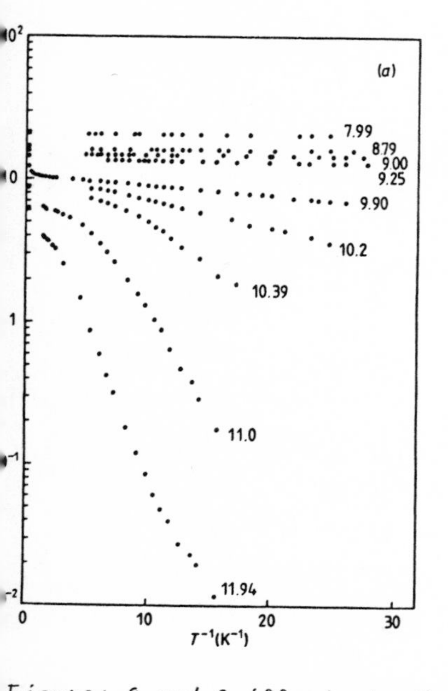

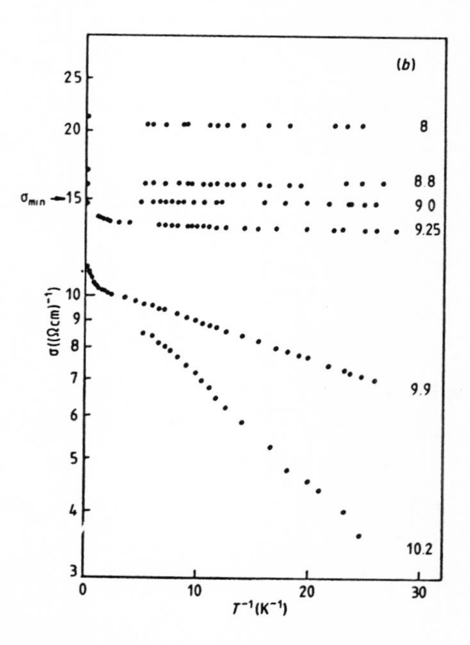

Figures 8 and 9 illustrate the magnetic field induced transition in InP doped at ≈ twice the critical concentration for the transition ~7 $10^{16} cm^{-3}$. The values of magnetic field are indicated.

in a magnetic field without observation of σ_{min}, similarly GaAs and InSb show a continuous transition[44,47], Figures 10 and 11. As the change in the density of states is proportional to the change in the conductivity it is not clear what happens when this becomes large. Anomalous effects have been found in Sb doped Si which have been attributed to an electronic phase transition driven by the interaction[16].

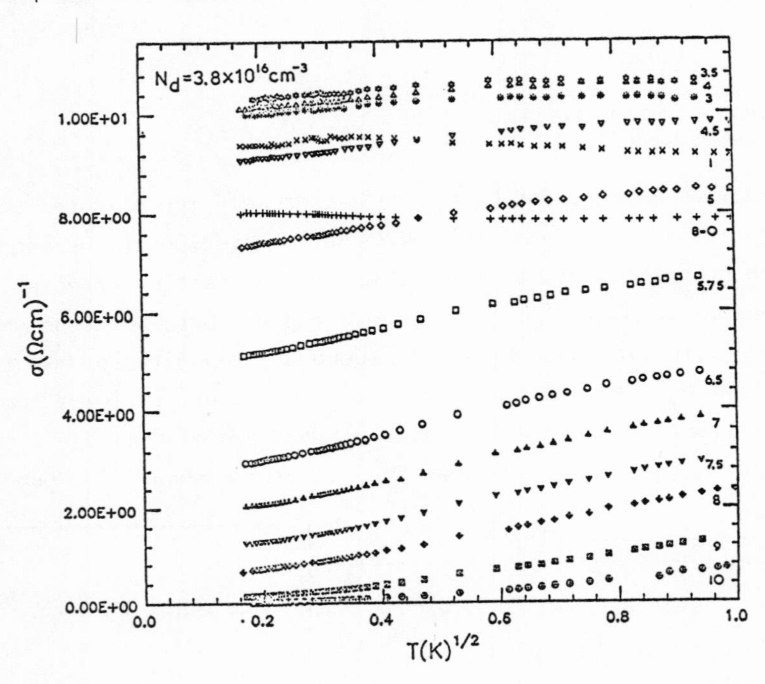

Fig. 10. Conductivity plotted against $T^{1/2}$ for a n-type GaAs sample $N_D=3.8\times10^{16}cm^{-3}$. The values of magnetic field are indicated.

As the temperature is reduced so interactions can dominate over quantum interference. We will discuss the temperature dependence of conductivity of 3D metallic InSb; at higher temperatures the conductivity rises but then as $(1/\ell-1/L_{IN})$ tends to a constant value so the interaction correction can be observed following a $T^{1/2}$ law. Figures 10 and 11 show the behaviour for GaAs doped on the metallic side of the transition. At low values of field the $T^{1/2}$ correction causes a drop in resistance with decreasing temperature but at higher fields the interaction correction changes sign and causes the resistance to increase.

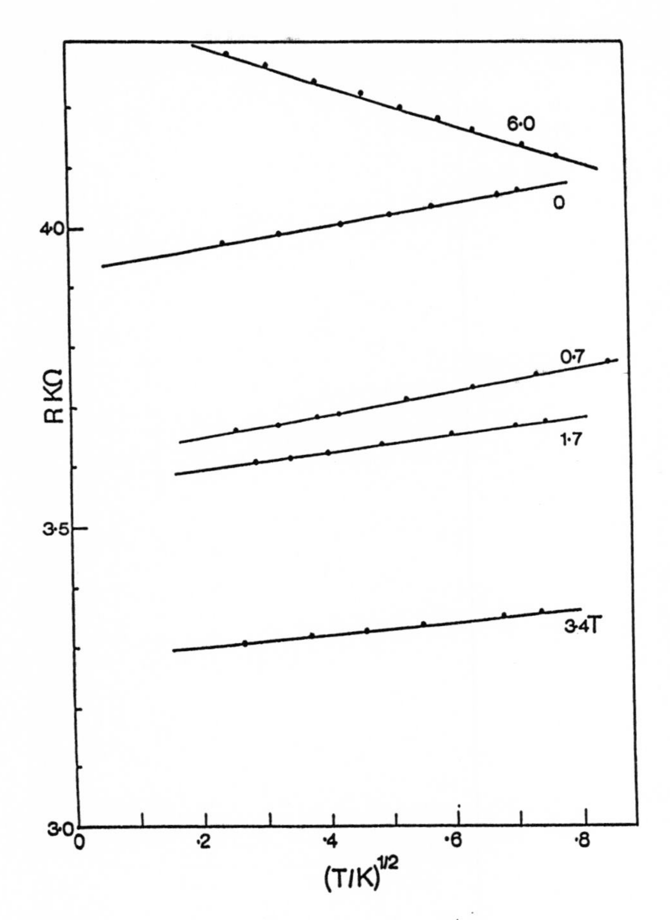

Fig. 11. Resistance of GaAs sample vs. $T^{1/2}$ at higher fields. (Newson Ph.D Thesis).

Castellani[48] et al have suggested that in the $T^{1/2}$ regime m varies as $(\sigma(0))^{-3/2}$. This was tested for InSb[49] and good agreement found except very near the transition where the relation broke down. Theory suggests that very close to the transition the $T^{1/2}$ behaviour turns into $T^{1/3}$. Figures 12 and 13 show results of Newson[49] for InSb suggesting that this may be the case.

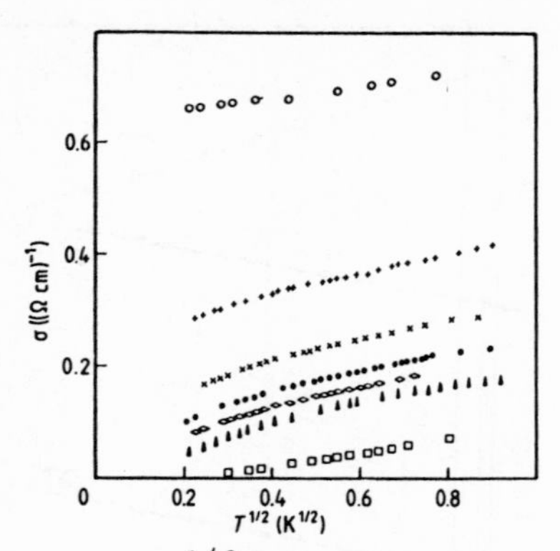

Fig. 12. Conductivity vs $T^{1/2}$ for a metallic InSb sample. The points are for magnetic fields between 0.6 and 1.0 Tesla. (Newson Ph.D Thesis).

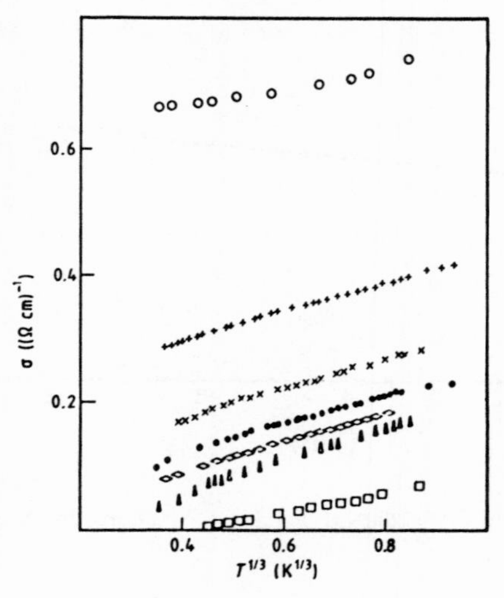

Fig. 13. The results of Figure 12 plotted as σ versus $T^{1/3}$.

10.

INTERACTION EFFECTS IN HIGH MAGNETIC FIELDS IN 2D

A magnetic field is an excellent method of separating quantum interference and interaction corrections[50,51]. In strong fields such that $\omega>1$, the density of states of a 2D electron gas splits into Landau levels and now the corrected predictions to the resistance R^0 and Hall constant R_H are as follows[52,53]

$$\frac{\Delta R}{R^{O}} = - \frac{R^{O}_{\square} g' e^{2}}{2\pi^{2}\hbar} \quad (1 - \omega^{2}\tau^{2})\ell n(T/T_{o})$$

$$\frac{\Delta R_{H}}{R_{H}} = - 2R^{O}_{\square} \frac{g' e^{2}}{2\pi^{2}\hbar} \ell n(T/T_{o})$$

Here g´ is the interaction coupling constant considered in detail by Fukuyama[54]. The theory predicts a ratio

$$\frac{SR_{H}/R_{H}}{SR/R} = 2/1 - \omega^{2}\tau^{2}$$

when $\omega\tau$ is small this is the ratio of 2 which has been confirmed.

The validity of this relation has been investigated in Silicon inversion layers with high impurity scattering[55]. This broadened the Landau levels reducing the role of thermal excitation. Figure 14 shows the

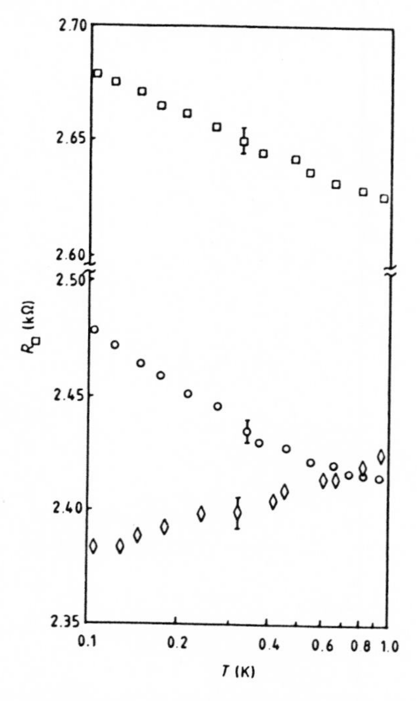

Fig. 14. The resistivity as a function of temperature for the values of magnetic field indicated. ○B=0.61T; □B=3.8T; ◇B=4.5T.

behaviour of the resistivity (R_{o}) illustrating the logarithmic behaviour and the change in sign which occurs very near the $\omega\tau$=1 condition. The

Hall constant shows little change in magnitude as, Figure 15, the magnetic field increases; the small effect could be due to a change in g, the coupling constant. Good agreement with theory is revealed in Figure 16, where the difference in behaviour is clearly seen.

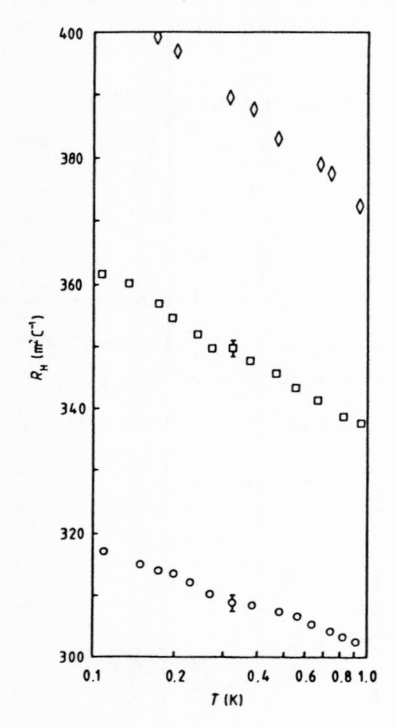

Fig. 15. The Hall constant as a function of temperature for the value of magnetic field indicated. ○$B=0.61T$, □$B=3.8T$; ◇$B=4.5T$.

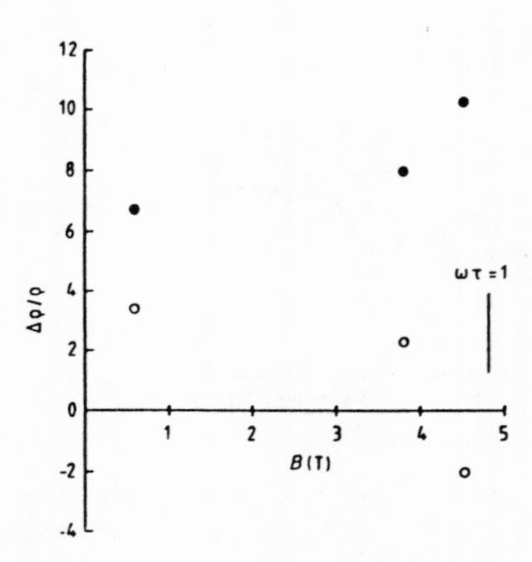

Fig. 16. The percentage changes of resistivity (○) and Hall resistivity (●) per decade change of temperature are plotted against magnetic field.

Figure 17 illustrates the effects of parallel and transverse fields on the Si inversion layer. As seen in the transverse field the quantum interference is first suppressed and then the enhancement of the interaction effect is observed. In a parallel field the quantum interference is not significantly affected.

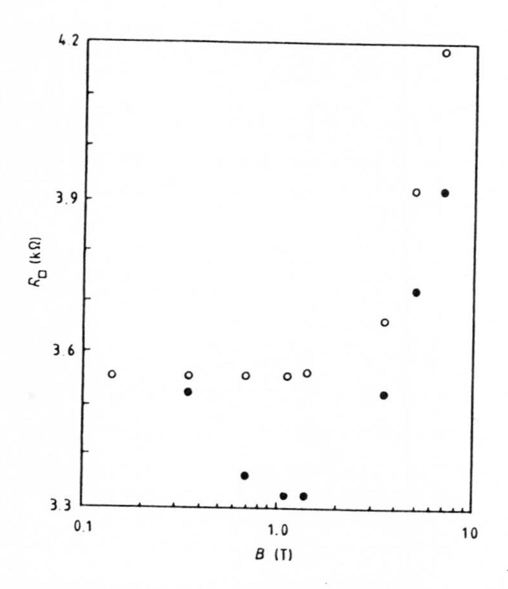

Fig. 17. The resistivity is plotted against magnetic field for normal (●) and parallel (o) directions. The carrier concentration is 1.1x10^{12} electrons/cm^2, temperature is 1.35K.

It is clear that similar experiments on a structure with a wider channel could give rise to an observable suppression of quantum interference by a parallel field. This has been achieved by the use of the Gallium Arsenide Field Effect Transistor[41,56], a structure in which the thickness of the conducting channel can be varied continuously from a few thousand Angstroms to zero. Altshuler and Aronov[6] have predicted that when $L_c \gg aL_{IN}$ and $L_{IN} > a$, the negative magnetoresistance $\Delta\sigma(B)$ is given by

$$\Delta\sigma(B) = \frac{e^2}{2\pi^2\hbar} \ln\left(\frac{a^2 L_{IN} B^2 e^2}{3\hbar^2} + 1\right)$$

The GaAs device has been used to confirm the validity of this expression and L_{IN} has been extracted by the use of both transverse and parallel magnetic fields. Figure 18 shows the temperature dependence of L_{IN} obtained by the two methods and, as seen, the agreement is extremely good.

When L_c<<a the negative magnetoresistance is 3D but an intermediate regime is found to be present between the 3D and low field (B^2) regimes. (The B_2 regime occurs when $L_c >> (L_{IN}a)^{\frac{1}{2}}$. The intermediate regime arises because diffuse scattering from the enclosing potential of the conductin channel imposes a momentum indeterminacy. This problem has been considered using the technique of Bergmann which considers the geometry of the region in which quantum interference occurs[57]. The result is that the intermediate parallel field regime is predicted to show a behaviour

$$\delta\sigma = \frac{e^2}{2\pi^2\hbar}(1 - 2a/L_c)\ell nT.$$

This equation is in good agreement with experiment and possesses the characteristic lnT dependence of a 2D system and the L_c^{-1} ($B^{1/2}$) dependence of 3D. This device has been used for a variety of experiments in which the transition from 3D to 2D is clearly seen in the quantum interference.

The 2D to 1D transition has been investigated in modulation doped GaAs-AlGaAs heterojunctions[58]. Here the 2D channel is electrostatically squeezed and a transition to 1D behaviour occurs. Negative magnetoresistance is found in agreement with the Altshuler-Aronov formula[59].

$$\delta G = -\frac{e^2}{\pi\hbar}\left[\frac{1}{L_{IN}^2} + \frac{W^2}{3L_c^4}\right]^{-1/2}$$

where W is the width of the channel and G is the conductance per unit length. As the length scales are long the negative magnetoresistance ca be explored without enhancement of interaction effects. These experiments showed that the inelastic (phase relaxation length) varied as $T^{-1/}$ in agreement with theoretical predictions for the scattering of electrons by fluctuations in electronic charge density[60]. Correcting the conductance for the interference correction

$$\delta G = -\frac{e^2 L_{IN}}{\pi\hbar}$$

allows observation of the interaction effects. Here the predicted variation is

$$\delta G = \left(\frac{e^2 g_{1D}}{\pi\hbar}\right)\left(\frac{\hbar D}{2kT}\right)^{1/2}$$

where g_{1D} is a 1D coupling constant having value 1.35. Figure 19 shows the experimental results, a satisfactory feature was that extrapolating the straight line to $T^{-1/2} = 0$ gives the Boltzmann conductance. Assuming a 2D density of states gives the same value of diffusion coefficient, D, as the slope. A combination of magnetic fields and semiconductor technology allows investigation of a number of aspects of quantum interference and interaction effects.

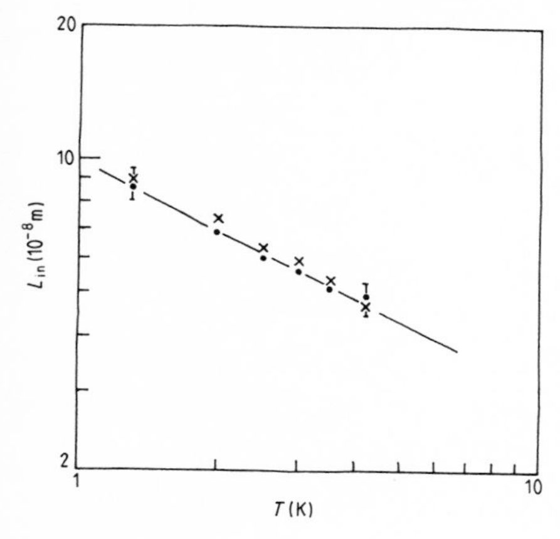

Fig. 18. Inelastic length L_{in}, obtained from the two methods for a channel thickness of 440Å ● Parallel field; × perpendicular field.

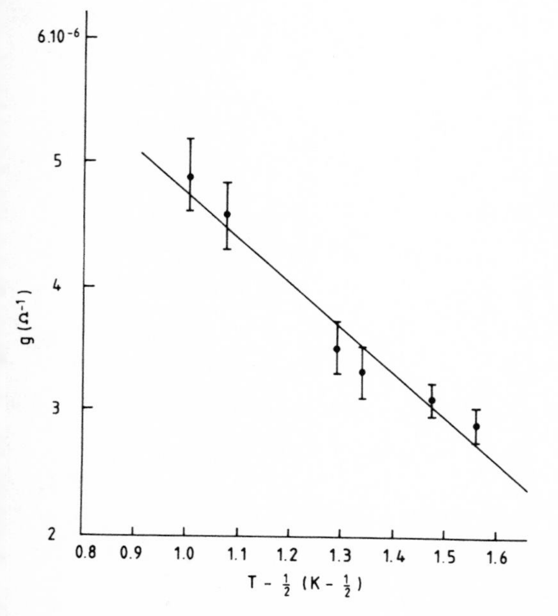

Fig. 19. The conductance after addition of the quantum-interference correction, plotted against $T^{-1/2}$.

REFERENCES

1. Mott, N.F., Proc. Phys. Soc. A 62, 416,(1949).
2. Mott, N.F., Can. J. Phys. 34, 1356 (1956).
3. Mott, N.F., "Metal-Insulator" Transitions, Taylor and Francis, London,(1974).
4. Anderson, P.W., Phys, Rev. 86, 694 (1952).
5. Mott, N.F. and Davis, E.A., "Electronic Processes in Non-Crystalline Materials", Oxford University Press, 2nd Edition, 1979.
6. Altshuler, B.L. and Aronov, A.G., in "Electron-Electron Interactions in Disordered Systems", Ed. Efros, A.L. and Pollak, M., North Holland, 1985.
7. Wigner, E., Trans. Far. Soc. 34, 678,(1938).
8. Care C.M. and March, N.H., Adv. Phys. 24, 101 (1975).
9. Care C.M. and March, N.H., J. Phys. C. 4, L372 (1971).
10. Rosenbaum, T.F., Field, S.B., Nelson, D.A. and Littlewood, P.B., Phys. Rev. Lett. 54, 241 (1985).
11. Tsui, D.C., Stormer, H.L., Hwang, J.C.M., Brooks, J.S. and Naughton, M.J., Phys. Rev. B. 28, 2274 (1983).
12. Mott, N.F., Pepper, M., Pollitt, S., Wallis, R.H. and Adkins, C.J., Proc. Roy. Soc. A 345, 169 (1975).
13. Pepper, M., Pollitt, S., Adkins C.J. and Oakley, R.E., Phys. Lett. 47A, 71, (1974).
14. Yafet, Y., Keyes, R.W. and Adams, E.N., J. Phys. Chem. Solids. 1, 137 (1956).
15. Sladek, R.J., J. Phys. Chem. Solids 5, 157 (1958).
16. Long, A.P. and Pepper M., J. Phys. C. 17, L425 (1984).
17. Shafarman, W.N., Castner, T.G., Brooks, J.S., Martin, K.P. and Naughton, M.J., Phys. Rev. Lett. 56, 980 (1986).
18. Shafarman, W.N. and Castner, T.G., in "Proceedings of the Seventeenth International Conference on the Physics of Semiconductors 1984 Ed. Chadi, J.D. and Harrison, W.A. Springer 1985, 1079.
19. Shklovskii, B.I. and Efros, A.L., Sov. Phys. Semicond. 14 (5) 487, (1980).
20. Newson, D.J., McFadden, C. and Pepper, M., Phil. Mag. B 52, 437, (1985).

21. Pepper, M., Phil. Mag. B 38, 515 (1978).
22. Robert, J.L., Raymond, A., Aulombard, R.L. and Bousquet, C., Phil. Mag. B 42, 1003 (1980).
23. Mansfield, R., Abdul-Gader, M. and Fozooni, P., Solid State Electron 28, 109 (1985).
24. Dubois, H.G., Biskupski, G., Wojkiewicz, J.L., Briggs, H. and Remenyi, G., J. Phys. C. 17, L411 (1984).
25. Von Molnar, S., Briggs, A., Flouquet, J. and Remenyi, G., Phys. Rev. Lett. 51, 706 (1983).
26. Myczielski, A. and Myczielski, J., Proc. 15th Intern. Conf. on Semiconductor Physics (Proc.Phys.Soc. Japan), 807 (1980).
27. Wojtowicz, T., Dietl, T., Sawicki, M., Plesiewicz W. and Jaroszynski,J. Phys. Rev. Lett. 56, 2419 (1986).
28. Field, S.B., Reich, D.H., Shivaram, B.S., Rosenbaum, T.F., Nelson, D.A., and Littlewood, P.B., Phys. Rev. B 33, 5082 (1986).
29. Aronzon, B.A., Kopylov, A.V., Meilikhov, E.Z., Gorbatyuk, I.N., Rarenko, I.M. and Talyanskii, E.B., Sov. Phys. Jetp. 62, 71 (1985).
30. Miller, A. and Abrahams, E., Phys. Rev. 120, 745 (1960).
31. Gershenzon, E.M., Kurilenko, J.N. and Litvak-Gorskaya, L.B., Sov. Phys. Semicond. 8, 689 (1974).
32. Ferre, D., Dubois, H. and Biskupski, G., Phys. Stat. Sol. B70, 81, (1975).
33. Pepper, M., Communcations on Physics, 1, 147 (1976).
34. Pollak, M. and Ortuno, M., in "Electron-Electron Interactions in Disordered Systems", Ed. Efros, A.L. and Pollak, M. 287, (1985).
35. Dietl, T., Antoszewski, J. and Swierkovski, L., Physica 117 B, 491 (1983).
36. Khmelnitskii, D.E. and Larkin, A.J., Solid State Comm. 39,1069 (1981).
37. Berggren, K-F., J. Phys. C. 15, L843 (1982).
38. Kawabata, A., Solid State Comm. 38, 823 (1981).
39. Shapiro, B., Phil. Mag. B 50, 241 (1984).
40. Spriet, J.P., Biskupski, G., Duvois, H. and Briggs, A., Phil. Mag. Lett., In the Press.
41. Newson, D.J. and Pepper, M., J. Phys. C. 18, L1049 (1985).
42. Reference 38 and Kawabata, A.J., Phys. Sol. Japan 49, 628 (1980).
43. Reference 20.

44. Reference 23, Castner, T.G. and Shafarman, W.N. in "Localisation and Metal-Insulator Transitions", Ed. Fritzsche, H. and Adler, D., 9, (1985).
45. Reference 24.
46. Long, A.P. and Pepper, M., J. Phys. C 17, 3391 (1984).
47. Newson, D.J. et al, to be published.
48. Castellani, C., Di Castro, C., Lee, P.A. and Ma, M., Phys. Rev. B 30, 527 (1984).
49. Newson, D.J. et al, to be published.
50. Uren, M.J., Davies, R.A. and Pepper, M., J. Phys. C 13, L985 (1980).
51. Uren, M.J., Davies, R.A., Kaveh, M. and Pepper, M., J. Phys. C 14, 5737 (1981).
52. Houghton, A., Senna, J.R. and Ying, S.C., Phys. Rev. B 25, 2169 (198
53. Houghton, A., Senna, J.R. and Ying, S.C., Phys. Rev. B 25, 6468 (198
54. Fukuyama, H. in "Electron-Electron Interactions in Disordered Systems", Ed. Efros, A.L. and Pollak, M., 155 (1985).
55. Davies, R.A. and Pepper, M., J. Phys. C 16, L679 (1983).
56. McFadden, C., Newson, D.J., Pepper, M. and Mason, N.J., J. Phys. C 1 L383 (1985).
57. Bergmann, G.D., Phys. Rev. B28, 2914 (1983).
58. Thornton, T.J., Pepper, M., Ahmed, H., Andrews, D. and Davies, G.J., Phys. Rev. Lett., 56, 1198 (1986).
59. Altshuler, B.L. and Aronov, A.G., JETP Lett. 33, 499 (1981).
60. Altshuler, B.L. and Aronov, A.G., Solid State Comm. 54, 617 (1985).

STATIC AND DYNAMIC MAGNETIC PROBES OF THE M-I TRANSITION

D. F. Holcomb

Cornell University, Ithaca, NY 14853 USA

1.
INTRODUCTION

The study of localization transitions naturally focuses on the electrical transport properties. However, important characteristics of the system of charge carriers can also be determined by measurement and interpretation of magnetic properties. These notes will describe the nature of the information which can be and has been obtained from measurements of the magnetic susceptibility, χ, of the magnetization per unit volume, M, and of various characteristics of the magnetic resonance signals obtained both from electron spin resonance (ESR) and nuclear magnetic resonance (NMR) measurements. The largest block of experimental data comes from the bellwether system, Si:P. Some measurements on the other most common silicon donor system, Si:As, and on the germanium companion, Ge:As, will be described. I will also describe extensive NMR measurements which have been made on the sodium bronze systems, Na_xWO_3 and $Na_xTa_yW_{1-y}O_3$, and relate those to the NMR observations in the silicon materials.

A similar review was given at an earlier session of this School.(1) I shall concentrate on information and interpretation which has been added since that time, referring to the work before

1978 only as necessary to give background and continuity.

It is useful to remember that the magnetic susceptibility and magnetization are bulk properties and, like electrical transport data, require the intermediary of a theoretical model in order to relate the experimental data to microscopic properties. The task of developing this model is particularly difficult in the strong disorder of the doped semiconductors, where local impurity densities differ greatly from one point to another. An advantage to these bulk magnetic properties, however, is that they are thermodynamic properties and, thus, theoretical expressions can often be directly derived from expressions for partition function or free energy. Contact with specific heat measurements will also occasionally be made.

The ESR and NMR data, under favorable conditions, can give quite direct information about microscopic interactions among electrons and nuclei. Unfortunately, in disordered systems the observed resonance signal frequently reflects the sum of resonance signals from groups of spins in quite different local environments. While the observed signal contains, in principle, interesting information about the details of the distribution of local environments seen by the resonating spins, it is frequently difficult to deconvolute the absorption line in an unambiguous way.

Since the largest part of these lectures will deal with the Si:P system, it is appropriate to give a brief, summary overview of the state of knowledge of magnetic properties of this system, over the donor concentration range from 1E17 to 1E20*. This range covers a factor of approximately 30 on either side of n_c at 3.7E18. Figure 1 displays six regions of concentration, labeled I to VI, and gives a very brief description of the characteristic features of each region, as revealed in either static or magnetic resonance experiments.

* *We shall use a notation for concentration in which 1E17 ≡ 1.0 × 10^{17} cm^{-3}.*

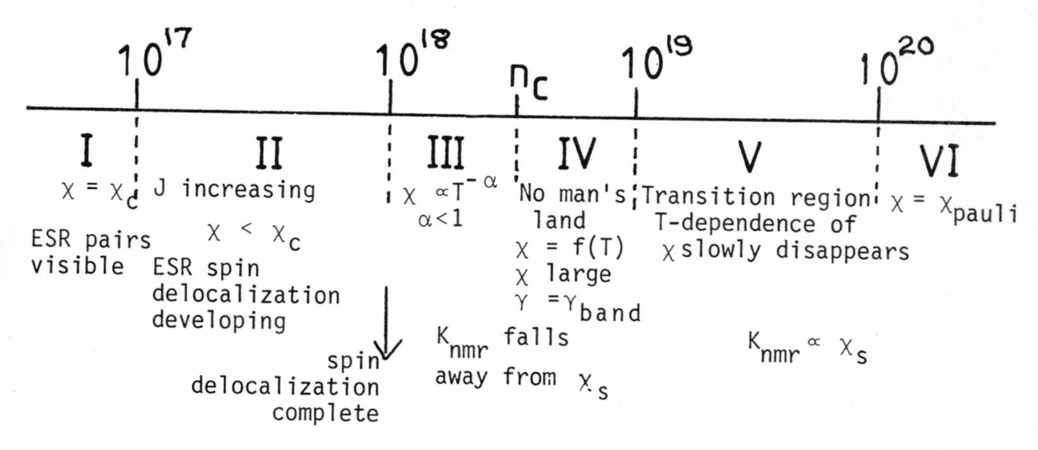

Fig. 1. Characteristics of Magnetic Properties of Si:P

For the other system considered in these notes, the sodium tungsten bronzes, information on the magnetic properties is not so nearly complete. That information is considered in detail in Section 3.4.

2.
STATIC MAGNETIC AND ESR PROPERTIES OF DONOR ELECTRONS

2.1 Simple Models

We shall be concerned primarily with the spin component of susceptibility and magnetization, with orbital or diamagnetic contributions being viewed as unwanted interlopers. For simplicity, we shall also restrict ourselves to single electron atoms, with $S = 1/2$. (Fortunately, most of the experimental data which we examine comes from single electron, $S = 1/2$ systems.) The susceptibility of a set of noninteracting electrons is given by the Curie law relationship,

$$\chi_c = NS(S + 1)\, g^2\beta^2/3kT \qquad (2.1)$$

for high temperatures. At low temperatures, we must substitute the expression, B_J/H for the quantity $(S + 1)g\beta/3kT$, where B_J is the Brillouin function. Substitution of the quantity $(T + \Theta)$ for T gives the Curie-Weiss form, and allows for the presence of spin-spin interactions. The value of Θ is negative for ferromagnetic alignment of interacting spins and positive for antiferromagnetic alignment.

In the limit of completely delocalized, independent electrons, we will have the Pauli form for the spin susceptibility,

$$\chi_S = 3N\beta^2/2kT_F\ [1 - \pi^2/12(kT/E_F)^2], \qquad (2.2)$$

where I have included the first correction term for non-zero temperature.

In the same model, the specific heat $c = \gamma T$, and we have the simple relationship, $\chi/\gamma = 3\beta^2/\pi^2 k$.

2.2 Experimental Data on χ_S and M

Much of the experimental data which we examine comes from measurements

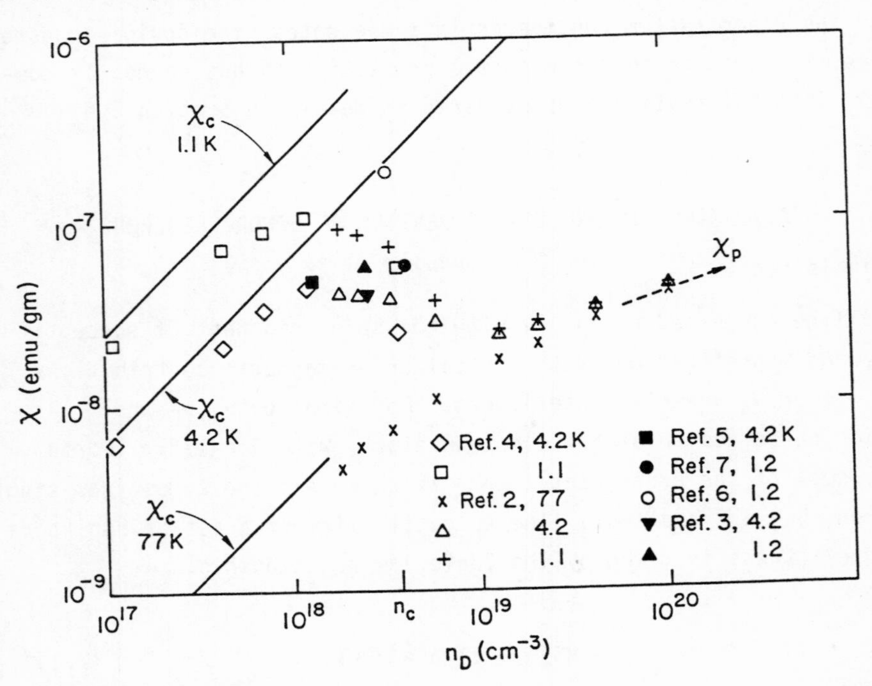

Fig. 2: Spin susceptibility data for Si:P from sources as indicated. The straight lines are drawn for χ_c at the temperatures noted. The line marked χ_p is calculated from Eq. (2.2) with T_F calculated from the expression, $T_F = (3\pi^2 n)^{2/3}\hbar^2/2km_{de}$. The value of m_{de}, the "density of states effective mass" is $(m_l^* m_t^{*2})^{1/3} = 0.33\ m_e$ for Si. The value of T_F is 830K at 1E20, 179K at 1E19, 113K at 5E18 and 39K at 1E18.

on Si:P, the "hydrogen atom" of M-I transitions in impurity systems. Figure 2 displays data for spin susceptibility, χ_S, as a function of donor concentration, n_D, taken from a variety of sources.

Data of Fig. 2 were obtained by several different experimental techniques. The most important differentiation is between data derived from measurements of total susceptibility (by Andres, et al.,[4] using a SQUID magnetometer and by Sarachik, et al.[5] using a Faraday balance) and from integration of the area under an ESR signal (the remainder of the measurements.) Total susceptibility can be rather reliably corrected for host diamagnetism, but is subject to uncertainty with respect to any diamagnetic contribution from mobile donor electrons. One would expect this diamagnetic contribution to become significant only quite near the M-I transition. The ESR measurements all rely on the fact that one can extract values of χ_S by tracing out the RF susceptibility $\chi''(\omega)$ and writing

$$\chi_S = \frac{2}{\pi\omega_0} \int_0^\infty \chi''(\omega)d\omega \qquad (2.3)$$

(See pp. 49-50 of Ref. 9 for a discussion of this relationship.) It is virtually impossible to calculate or measure apparatus parameters with sufficient accuracy to determine the absolute value of $\chi''(\omega)$ reliably. Thus, one must obtain the absolute value of χ_S by a calibration scheme, using a sample with a known value of χ_S as a reference sample. The various ESR measurements given in Fig. 2 fall into two classes. For the most extensive set of data, that of Quirt and Marko,[2] as well as for the measurements of Jerome, et al.[3], the calibration reference is a sample of $CuSO_4 \cdot 5H_2O$, for which a Curie law calculation is believed to be correct. Two data points, those of Paalanen, et al.[7] and Ikehata and Kobayashi[6], rely on the Schumacher-Slichter[8] technique, in which the calibration is made against the nuclear resonance signal from ^{29}Si nuclei measured in the same sample, with the same apparatus, operating at the same frequency. (See pp. 117-118 of Ref. 9 for a discussion of this experimental scheme.) Unfortunately, those two data points differ by a factor of 3, in spite of the fact that the sample composition and measurement temperature are nearly identical. The accuracy of the elegant Schumacher-Slichter technique is subject to two conditions. (1) For neither ESR nor NMR absorption signals can there be any saturation effect. (2) One must be certain that the wings of the line are accurately taken into account. Establishment of the absolute baseline is important. One or the other of the two measurements may have fallen victim to one of these two sources of error.

There are two other sets of ESR-derived data in the literature which I have not plotted, in order to avoid an excessively cluttered plot. These are an early set of measurements by Ue and Maekawa(10) and a more recent set of measurements by Wagner and Schwerdtfeger.(11) Both of these sets of data fall in the same range of concentration and temperature as those of Quirt and Marko. I have arbitrarily decided to downplay them, partly because of the rather high degree of consistency between the data of Quirt and Marko and that of Andres, et al., and partly because the Quirt and Marko value at $n_D = 10^{20}$, falling very close to the calculated value of Pauli susceptibility, seems more consistent with the implications of specific heat(12) and NMR [Sundfors and Holcomb,(13) Sasaki, et al.,(14)] data.

For those who have in mind the dramatic drop in dc conductivity, σ, at the M-I transition in Si:P,(15) the most striking feature of the data of Fig. 2 may be the fact that there is neither a sharp jump nor even a sharp change in $d\chi/dn$ at the critical concentration n_C. That feature is also characteristic of the specific heat data(12) shown in Fig. 3. It emphasizes the fact that the localization phenomenon in systems such as this is associated with the specific nature of the electron wave functions. The thermodynamic density of electron states at the Fermi energy suffers no dramatic change whatsoever, and obviously continues to be non-zero for a substantial range of concentration below n_C.

Two substantial studies of magnetization M(H,T) in Si:P are found in the literature, one by Ikehata, Ema, Kobayashi, and Sasaki(16) and a more recent one by Sarachik, Roy, Turner, Levy, Isaacs, and He.[5] The measurements of Ikehata, et al. cover a concentration range from 5.3E17 to 6.5E18, a temperature range from 1.5K to room temperature, and values of field to 95 kG. Because the magnetization is dominated by the diamagnetic contributions, it seems difficult to extract very much information beyond that given by the susceptibility measurements shown in Fig. 2. Samples of Si:P measured by Sarachik, et al. for temperatures at and above 1.25K, and for fields to 50 kG all have donor concentrations less than n_C. Their results bear primarily on the model of Bhatt and Lee[17,18] to be discussed in the next section.

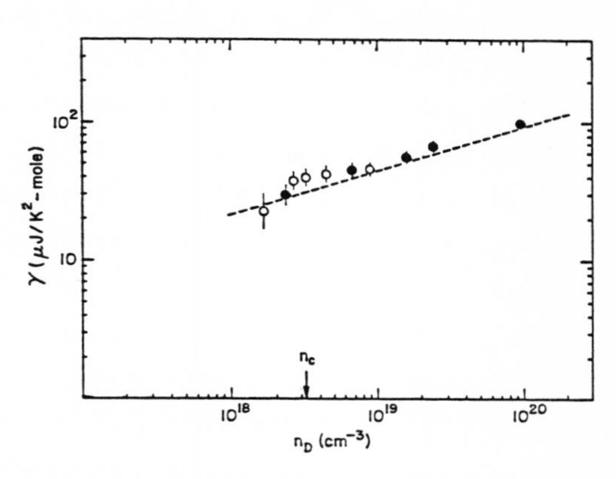

Fig. 3: Measurements of the electronic specific heat coefficient γ *reported by Kobayashi, et al.*[12] *Values of donor concentration are derived from Hall constant measurements at room temperature. The broken line corresponds to values calculated on the basis of the band structure of pure Si.*

Measurements in Si:B by Sarachik, He, Li, Levy and Brooks[19] have also been reported. The results are not qualitatively different from those in Si:P, although it is clear that there are effects associated with the fact that the hole wave functions are derived from the valence band rather than the conduction band and with the fact that the acceptor wave functions have J = 3/2, rather than the J = S = 1/2 state for the donor electrons.

Referring to the regions of concentration defined in Fig. 1, we note that Regions I and VI can be successfully described by simple models. Included in Fig. 2 are three solid lines which show values of χ given by Eq. (2.1) for three temperatures, 1.1K, 4.2K and 77K. These lines give us the information that the Andres, et al. measurements at n_D = 1E17, corrected for the silicon host diamagnetism, are very close to the appropriate Curie curves. Thus, the spin susceptibility is that of n_D non-interacting electrons with S = 1/2, g = 2. In Region VI, the measured χ is very close to the Pauli value, as indicated in Fig. 2.

ESR measurements permit us to sample the hyperfine coupling between bound electron and the central ^{31}P nucleus (nuclear spin I = 1/2). Observation of changes in this coupling gives valuable information about the magnetic interaction among the donor atoms which appears as the concentration increases. Slichter[20] gives a good description of the phenomenon. Good sketches of experimental results in Si:P are given by Maekawa and Kinoshita.[21] A modern re-examination of Si:P and a thorough investigation of the Si:As system is given by New and Castner.[22] In Fig. 4 are sketched the ESR absorption patterns which arise from (a) an isolated donor, from (b) two donors and (c) three donors with spin exchange, and finally (d) from a large cluster of donors with rapid spin exchange. The comment in Fig. 1 referring to Region I is a statement that for n_D = 1E17, one sees a small admixture of the pattern (b) in Fig. 3, resulting from two-spin clusters. New and Castner show that one sees this two-spin pattern when the separation between two donors is less than 110 Å.

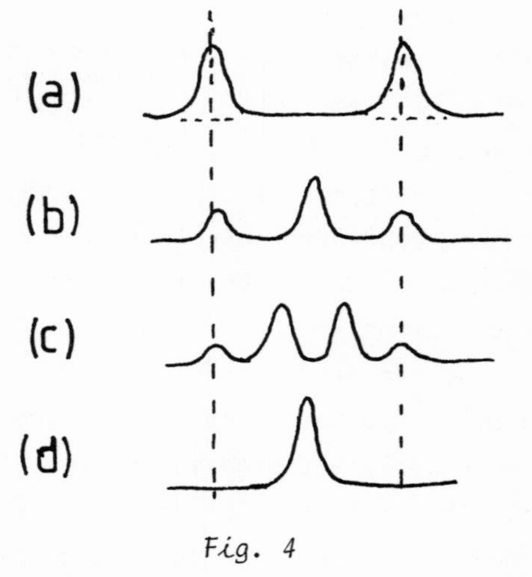

Fig. 4

2.3 Regions II and III--Models based on antiferromagnetic exchange

It has long been recognized [See, e.g. Sonder and Stevens[23]] that the drop of χ_s away from the Curie lines for n_D near 1E18 probably arises from antiferromagnetic coupling among the donor spins. Various cluster models (See Andres, et al.,[4] for references) have produced plausible explanations. However, it was clear to a number of those workers that the calculational difficulty of handling multi-spin interactions in the large clusters which develop near the M-I transition required a more elaborate treatment. The calculation of Bhatt and Lee is, to my knowledge, the first attempt to handle these multi-spin interactions in a way that has the ring of truth about it.

Bhatt and Lee calculate $\chi(T)$, the temperature-dependent susceptibility of a system of antiferromagnetically-coupled spins, located at random in a three-dimensional continuum. Their calculation is best described in Ref. 17. Interaction among the spins is given by the Hamiltonian

$$H = \frac{1}{2} \sum_{i \neq j} J_{ij} \; \mathbf{S}_i \cdot \mathbf{S}_j . \tag{2.4}$$

J_{ij} is the exchange coupling between spins $\mathbf{S}_i$ and $\mathbf{S}_j$. All values of J_{ij} are taken as positive (antiferromagnetic coupling). [It is also this exchange coupling which leads to the ESR cluster patterns displayed in Fig. 4. The criterion for observation of pattern (b) in Fig. 4 is that the value of J_{ij} for a particular pair be approximately equal in magnitude to the hyperfine splitting of pattern (a).] It is assumed the J_{ij} follows the form

$$J_{ij} = J_o e^{-2r_{ij}/a} \tag{2.5}$$

where J_o is the exchange coupling for nearest neighbors, r_{ij} is the distance between the two spins under consideration and a is the decay length.

The Bhatt and Lee computer calculation, with magnetic field $H \simeq 0$, proceeds along the following lines:

(1) Spins are sprinkled into space at random positions.

(2) Values of J_{ij} for all pairs are calculated according to Eq. (2.5).

(3) A decimation procedure, described in Refs. 17 and 18, is performed. At each step of this procedure, the spin pair with the largest value of J_{ij}, either raw or renormalized, is discarded. This step results from the fact that when an actual calculation of $\chi_s(T)$ is carried out, it will be at a temperature T such that $kT \ll J_{ij}$. Any pair with such a large exchange coupling will remain in the singlet state, since there isn't enough thermal energy to excite the triplet. Thus, it will not contribute to the susceptibility. Before the pair in question is discarded, its effect on energy levels of nearby spins is preserved through the renormalization procedure.

(4) The next step in the calculation can be described by reference to Fig. 5. We have sketched a ladder of values of J_{ij}. Suppose that by the iterative process referred to in step (3) and described in some detail in Refs. 17 and 18, all values of J_{ij} larger than some value J_M have been discarded. The value of J_M is established by the temperature at which we wish to calculate the value of χ_s. $J_M = kT$. We now approach the key approximation of the Bhatt and Lee calculation which makes it tractable. Because of the random location of spins, there will be a very wide distribution in values of J_{ij}. We are calculating χ_s at temperature $T = J_M/k$. It is assumed that for all spin pairs with $J_{ij}<J_M$, any dimunition in polarization caused by the exchange coupling is negligible and the susceptibility (for S = 1/2) will be given by

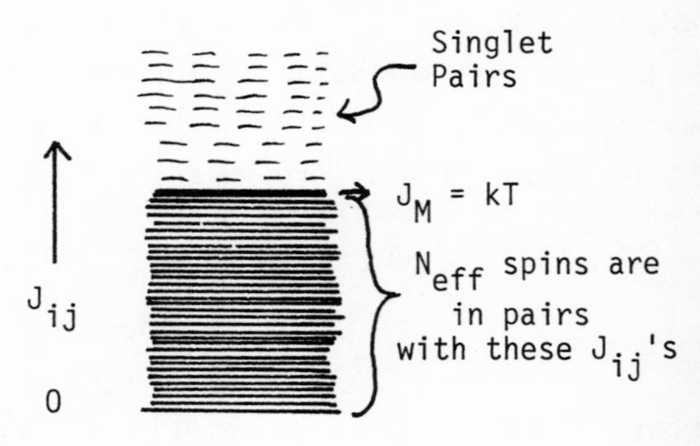

Fig. 5

$$\chi_s = N_{eff}\, g^2\beta^2/4kT \qquad (2.6)$$

where N_{eff} is the number of spins in the pairs with $J_{ij} < J_M$. The consequence of this procedure is that, as T drops, χ_s falls farther and farther below the Curie value, because more and more pairs have been discarded as $J_M = kT$ is reduced. Figure 6 reproduces the

Bhatt and Lee result. It gives a rather good description of the temperature dependence of the Andres, et al. curves shown in Fig. 7, at least for T > 0.1K. (It will be noted that the measurements of Andres, et al. for very low temperatures flatten out more than the Bhatt and Lee calculation would suggest. One must be cautious about attaching great significance to this flattening. If the sample drifts out of equilibrium with the thermometer, which is an ever-present problem at very low temperature, the effect will often be just that seen -- an apparent leveling off of the experimental property of interest.)

The Bhatt and Lee calculation is based on ideas suggested earlier by a number of authors, identified in the references given in their papers. I found one of those references, that by Clark and Tippie[24], to be particularly useful as a description of the procedure of earlier authors and how it relates to the computer model of Bhatt and Lee. In the earlier work, use of cluster models suggested that the distribution of values of J for nearest neighbor pairs would follow the distribution function,

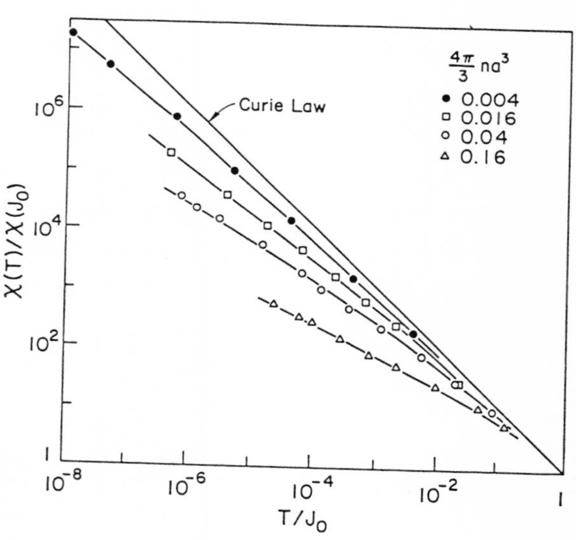

Fig. 6: Relative values of $\chi(T)$ calculated by Bhatt and Lee.[17] If we take the constant a to have the value of the Bohr radius for Si:P, 17Å, the curve at highest concentration (open triangles) corresponds to n_D = 7.8E18.

$$P(J) \propto J^{-\alpha} , \tag{2.7}$$

where α is a number less than one. With this distribution function, Clark and Tippie show that the temperature dependence of χ is given by $\chi \propto T^{-\alpha}$. The Bhatt and Lee curves of Fig. 6 show a similar temperature dependence, with α a function of density of spins.

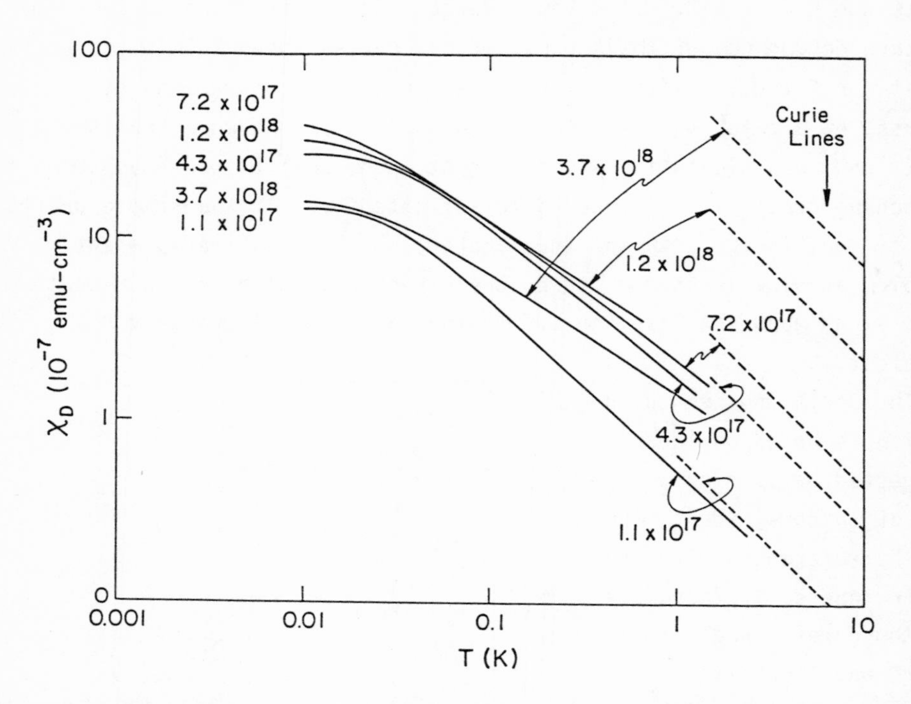

Fig. 7: A compilation of data for total donor susceptibility in Si:P, taken from Figs. 4 and 5 of Ref. 4. Values of n_D are noted for the five curves. Raw experimental data was corrected for the Si host diamagnetism. The solid curves represent the data points of Ref. 4 taken at low field, either 10G or 50G. The five short, dashed lines at 45° represent Curie susceptibility lines, calculated on the basis of a contribution from n_D free spins in the five samples of interest.

Taking this same form for the distribution function for <u>renormalized</u> J's [presumably with a value of α different from that which would be appropriate to the simpler picture for which P(J) is the distribution function for the raw, pairwise values of J], Sarachik, et al.[5] show that their magnetization results for Si:P at fields up to 50 kG and T>1.3K, for samples with n_D = 0.67E18, 1.3E18, 2.4E18, and 2.8E18 can be fully represented by this distribution function, with values of α ranging from 0.78 to 0.64 for the samples noted. The expression for M(H,T) which they use is

$$M(H,T) = (kT/g\beta)\chi(T)f_{\alpha}\ (g\beta H/kT), \qquad (2.8)$$

where χ(T) is the measured susceptibility and f_α is a function given in their paper which can be derived directly from the thermodynamics of the coupled spin pair model, as outlined in Ref. 24.

In a certain sense, the primary value of the Bhatt and Lee calculation may have been to give firm justification to the model used by Clark and Tippie and others, based on the form of P(J) given above. In any event, there seems a reasonably firm basis for the assumption that the Bhatt and Lee model, based on the assumption that effects of the pair-wise antiferromagnetic exchange coupling govern χ for n_D just below n_c, is close to the truth.

Kamimura[25] has suggested an alternate picture in which the deviations from Curie law are governed by the appearance of effects of the intrastate Coulomb repulsion, U. U is a measure of the energy required to place an electron of opposite spin on an already occupied site. If one neglects effects of the intersite exchange coupling, J, upon which the Bhatt and Lee model is based, this model gives a form for χ which approaches the Curie law as T approaches zero, rather than falling away from the Curie law as does the Bhatt and Lee result. Kamimura seems to suggest that this result fits the Andres, et al. results as plotted in the form shown in Fig. 8. It seems to us that a look at the original data plot shown in Fig. 7 shows that such a conclusion is not justified. The trouble seems to lie in the fact that the 1/χ plot vs. T of Fig. 8 is particularly insensitive to what's happening at low T, because of the crowding of all experimental points into a small region.

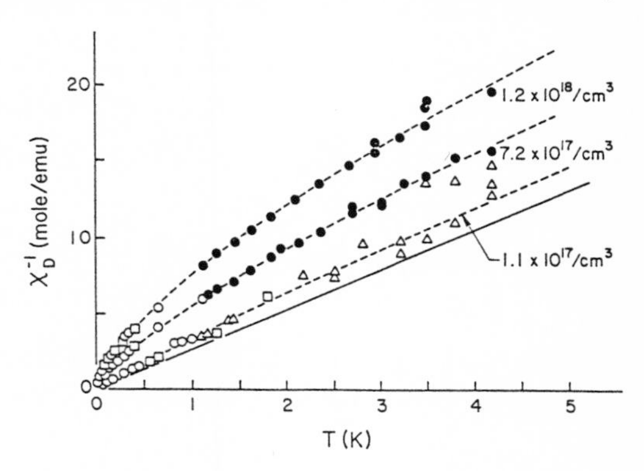

Fig. 8: Data from Ref. 4 plotted in the form, 1/χ vs. T. The data near T = OK, indicated by open circles, was measured in a field of 50G.

Region II of Fig. 1 should be the primary domain of the Bhatt and Lee model. Since the Bhatt and Lee model neglects any effects of the Coulomb repulsion energy, U, it is uncertain how far into Region III one would expect it to be valid. The treatment by Sarachik, et al. for their samples with n_D up to 2.8E18 suggests that the model is rather good throughout most of this region.

2.4 Charge and Spin Delocalization

The single ESR absorption line sketched in Fig. 4(d) is seen in Si:P for all values of $n_D > 1E18$. By this criterion, we would say that the donor spins are completely delocalized -- the exchange coupling is sufficiently strong that all spins are in magnetic communication with one another. Does this observation contradict the fact that the conductivity transition occurs at $n_c = 3.7E18$? The answer is "No." This spin delocalization can and does occur in the ground state of the system -- no energy excitation is necessary other than the gentle probing of the RF field of the ESR experiment. For DC charge transport, we must have the ability to put a second electron on a donor site. Screening must be sufficiently large that the Coulomb energy, U, is reduced to a value less than kT. This necessary reduction in U is reached only at the higher concentration, n_c.

New and Castner[22] have carefully investigated the ESR cluster patterns for both Si:P and Si:As. They define a characteristic volume for cluster formation, V_c, by the equation $\bar{n} \equiv n_D V_c$, where $\bar{n}$ is the mean cluster size in a Poisson distribution. They find the value of V_c to be about 8.2E6 $Å^3$ for Si:P and 5.6E6 $Å^3$ for Si:As. These values correspond to a critical distance for spin exchange of 110Å for Si:P and 126Å for Si:As. The value for Si:P is almost exactly a factor of 2 larger than the critical distance for charge delocalization which can be estimated on the basis of a percolation calculation. Using a model of randomly placed spheres of radius r_c, Pike & Seeger[26] show that the percolation threshold concentration, n, satisfies the relationship,

$$(4\pi/3)\ r_c^{\ 3}\ n = 0.305. \tag{2.9}$$

If we identify the value of n with the M-I transition concentration, n_c, 3.7E18, we find the value of critical separation, $2r_c$, to be 54 Å.

2.5 Susceptibility in the range, $3E18<n_D<1E19$

Examination of the data in Fig. 2 shows that in Region IV, values of χ_s lie substantially above the Pauli susceptibility line. Values of the Fermi temperature T_F for various electron concentrations are given in the caption for Fig. 2. It is clear that we are far below T_F for samples in this range, yet χ_s shows the strong temperature dependence shown in Fig. 2. Quirt and Marko were able to give a reasonable representation of their data by a model which assumed χ to be a sum of two terms, a Curie-Weiss term from localized spins and a pure Pauli term from delocalized spins. While it is indeed possible to produce a reasonable fit of the data by this scheme, there is no experimental evidence to support the idea that spins are divided into two well-defined camps. ESR data shows a single, narrow ESR line (See Ref. 21, for example) and the NMR data also implies that all electron spins are in intimate communication with one another.

A number of authors have suggested that this enhancement and strong temperature dependence of χ may arise from a situation in which there are sites of low impurity density at which a nearly local moment exists. While all electrons are indeed part of a communicating system, the local magnetization envelope displays a polarization characteristic which mimics a truly localized moment. The picture is reminiscent of spin glass systems such as Au:Fe at relatively low Au concentrations. However, the data of Fig. 7 show that the cusp in $\chi(T)$ characteristic of the spin glass systems [Cf., e.g., Cannella and Mydosh[27]] is absent.

With an enhanced susceptibility, one is naturally led to examine experimental values of specific heat to see if related effects are seen there. The fact that there is only a very modest enhancement of γ (See Fig. 3) must be accommodated into any complete picture of the enhancement of χ.

Further evidence for the existence of unusual magnetic properties in Region IV (cf. Fig. 1) is to be found in an observed dependence of the NMR spin-lattice relaxation time T_1 on magnetic field strength. In the basic Korringa model (developed in the next section) which describes the effect of the mobile electron system on the NMR properties, the value of T_1 is independent of magnetic field strength. Ikehata, Sasaki, and Kobayashi[28] observed that while T_1 is indeed independent of H in the upper part of Region V, (cf. Fig. 1) a field dependence appears as n_D decreases through 1E19 and becomes stronger in Region IV. They point out that such an effect could be expected if the electron spin system has magnetic couplings which generate spin waves or other coherent magnetic phenomena. Tunstall and Deshmukh[29] observed a magnetic field dependence for T_1 in Ge:As of a similar magnitude in the region $n_D > n_C$. Palaanen, et al.[30] in their study of Si:P with n_D near n_C at very low temperatures followed this field dependence of T_1 more completely, to very low field. They suggested that, again by analogy with other local moment systems, there will be an inhibition of electron spin diffusion associated with the fact that the conduction electrons must carry the large magnetization of the local moment system from site to site. The longer correlation time for the electron-nucleus interaction leads to enhanced nuclear relaxation. Although no quantitative calculation of the field dependence of T_1 seen in these three studies is yet available, there is certainly strong evidence that a correlation time substantially longer than that which appears in the Korringa theory has indeed made its appearance.

In this concentration range, we seem to be particularly bedeviled by our inability to theoretically handle the wide distribution of local properties which occurs in the rather dilute impurity system. Any model which relies on analogy to either a periodic system or to one in which all impurities have essentially identical local nature, such as Au:Fe or Cu:Mn, is unlikely to produce the kind of refinement which appears to be required.

Just recently, a new approach to the problem has appeared, growing out of the work of Finkel'shtein[31]. Castellani, et al[32] have added effects of disorder to the Fermi liquid theory approach used by Finkel'shtein, and show that within this model, a temperature-dependent, enhanced spin susceptibility and a slowing of spin diffusion appear. While this model shows promise, its rather high level of abstraction makes it difficult to know whether it is describing any particular real material.

In a contribution to the Mott Festschrift volume last year in which he discussed the thermodynamic properties of Si:P, Rice [33] commented, "This all goes to show that the problem introduced by Sir Nevill Mott 35 years ago still resists a definitive solution but still exerts a definite attraction." This seems as good a place to leave the discussion of Region IV as any. Many features of the data point to some kind of local moment system. But it also seems clear that none of the models developed for other magnetic systems can yet give a quantitative picture of the Si:P system.

3. NMR EXPERIMENTS -- SILICON, GERMANIUM, AND TUNGSTEN BRONZES

3.1 Hyperfine Interaction and the Korringa Theory

The reader who wishes a more complete treatment of the NMR background to the discussion in this section is referred to the book by C.P. Slichter.[9] Pp. 106-121 and 144-150 of that reference are particularly germane.

We shall be primarily interested in situations in which important NMR properties are determined by interaction of the nuclear moments with the conduction electrons or holes. The two properties to which attention is most commonly directed are the Knight shift K and the spin-lattice relaxation time T_1.

The primary source of static local field experienced by the nuclear moments in a metallic system is the paramagnetism of the mobile electrons. The relative shift in resonance frequency, $\Delta\omega/\omega_0$, brought about by these local fields is called the Knight shift K. We call immediate attention to the fact that, in a disordered material, the Knight shift will typically vary from site to site.

Time-dependent magnetic fields generated by motion of the charge carriers will stimulate transitions of the nuclei between the Zeeman levels which are separated by an energy $\Delta E = \gamma h H_0$, and, thus, will frequently determine the value of T_1.

The value of K is a measure of the time-average local magnetic field at the nuclear site, and thus is related to the magnitude of the electron spin susceptibility, χ_s. In the simplest situation, the nucleus-electron coupling is generated by the hyperfine interaction, A **I**•**S**. **I** is the nuclear spin vector and **S** the electron spin vector. In the basic, noninteracting Korringa model, described at length in Ref. 9, the resonance field shift can be written as

$$K = \frac{\Delta\omega}{\omega_o} = \frac{8\pi}{3}\left\langle |u(o)|^2 \right\rangle_{E_F} \chi_s, \tag{3.1}$$

where $\left\langle |u(o)|^2 \right\rangle_{E_F}$ is the square of the electron wave function at the nuclear site averaged over electron states on the Fermi surface.

In the systems we consider, there is a broad distribution of local environments for the resonant nuclei and we need to modify the homogeneous form of Eq. (3.1). A useful form is given as

$$K_i = 8\pi/3 < |u(r_i)|^2 >_{E_F} N_{loc} \frac{\gamma_e \hbar}{H_o} <S_z>. \tag{3.2}$$

N_{loc} is the local electron density averaged over a suitably chosen sampling volume. $<S_z>$ is the value of the component of S parallel to the external field, again averaged over a suitably chosen sampling volume. This equation gives an expression for the value of K at a particular nuclear position r_i. We have broken up the expression for χ_s by writing it in the form

$$\chi_s = N_{loc} \frac{\gamma_e \hbar}{H_o} <S_z>. \tag{3.3}$$

As noted earlier, the ESR experiments show strong exchange interactions between impurity electron spins for a certain range of values of $n < n_c$. Thus, we believe it reasonable to assume that $<S_z>$ is not a local property, but has the same value at all points because of the rapid spin exchange. Our model requires that the characteristic diameter of the sampling volume used to determine N_{loc} be large compared to the interimpurity spacing, so that many impurity sites contribute an electron to the electron distribution sampled by a given nucleus. The wavefunction factor, $<|u(r_i)|^2>$, which we represent subsequently by the simplified symbol P_i, is expected to vary from site to site as determined by the degree of proximity of that site to one or more donor ions.

According to the model represented by Eq. (3.3), typical NMR absorption line shapes, such as those sketched in Fig. 9, are plots of the probability of various values of the product P_iN_{loc} for a sample with a given value of $<S_z>$.

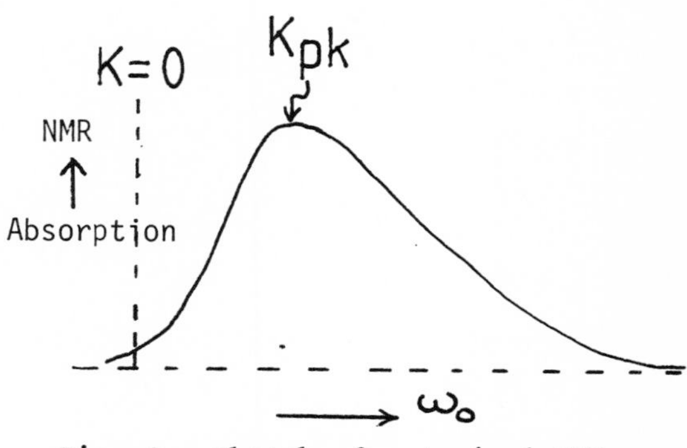

Fig. 9: Sketch of a typical NMR absorption line shape seen for nuclei in a disordered metal.

3.2 Knight Shift Measurements in Si:P, Si:As, and Ge:As

Figure 10 shows a collection of measurements of K_{pk} (See Fig. 9 for definition of K_{pk}) in Si:P, all taken at temperatures of 4.2 or 4.3K. For comparison the Quirt and Marko data for χ_s are also plotted, using the scale at the right and normalized so that values of K and χ coincide at n_D = 1E20. The most complete data is that given by Kobayashi, Fukugawa, Ikehata and Sasaki,(34) measured at a field of 9.1 kG. The data by Hirsch and Holcomb,(35) measured at 58.5 kG, is in good agreement with the earlier measurements at low and high concentrations, but differ somewhat in the region of n_D near 1E19. Possible reasons for the differences are discussed in Ref. 35.

One sees two prominent features of the Knight shift data of Fig. 10. (1) For values of n_D near 1E20, values of K follow the spin susceptibility, as predicted by Eq. (3.1), but fall away as n_D is decreased through a value of 1E19. (2) As with χ_s and the specific heat, there is no sharp change in values of K at the transition concentration n_c. One can perhaps summarize the message from these two features in the following fashion. As the donor concentration decreases through the M-I transition, the spatial distribution of electron probability density at the ^{29}Si sites changes. A much larger fraction of the ^{29}Si nuclei find themselves in regions far from a donor, with suitably attenuated values of P_i. Whether this effect is largely a statistical one, as argued in Ref. 35 or is intrinsically associated with the electron localization signaled by the sharp transition in dc conductivity at the transition, as argued in Ref. 34, remains unclear. More limited NMR data for Si:As[35] indicate that the knee of the K vs. n_D curve shifts to higher concentation. This effect is naturally explained on the basis of the tighter donor wave functions of the As impurity -- various effects are shifted to higher concentration than for Si:P.

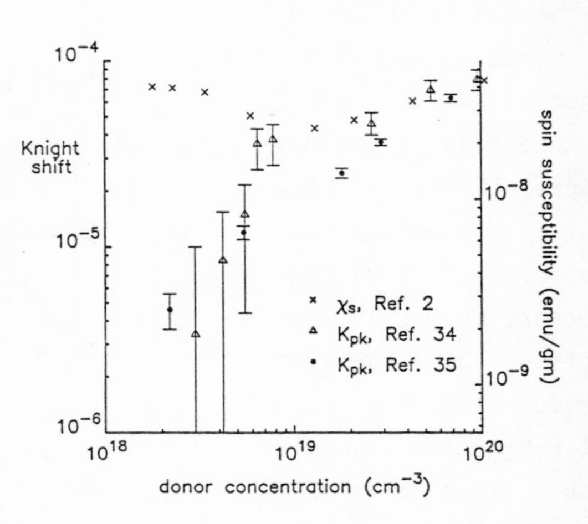

Fig. 10: A compilation of experimental values of K_{pk}, the Knight shift at peak amplitude, for the ^{29}Si NMR in Si:P, taken from Refs. 34 (open triangles) and 35 (solid circles).

Data for Ge:As by Deshmukh and Tunstall[36] suggest that the falloff of the magnitude of K in comparison with χ_s occurs more precipitously in this material, and occurs more nearly centered at n_c for Ge:As than for Si:P or Si:As. But there is no data for $n_D < n_c$ in the Ge:As system, and a sharp description of differences between Ge:As and Si:P or Si:As is not possible.

3.3 Relaxation Time Measurements in Si:P, Si:As, and Ge:As

If the dominant contribution to the nuclear spin-lattice relaxation time T_1 comes from hyperfine interactions of the nuclei with the conduction electrons, then the Korringa relation[37] is valid. This relation states that

$$K^2T_1T = \hbar/4\pi k(\gamma_e/\gamma_n)^2, \qquad (3.4)$$

where γ_e and γ_n are the electronic and nuclear gyromagnetic ratios, respectively. The right side of Eq. (3.4) is called the Korringa constant and has an invariant value for a given nuclear species.

Kobayashi, et al.(34) showed that, at least over certain parts of the range of values of n_D, the ^{29}Si nuclear magnetization in Si:P does not relax in time along a single exponential. Rather, there is a local value of T_1 which matches a local value of K. Kobayashi, et al., note that this lack of thermal equilibrium within the nuclear spin system probably arises because spin diffusion is inhibited by the differing values of local magnetic field from site to site. Hirsch and Holcomb(35) found a similar relaxation behavior for the ^{29}Si spin system in Si:As. It is notable that measureable local Knight shifts and local relaxation behavior resembling those expected from the Korringa model occur in samples well below the carrier localization transition at n_C. Korringa-like effects extend down(3,35) to values $n_D/n_C = 0.44$.

In their study of the relaxation behavior of the ^{73}Ge spin system in Ge:As, Tunstall and Deshmukh[29] found a concentration dependence for T_1 roughly matching the simple Korringa picture, with $T_1 \propto n_D^{-2/3}$. This concentration dependence is the same on either side of the M-I transition at $n_D \approx$ 3E17, with no noticeable jog at n_C. Evidently, in samples of both Si:P and Ge:As with $n_D < n_C$ a local density of states is sufficiently well developed to sustain a Korringa-like relaxation. As we shall see in the next section, that effect can be studied in more detail in the sodium tungsten bronze system.

3.4 K and T_1 Measurements in Na_xWO_3 and $Na_xTa_yW_{1-y}O_3$

Study of the M-I transition in Na_xWO_3(1) was inhibited by the difficulty in maintaining any given crystal structure for compositions on both sides of the transition. In the 19th SUSSP, held at St. Andrews in 1978, J.P. Doumerc described experiments which showed that by replacing a certain fraction of W ions with Ta, it is possible to maintain the pseudo-cubic crystal structure on both sides of the transition. Subsequent publications by Doumerc and other members of the Bordeaux group [38,39,40] described results of subsequent experiments. Encouraged by this success of the Bordeaux group, we set out in our laboratory to carry the earlier NMR studies on Na_xWO_3 into this modified material. While there are no published data on spin susceptibility in either the uncompensated or compensated version of sodium tungsten bronze, the NMR properties by themselves highlight certain features of the magnetic system which are likely to be applicable to other impurity systems such as heavily doped silicon and germanium.

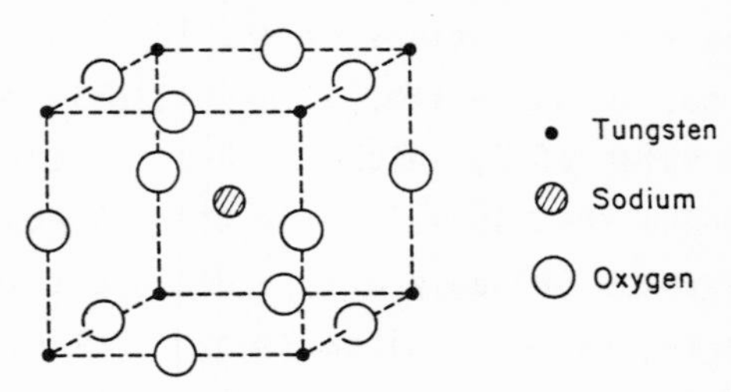

Fig. 11: A lattice cell of NaWO3. In the material, $Na_xTa_yW_{1-y}O_3$, Na occupies the Na site in the figure with probability x, and Ta occupies a W site with probability y.

Figure 11 shows the perovskite structure of this material, with the center of the WO_3 cube occupied by Na ions with probability x, and W ions replaced by Ta ions with probability y. Through the work of the Bordeaux group and work in our laboratory[41,42] the transition in dc conductivity is fairly well mapped out for temperatures above 1.5K. Figure 12 collects the experimental data. We place the M-I transition at a composition x - y = 0.18. Our interpretation of the conductivity data is somewhat different from that of Dordor, et al.[40] who would extend the range of the non-metallic phase to higher values of x - y.

Using the ^{183}W nuclear spin system, measurements have been made of absorption line shape, Knight shift distribution, spin-lattice relaxation properties, and spin phase memory time T_2[41,43] Figure 13 displays the absorption line shapes for samples with various compositions. Nuclear dipolar broadening in the ^{183}W spin system is extraordinarily weak, and the broad absorption lines seen result entirely from the wide distribution in local values of Knight shift (or chemical shift) generated by the electron distribution. As for uncompensated Na_xWO_3,[44,45] the Knight shift in the metallic phase is negative, apparently arising from core polarization effects.

In broad terms, the interpretation of Fig. 13 is straightforward. As the value of x - y is lowered toward that for the M-I transition, the width of the distribution of local values of Knight shifts shrinks and seems to be settling in on the value of zero. However, it is well to note that even for the sample with x-y = 0.10, which we believe to be well on the insulating side of the transition, the linewidth is about 30 G, far larger than the nuclear dipolar linewidth. This linewidth suggests that there remains a rather wide distribution of local magnetic fields caused by variation in local electron spin density. Because there are several different mechanisms contributing to the Knight shift[43,40] it is not possible to do a quantitative analysis of this distribution of local fields in order to obtain information about the nature of the local clusters of uncompensated Na ions.

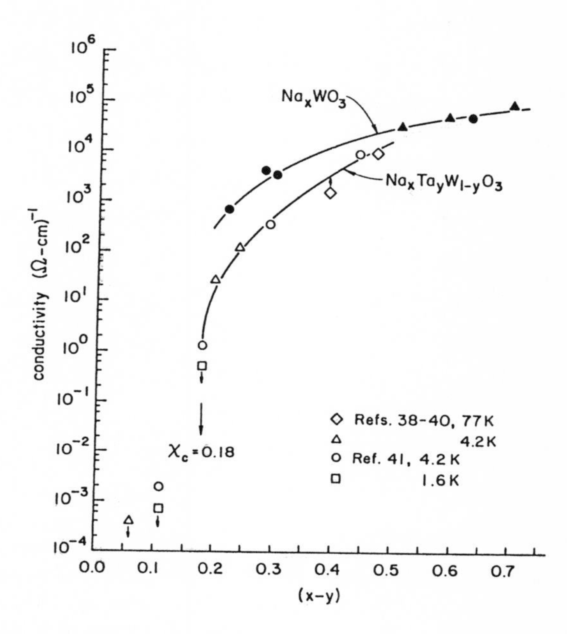

Fig. 12: DC conductivity, σ, as a function of the net electron concentration, x-y, for the sodium tantalum bronze. Data is taken from Refs. 38-41, as indicated.

The nuclear spin relaxation properties of the tantalated bronze are more revealing. Figure 14 shows relaxation curves for samples of four different compositions. The plots are made in such a fashion as to call attention to the change in the character of the relaxation curve. For the metallic sample at x - y = 0.44, the height of the spin echo measured at time t following a saturating RF pulse recovers along a simple exponential. That is, it has a well defined value of relaxation time T_1. For the other three samples, the relaxation

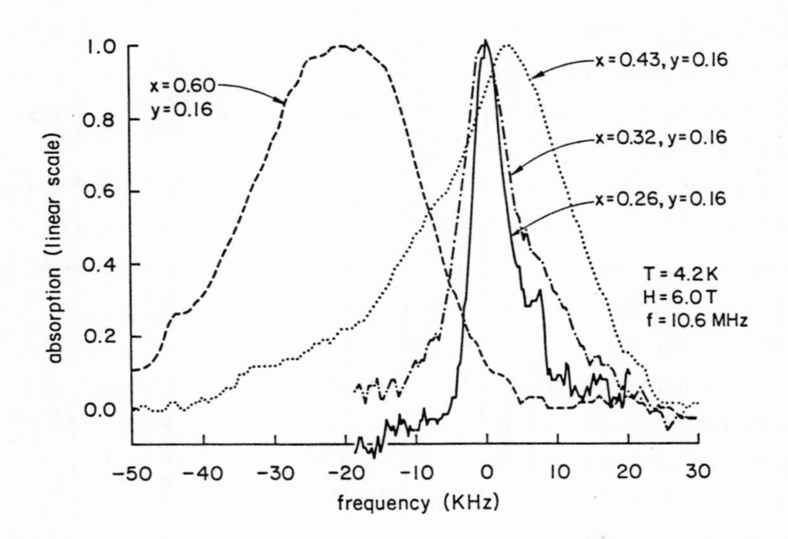

Fig. 13: NMR absorption linshapes in $Na_xTa_yW_{1-y}O_3$. Data were obtained by the spin echo integration technique first suggested by W.G. Clark, Rev. Sci. Inst. 35, 316 (1964). Further discussion of the technique is given by A. Avogadro, G. Bonera, and M. Villa, J. Mag. Reson. 35, 387 (1979).

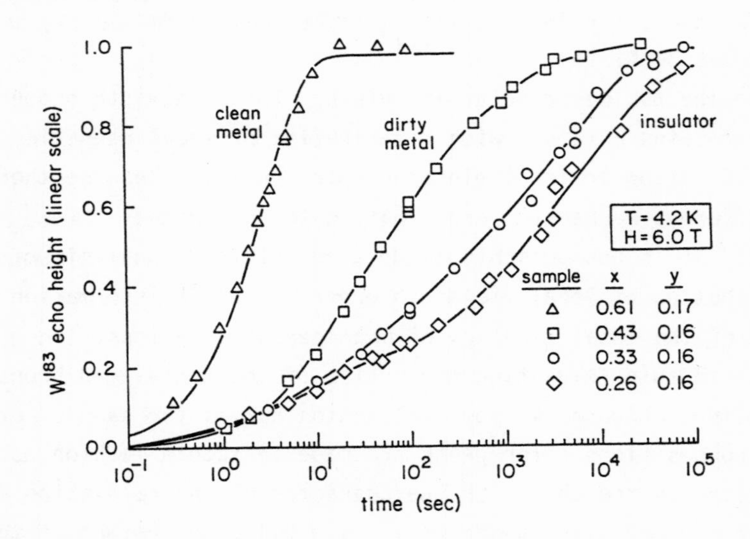

Fig. 14: ^{183}W magnetization recovery curves for $Na_xTa_yW_{1-y}O_3$. The solid lines are multiple exponential fits, described in Rev. 43. The curve for the "clean metal" is a single exponential, with T_1 = 3.0 sec.

spreads out over progressively longer times. The most striking result is that for the insulating sample at x - y = 0.10. With some portion of the nuclear spin system still relaxing after 24 hours, there are many nuclei which relax on a very much shorter time scale, of the order of 10 seconds.

The particular interest in the relaxation properties of this system lies in the fact that the very weak W-W dipolar coupling leads to a virtual absence of spin diffusion. Thus, the observed distribution of relaxation rates senses the distribution of local electronic environments. In some sense, the fast relaxers are in metal-like locales, while the slow relaxers see insulating environments. We need a model which connects the local relaxation rate to some feature of the local electronic structure.

As with the relaxation results in Si:P and Si:As noted in the previous section, it is tempting to try to develop some kind of local Korringa relation to relate local value of K to local value of T_1. Unfortunately, there exists no general theory of NMR in solids near a M-I transition. The Warren-Götze-Ketterle model[47,48] treats enhanced relaxation in metals with very strong electron scattering. But Ref. 40 shows that the results in $Na_xTa_yW_{1-y}WO_3$ do not fit this model very well.

At this point, I quote from Ref. 43:

"The suggestion that electron-nucleus spin-flip collisions remain the dominant nuclear relaxation mechanism on both sides of the M-I transition seems in conflict with the picture of disorder-induced localization of electronic states. Once the electronic eigenstates are localized, each nucleus sees a discrete spectrum of states. One might expect this to drastically inhibit nuclear relaxation since empty states at the Fermi level are needed in order for energy-conserving spin-flip collisions to occur. Using a density of states derived from Zumsteg's[49] specific heat measurements for Na_xWO_3, one can estimate an energy level spacing of about 20 meV for a region containing 100 atomic cells. The electron spin flip energy, $2\beta H$, is only 0.7 meV at H = 6.0 T. Thus, the conditions seem ripe for inhibition of the nuclear relaxation rate. ...However, NMR studies of small Pt particles[50,51] (in which the localization length is clearly the particle size) indicate no such inhibition of the relaxation rate. So far as we know, the reason for this behavior in the Pt particles is not certainly known. It may be that for both that system and the bronzes lifetime broadening of the electronic levels produces the necessary quasi-continuum.

4.
SUMMARY OF INFORMATION OBTAINED FROM MAGNETIC MEASUREMENTS

Two key features of magnetic properties stand out. (1) There are no discontinuous or sharp changes at the M-I transition in any of the magnetic properties of either heavily-doped semiconductors such as Si:P or in related impurity systems such as $Na_xTa_yW_{1-y}O_3$. (2) Satisfactory theoretical models for these properties will require that we take full cognizance of the very wide and continuous distribution of local properties in the relatively dilute systems which form the subject of these notes.

4.1 Spin Susceptibility and Magnetization

It is my judgement that we have a sound understanding of the behavior of spin susceptibility and magnetization in Si:P for $n_D < 3E18$. The experiments by ESR[2] and static methods[4,5] and the theory[17,18] give a rather satisfying picture. By implication, we expect similar effects in other impurity systems characterized by the fully random placement of the impurities. For $n_D > 3E19$ in Si:P, a similar situation prevails -- simple metallic models do a good job.

For the region with $3E18 < n_D < 3E19$, many unsolved features remain. A model invoking strong correlations in the spin system seems to be required in order to produce the strong temperature and concentration dependence of χ_s and the ESR and NMR relaxation behavior. Although promising theoretical models are being currently explored, there is as yet no clear idea how to apply these models in any detail to real materials.

4.2 NMR Properties

NMR Knight shifts for the host system (i.e., ^{29}Si in Si:P, ^{73}Ge in Ge:As, or ^{183}W in the sodium tungsten bronzes) show a characteristic falloff at the M-I transition -- a falloff for which a fully satisfying explanation remains elusive.

Recent relaxation time measurements have shown promise -- either as a way to explore correlation effects in the puzzling Region IV (Fig. 1) or as a probe of the local energy level distribution in the insulating state. Whether these advances can lead us to a sharp and quantitative atomic description of the electron system near the M-I transition remains to be seen.

ACKNOWLEDGEMENTS

I am grateful to M.P. Sarachik and M. Paalenen for communicating results from Refs. 5 and 7, respectively, prior to publication. Conversations with them, as well as with R.N. Bhatt were useful. Most particularly, I have profited greatly from many conversations with recent Cornell colleagues, M.A. Dubson and M.J. Hirsch.

REFERENCES

1. D.F. Holcomb, "Magnetic Properties of Doped Semiconductors and Tungsten Bronzes," in The Metal Non-Metal Transition in Disordered Systems, Proceedings of the 19th Scottish Universities Summer School in Physics, 1978.
2. J.D. Quirt and J.R. Marko, Phys. Rev. Letters 26, 318 (1971), Phys. Rev. B 5, 1716 (1982), Phys. Rev. B 7, 3842 (1973).
3. D. Jerome, C. Ryter, and J.M. Winter, Physics 2, 81 (1965); D. Jerome, C. Ryter, H.J. Schulz, and J. Friedel, Phil. Mag, B 52, 403 (1985).
4. K. Andres, R.N. Bhatt, P. Goalwin, T.M. Rice, and R.E. Walstedt, Phys. Rev. B 24, 244 (1981).
5. M.P. Sarachik, A. Roy, M. Turner, M. Levy, D. He, L.L. Isaacs, and R.N. Bhatt, B34, 387 (1986).
6. S. Ikehata and S. Kobayashi, Solid State Comm. 56, 607 (1985).
7. M.A. Palaanen, S. Sachdev, and R.N. Bhatt, to be presented at 18th International Conference on the Physics of Semiconductors, Stockholm, 1986.
8. R.T. Schumacher and C.P. Slichter, Phys. Rev. 101, 58 (1956).
9. C.P. Slichter, Principles of Magnetic Resonance, 2nd Edition; Springer-Verlag, Berlin-Heidelberg-New York, 1980.
10. H. Ue and S. Maekawa, Phys. Rev. B. 3, 4232 (1971).
11. H.J. Wagner and C.F. Schwerdtfeger, Solid State Comm. 30, 597 (1979).
12. N. Kobayashi, S. Ikehata, S. Kobayashi, and W. Sasaki, Solid State Comm. 32, 1147 (1979).
13. R.K. Sundfors and D.F. Holcomb, Phys. Rev. 136, A810 (1964).
14. W. Sasaki, S. Ikehata, and S. Kobayashi, J. Phys. Soc. Jpn. 36, 1377 (1974).
15. T.F. Rosenbaum, K. Andres, G.A. Thomas, and R.N. Bhatt, Phys. Rev. Lett. 45, 1723 (1980).
16. S. Ikehata, T. Ema, S. Kobayashi, and W. Sasaki, J. Phys. Soc. Japan 50, 3655 (1981).
17. R.N. Bhatt and P.A. Lee, J. Appl. Phys. 52, 1703 (1981).

18. R.N. Bhatt and P.A. Lee, Phys. Rev. Lett. 48, 344 (1982).
19. M.P. Sarachik, D.R. He, W. Li, M. Levy, and J.S. Brooks, Phys. Rev. B 31, 1469 (1985).
20. C.P. Slichter, Phys. Rev. 99, 479 (1955).
21. S. Maekawa and N. Kinoshita, J. Phys. Soc. Jpn. 20, 1447 (1965).
22. D. New and T.G. Castner, Phys. Rev. B. 29, 2077 (1984).
23. E. Sonder and D.K. Stevens, Phys. Rev. 110, 1027 (1958).
24. W.G. Clark and L.C. Tippie, Phys. Rev. B 20, 2914 (1979).
25. H. Kamimura, Phil. Mag. B 52, 541 (1985).
26. G.E. Pike and C.H. Seager, Phys. Rev. B 10, 1421 (1974).
27. V. Cannella and J.A. Mydosh, Phys. Rev. B 6, 4220 (1972).
28. S. Ikehata, W. Sasaki, and S. Kobayashi, Solid State Comm. 19, 655 (1976).
29. D.P. Tunstall and V.G.I. Deshmukh, J. Phys. C 12, 2295 (1979).
30. M.A. Palaanen, A.E. Ruckenstein, and G.A. Thomas, Phys. Rev. Lett. 54, 1295, (1985).
31. A.M. Finkel'shtein, Sov. Phys. JETP 57, 97 (1983) and Z. Phys. B. 56, 189 (1984).
32. C. Castellani, C. DiCastro, P.A. Lee, M. Ma, S. Sorella, and E. Tabet, Phys. Rev. B. 30, 1596 (1984) and Phys. Rev. B 33, 6169 (1986).
33. T.M. Rice, Phil. Mag. B 52, 419 (1986).
34. S. Kobayashi, Y. Fukagawa, S. Ikehata, and W. Sasaki, J. Phys. Soc. Jpn. 45, 1276 (1978).
35. M.J. Hirsch and D.F. Holcomb, Phys. Rev. B 33, 2520 (1986).
36. V.G.I. Deshmukh and D.P. Tunstall, J. de Phys. Colloque C-4, 329 (1976).
37. J. Korringa, Physica 16, 601 (1950).
38. J.P. Doumerc, J. Marcus, M. Pouchard, and P. Hagenmuller, Mat. Res. Bull. 14, 201, (1979).
39. J.P. Doumerc, P. Dordor, E. Marquestaut, M. Pouchard, and P. Hagenmuller, Phil. Mag B 42, 487, (1980).
40. P. Dordor, J.P. Doumerc, and G. Villeneuve, Phil. Mag. B 47, 315 (1983).
41. M.A. Dubson, Ph.D. thesis, Cornell University, 1984 (unpublished).

42. M.A. Dubson and D.F. Holcomb, Phys. Rev. B 32, 1955 (1985).
43. M.A. Dubson and D.F. Holcomb, Phys. Rev. B 34, 25 (1986).
44. A. Narath and D.C. Wallace, Phys. Rev. 127, 724 (1962).
45. B.R. Weinberger, Phys. Rev. B 17, 566 (1978).
46. D.P. Tunstall and W. Ramage, J. Phys. C 13, 725 (1980).
47. W.W. Warren Jr., Phys. Rev. B 3, 3708 (1971).
48. W. Gotze and W. Ketterle, Z. Phys. B 54, 49 (1983).
49. F.C. Zumsteg Jr., Phys. Rev. B 14, 1406 (1976).
50. I. Yu, A.A.V. Gibson, E.R. Hunt, and W.P. Halperin, Phys. Rev. Lett. 44, 348 (1980).
51. H.E. Rhodes, P.-K. Wang, C.D. Makowka, S.L. Rudaz, H.T. Stokes, and C.P. Slichter, Phys. Rev. B 26, 3569 (1982).

BAND GAP NARROWING DUE TO HEAVY DOPING IN SEMICONDUCTORS

Joachim Wagner
Fraunhofer-Institut für Angewandte Festkörperphysik
Eckerstr. 4, 7800 Freiburg, West Germany

1. INTRODUCTION

Heavy doping of semiconductors is known to modify the electronic band-structure[1,2]. This modification can be described in terms of a filling of the conduction (valence) band due to the large number of free electrons (holes), of a renormalization of the energy band caused by many body interactions and of a band tailing resulting from the random distribution of the impurity atoms. The two latter effects lead to a narrowing of the fundamental band gap of the semiconductor.

In heavily doped silicon and germanium, for example, this band gap narrowing was observed experimentally in optical absorption[3-7] and by photoluminescence experiments[8-16]. Further evidence for this gap shrinkage was obtained from electrical transport measurements on bipolar transistors and diode structures [17,18], which revealed a significant reduction in band gap energy with increasing doping level.

Comparing these experiments, considerable discrepancies are found between luminescence and absorption [19] as well as absorption and transport data[20]. Recent experimental work using photoluminescence excitation spectroscopy[12,15] instead of conventional absorption measurements[5-7] could remove the discrepancy between luminescence and absorption data. The relationship between the gap shrinkage determined by optical measurements and the shrinkage deduced from transport data, however, is still not known in detail[21].

All the experimental work cited so far was concentrated on the renormalization of the fundamental band gap. Heavy doping, however, affects also higher lying band edges. This was shown experimentally, using ellipsometry, for the E_1 and E_2 edge in heavily doped silicon and germanium[22,23]. The shift of the E_0 gap in heavily doped p-type germanium was studied by photoluminescence spectroscopy[14].

Much theoretical work has been done to describe the band gap reduction in heavily doped semiconductors. Using many body theory with various approximations numerical calculations were performed for the low temperature limit $T = 0$[1,24-27] as well as for finite temperatures[28]. Recently attempts have also been made to formulate a theory which gives a simple analytical expression for the band gap shrinkage at the expense of some of the accuracy of the many body treatment[29].

Besides the basic aspects discussed above, the band gap reduction upon heavy doping has also implications on the physics of semiconductor devices. Band gap narrowing affects e.g. the gain of bipolar transistors[30,31] as well as the open circuit voltage of solar cells[32]. Very large scale integration, on the other hand, requires increasingly higher doping levels[21]. This makes the detailed understanding of the doping induced gap shrinkage essential for the design of high performance devices.

In the following Section 2 a short review of the theoretical understanding of the band gap renormalization is given. Section 3 describes optical measurements of this gap renormalization with special emphasis on silicon. In Section 4 the implications of the band gap narrowing on semiconductor devices are discussed and in Section 5 the conclusions of the present paper are summarized.

2.
THEORETICAL APPROACH

For a heavily doped n-type (p-type) semiconductor with the charge carrier concentration far above the Mott critical density for metallization the conduction (valence) band is partially filled with the extra electrons (holes) introduced by the doping. This filling of the conduction or valence band leads to a high energy shift of the band gap measured by absorption. This is the so-called Burstein-Moss shift. The interaction of the electrons among themselves and with the ionized dopant impurities, on the other hand, causes a down-ward shift of the conduction and up-ward shift of the valence band, resulting in a narrowing of the band gap.

In heavily doped, degenerate semiconductors we have to distinguish two energy gaps. One, the so-called reduced band gap $E_{G,2}$, is the energy separation between the top of the valence and the bottom of the conduction band (see Fig.1). The other is the energy required to take an electron out to the valence band into an empty state of the conduction band, the optical gap energy $E_{G,1}$. These two gap energies differ by the band filling E_F.

According to Mahan[24], four contributions have to be considered for the computation of $E_{G,1}$ and $E_{G,2}$. These are the kinetic energy of the carriers, the exchange and correlation energy arising from the interaction of the carriers among themselves, the electron-donor interaction, and the electron-hole interaction. In the low-temperature limit T = 0 the kinetic energy or band filling E_F can be written as

$$E_F = \hbar^2 k_F^2/2m_d = (\hbar^2/2m_d)\cdot(3\pi^2/N)^{2/3}\cdot n^{2/3}$$
$$= 3.34\cdot(n/10^{18})^{2/3} \qquad (2.1)$$

with k_F as the Fermivector, m_d the effective density of states mass, n the electron density (n-type doping) and N the number of equivalent conduction band minima. The numerical value given in meV is valid for heavily doped n-type silicon with the carrier density given in units of 10^{18} cm^{-3}.

The next term to be considered is the exchange energy E_{ex}. Within the Hartree-Fock approximation this energy can be expressed as[24]

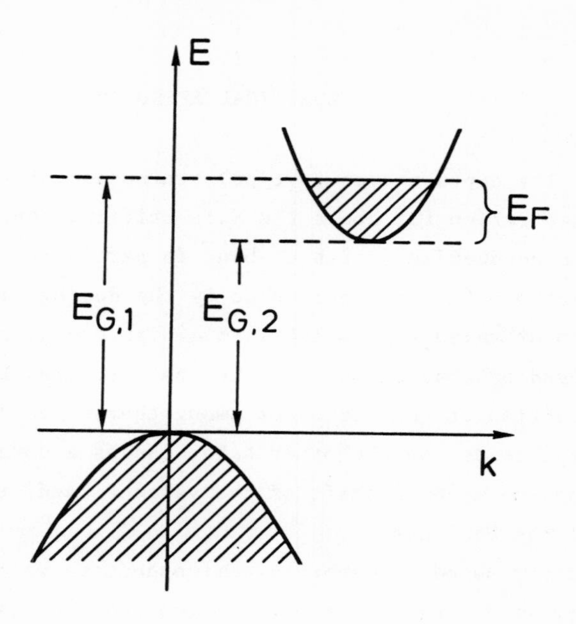

Fig.1 Schematic drawing of the band structure of a heavily n-type doped indirect semiconductor. Optical gap $E_{G,1}$*, reduced band gap* $E_{G,2}$*, and band filling* E_F *are indicated.*

$$E_{ex} = -(e^2/\pi\varepsilon_o)\cdot\Lambda\cdot(3\pi^2/N)^{1/3}\cdot n^{1/3}$$

$$= -6.47\cdot(n/10^{18})^{1/3} \qquad (2.2)$$

Λ is related to the longitudinal mass m_l and transverse mass m_t of the conduction band by

$$\Lambda = (m_l / m_t)^{1/3} \cdot (\tan^{-1}\delta / \delta)$$
$$\delta = \left[(m_l - m_t)/m_t\right]^{1/2}. \qquad (2.3)$$

The so-called correlation energy E_{cor} comprises all electron-electron interactions, which are not included in E_{ex}. It is smaller than E_{ex} and typically of the order of $-0.1\ E_{Rydberg}$ (24). The calculation of the energy arising from the electron donor interaction E_{ed} usually starts from a screened Coulomb potential to describe the ionized dopant impurities(24)

$$V(r) = (-e^2/\varepsilon_o \cdot r) \cdot \exp(-k_s \cdot r) \tag{2.4}$$

where k_s is the screening wavevector given in the Thomas-Fermi approximation by

$$k_s^2 \simeq \frac{6\pi e^2 \cdot n}{\varepsilon_o \cdot E_F} \sim n^{1/3}. \tag{2.5}$$

For the distribution of the free carriers around the ionized impurity atoms two extreme models can be considered. One assumes that all the carriers are acting as screening charges located around the impurities. For this model the interaction energy is

$$E_{ed} = -(e^2/8\varepsilon_o)k_s \sim n^{1/6}. \tag{2.6}$$

The other extreme is to assume a truly uniform carrier distribution, yielding

$$E_{ed} = -0.481 (e^2/\varepsilon_o)n^{1/3}. \tag{2.7}$$

The real situation is somewhere in between these two models. To account for this Mahan[24] performed a variational calculation for the ground state energy of the electron-donor system assuming a regular arrangement of the dopant impurities on a fcc lattice. The result is the electron-donor interaction term given in Equ.2.7 minus a constant energy. For heavily doped n-type silicon this can be written as

$$E_{ed} = -6.11 (n/10^{18})^{1/3} - 3.1. \tag{2.8}$$

The fourth term to be considered is the hole energy E_{eh} arising from the interaction of the holes among themselves and with the screened electron-donor system. A variational calculation gives for n-type silicon[24]

$$E_{eh} = - 13.1\ (n/10^{18})^{1/4} + 6.1\ (n/10^{18})^{1/3} \quad (2.9)$$

The minimum energy required to excite an electron out of the valence band into an empty state of the conduction band is then given by

$$E_{G,1} = E_F + E_{G,2} \quad (2.10)$$

and the reduced gap energy by

$$E_{G,2} = E_{ex} + E_{cor} + E_{ed} + E_{eh} + E_G \quad (2.11)$$

with E_G as the band gap energy of the undoped material. The terms $\pm\ 6.11(n/10^{18})^{1/3}$ in E_{ed} and E_{eh} cancel. So we are left with energy terms with negative sign except for E_F. Thus $E_{G,2}$ in heavily doped material is always smaller than E_G. This gap shrinkage contains terms which go as $n^{1/3}$ and $n^{1/4}$ with increasing carrier concentration plus a constant energy shift, which gives essentially the free exciton binding energy in the low density limit. The optical gap also contains E_F with a positive sign, which goes as $n^{2/3}$. This implies that for large enough carrier densities $E_{G,1}$ always shifts to higher energies. For intermediate carrier densities, however, $E_{G,1}$ might be smaller than E_G and might show only a very weak variation with n, depending on the density of states mass of the band occupied by the carriers.

Since the basic work by Mahan[24] the band gap narrowing in heavily doped silicon has been treated theoretically by many other groups. Their calculations differ mainly in the way they include the carrier impurity interaction. Berggren and Sernelius[25] used two different approaches for the arrangement of the donor impurities (n-type doping) - either on an fcc lattice or a random distribution. They also include the correlation energy explicitly in their calculation. The result for a regular arrangement of the impurities agrees with the one obtained by Mahan[24], whereas a random impurity distribution results in an increased band gap narrowing[25].

Selloni and Pantelides[26] also started from an ordered array of donor atoms. To calculate the electron donor interaction they used the single particle approach with the computational tools available as used for band structure calculations. These authors also included explicitly the effects of disorder and computed the luminescence emission line shape for degenerate n-type silicon.

Abram et al.[1,27] used two different approaches to describe the dielectric response of the charge carrier-impurity system. One is the Lindhard dielectric function and the other the plasmon pole approximation. They showed, that the latter gives results for n- and p -type doping with almost the same accuracy as the Lindhard function and facilitates the extension of the theory to finite temperatures[33].

Thuselt and Rösler[28] also calculated the band gap narrowing for n- and p-type material at finite temperatures. Besides silicon they also present theoretical data for heavily doped germanium and GaAs. A simple model for the calculation of the band gap shrinkage ΔE_G has been proposed by Landsberg et al.[29]. It starts from the Jellium model and uses Debye screening. In this model the band gap shift can be written in an analytic expression

$$\Delta E_G = (e^2/\varepsilon_o)\left[\frac{4\,e^2}{\varepsilon kT}\left[N_c\,F_{-1/2}\,(\gamma_c) + N_v\,F_{-1/2}\,(\gamma_v)\right]\right]^{1/2}$$

with $\gamma_c = (E_F - E_c)/kT, \quad \gamma_v = (E_v - E_F)/kT$

$$F_{-1/2}(\gamma) = \pi^{-1/2}\cdot\int_0^\infty x^{-1/2}\,/(1+\exp(x-\gamma))\,dx \qquad (2.12)$$

E_C and E_V are the energy of the conduction and valence band edge, respectively, E_F denotes the Fermi energy, and N_C and N_V are the effective densities of states at the conduction and valence band edges. This formula describes ΔE_G for n- and p-type doping in the low temperature limit as well as for finite temperatures. However, due to its simplicity this model is less accurate than the work based on many body theory discussed before.

All the theoretical calculations considered so far deal with the effect of heavy doping on the lowest, fundamental band gap. The dopant impurities are described in the effective mass approximation, which means that the ionized donor or acceptor is simply treated as a point charge. Any effects arising from the difference in the chemical nature of the dopant atom with respect to the host atoms are neglected. Allen[34] and Viña and Cardona[23] performed perturbation theoretical calculations of the change of the whole band structure in a heavily doped semiconductor. In these calculations which assume a random distribution of the impurities, terms of first and second order in the impurity potential have been included. For silicon a low energy shift of the E_1 and E_2 band edge is obtained which increases proportional to the square root of the impurity concentration. These calculations do not include exchange and correlation effects. To obtain the total shifts of band gaps which involve either the partially filled topmost valence band (p-type doping) or the conduction band (n-type doping), these terms have to be added.

All the theoretical work discussed in this Section, with the exception of the paper by Selloni and Pantelides[26], does not include disorder induced band tailing effects[35,36]. Numerical results from this work are shown in the following Section 3 and compared to experimental data.

3.
OPTICAL EXPERIMENTS

Optical absorption measurements have been widely used to study heavily doped semiconductors[3-7]. In the band-to-band absorption spectrum transitions are observed between filled states in the valence band and empty states in the conduction band.

For band-to-band absorption in an undoped indirect semiconductor as silicon or germanium, the low temperature absorption coefficient α_0 can be written as[6].

$$\alpha_o \sim (\hbar\omega - E_G - \hbar\omega_{Ph})^2 \tag{3.1}$$

with $\hbar\omega$ as the energy of the absorbed photon, E_G as the band gap energy and $\hbar\omega_{ph}$ as the energy of the emitted momentum conserving phonon. In degenerate material the absorption spectrum is modified due to the band filling. The absorption coefficient is then given by[4,7]

$$\alpha \sim \alpha_0 \cdot \int_0^1 \frac{x^{1/2}\,(1-x)^{1/2}\,dx}{1+\exp\left[(E_F - x\,(\hbar\omega - E_{G,2} - \hbar\omega_{Ph}))\,/kT\right]} \quad . \tag{3.2}$$

At low temperatures the absorption edge is shifted to higher energies by E_F relative to the reduced band gap $E_{G,2}$ (see Fig.1).

In heavily doped silicon, however, band-to-band absorption and absorption arising from the excitation of free electrons (holes) within the conduction (valence) band are strongly overlapping for photon energies of about the band gap energy(7) as shown in Fig.2. To deduce the band gap energy from such spectra, one has to subtract the free carrier absorption, which has an absorption coefficient proportional to the square of the wavelength of the absorbed light. The remaining absorption spectrum then has to be fitted using the expressions given in Equs.3.1 and 3.2 to extract $E_{G,2}$(7). Typical data for $E_{G,2}$ from such measurements are shown for low temperature and for 300 K in Figs. 6 and 7.

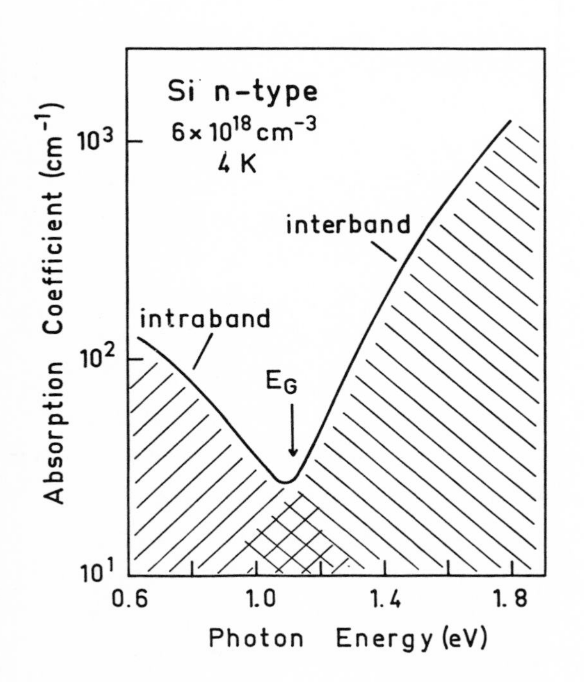

Fig.2 Typical low temperature absorption spectrum of heavily doped n-type silicon (see Ref.7). Intra- and interband absorption are indicated.

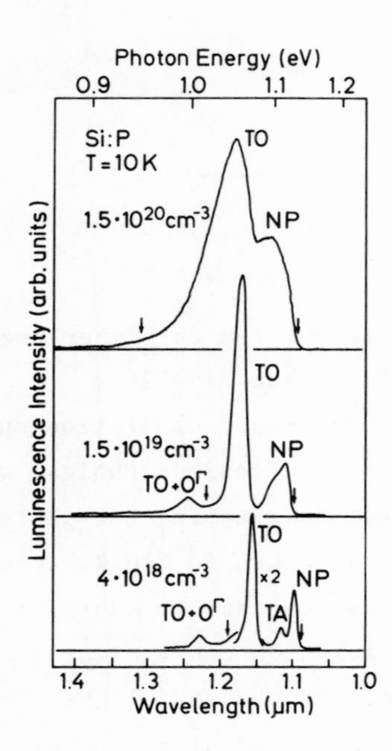

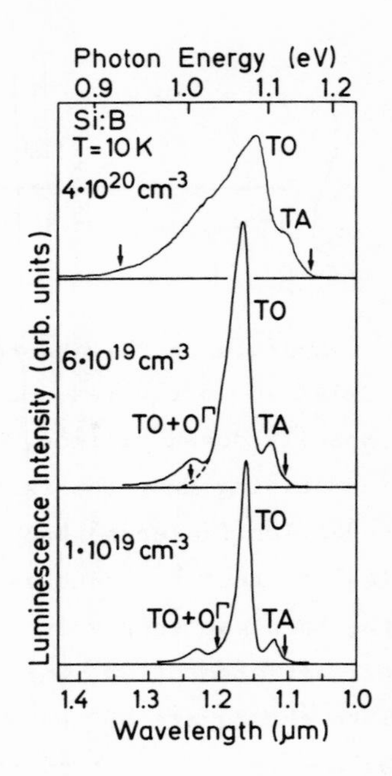

Fig.3 Photoluminescence spectra of heavily doped n-type (left) and p-type (right) Si for different carrier concentrations. Arrows indicate the high-energy cutoff of the NP line ($E_{G,1}$, left) and of the TA replica ($E_{G,1}-\hbar\omega_{TA}$, right), respectively, and the low-energy edge of the TO replica ($E_{G,2}-\hbar\omega_{TO}$). Spectral resolution varied between 8 Å for the more lightly doped samples and 24 Å for the most heavily doped ones (Ref.12).

Another optical technique which gives access to band filling and band gap narrowing in heavily doped material is the photoluminescence (PL) spectroscopy. Thereby the luminescence emitted by the radiative recombination of photocreated minority carriers is measured. The emission band extends from the reduced gap energy $E_{G,2}$ (low-energy edge) to the optical gap $E_{G,1}$ (high-energy cutoff). The width of the emission band represents the band filling E_F. In an indirect gap semiconductor zone boundary phonons have to participate in optical transitions in order to fulfil momentum conservation, leading to so-called phonon replicas. These replicas are shifted by the energy of the momentum-conserving phonon with respect to the electronic transition energy. At low temperatures, only the Stokes phonon replicas occur as the zone boundary phonons are not occupied thermally.

Fig.3 depicts low temperature PL spectra of heavily phosphorous (n-type) or boron (p-type) doped silicon. The emission spectra consist of three phonon replicas involving momentum-conserving transverse-acoustic (TA) and transverse-optical (TO) phonons and the combination of the TO phonon and the optical zone-center phonon O_{Γ}. For Si:P an additional no-phonon (NP) emission line is observed where the change in momentum is taken over by the donor impurities. The spectra show pure band-to-band emissions with the density of the photo excited carriers much smaller than the dopant concentration(12). With increasing doping level the luminescence bands become broader, and for the most heavily doped material only the two strongest transitions - TO and NP for Si:P and TO and TA for Si:B - are resolved.

The band gap energies $E_{G,1}$ and $E_{G,2}$, deduced from the PL spectra as indicated in Fig.3(12), are plotted in Fig.4 versus the dopant concentration. The reduced gap energy $E_{G,2}$ shows a pronounced reduction from 1.15 eV in pure silicon to $\sim$1.0 eV for a dopant concentration of $2\text{x}10^{20}$ cm^{-3}. This gap reduction is within the experimental accuracy the same for n- and p-type doping. The optical gap energy $E_{G,1}$ in contrast behaves differently for n- and p-type material. In n-type samples $E_{G,1}$ remains almost constant in the doping range $5\text{x}10^{18}$-$2\text{x}10^{20}\text{cm}^{-3}$, whereas for p-type doping a significant high-energy shift of $E_{G,1}$ is observed for dopant concentrations $\geq 10^{20}\text{cm}^{-3}$.

This difference in band filling reflects the difference in the density-of-states masses for electrons and holes. In pure material the mass of the electrons, m_{de} = 1.062 m_0, is nearly twice as large as the one for the holes, m_{dh} = 0.577 m_0(37,38). Upon heavy doping, these masses change somewhat with the hole mass becoming larger(39), but the band filling is always larger in n-type than in p-type material(15).

Fig.4 also displays theoretical data for $E_{G,1}$ and $E_{G,2}$. For n-type doping the results obtained by Berggren and Sernelius(25) as well as data extracted from the emission spectra calculated by Selloni and Pantelides(26) are shown. For p-type material, a comparison is made with theoretical data reported by Abram et al.(1). The overall agreement between experiment and theory is good and also finer details

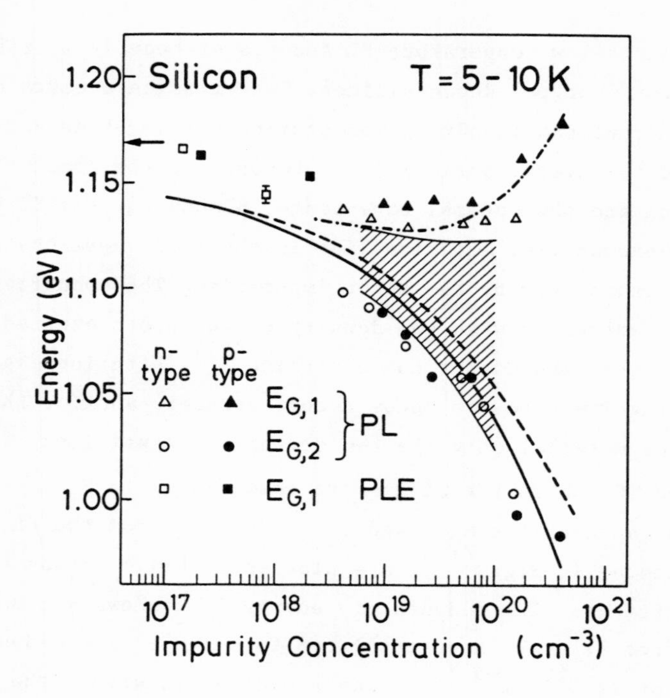

Fig.4 Optical ($E_{G,1}$) and reduced band gap ($E_{G,2}$) as determined by PL and PLE spectroscopy vs carrier concentration for n- and p-type material. Arrow on the energy scale indicates the band-gap energy of pure silicon. Solid line shows the result of the calculation of $E_{G,2}$ in n-type material by Berggren and Sernelius (Ref.25) and the dashed line refers to the calculation of $E_{G,2}$ for p-type doping by Abram et al. (Ref.1). Hatched area indicates the width and position of the luminescence spectra calculated by Selloni and Pantellides (Ref.26). Dashed-dotted line shows the optical gap for p-type silicon (Ref.12).

like the different behaviour of $E_{G,1}$ in n- and p-type material are reproduced by the theoretical data.

For a more accurate determination of the reduced gap energy $E_{G,2}$ from the PL spectra a detailed emission line shape analysis has been performed by several groups(9,11,15,16). The emission line shape $I(\hbar\omega)$ is expressed by the convolution of the densities of states D_e (E) and D_h (E) weighted by the distribution function $f_e(E)$ and $f_h(E)$ (the subscripts e and h stand for electrons and holes, respectively):

$$I\,(\hbar\omega) \sim \int_0^\infty f_e(E)D_e(E)\cdot f_h(\hbar\omega - E_{G,2} - E) \times D_h(\hbar\omega - E_{G,2} - E)\cdot dE\,. \qquad (3.3)$$

Thereby constant transition matrix elements are assumed. For n-type (p-type) doping $f_e\,(E)$ ($f_h\,(E)$) represents the Fermi distribution function and $f_h(E)(f_e(E))$ is described by a Boltzmann distribution. To account for band tailing effects and incomplete thermalization of the carriers, a Gaussian broadening of $I\,(\hbar\omega)$ is introduced(11,15,16)

$$\overline{I}\,(\hbar\omega) = \int_0^\infty I\,(\hbar\omega')\cdot\exp\left[-\,(\hbar\omega'-\hbar\omega)^2/E_s^2\right]\cdot d\hbar\omega' \qquad (3.4)$$

with E_s as an empirical broadening parameter.

Fig.5 displays PL spectra for heavily doped n- and p-type Si together with the fitted emission line shapes calculated using Equs. 3.3 and 3.4. The fit to the experimental spectra is reasonably good with the exception of the low energy edge. This, however, is due to the rather crude way used to introduce band tailing effects in the line shape calculation(15,16).

The shift ΔE_G of the reduced band gap energy $E_{G,2}$ at low temperatures, obtained from this line shape analysis, is plotted in Fig.6 versus the carrier concentration. For comparison theoretical results by Berggren and Sernelius(25) as well as low temperature absorption data by Schmid(7) are also shown. The agreement between the PL data and theory is remarkably good. The absorption data, however, give systematically a smaller gapshrinkage than the PL measurements. For carrier concentration of ~ 10^{19} cm^{-3} this difference amounts to up to 50 meV.

PL spectroscopy is also applicable to study the band gap narrowing at room temperature (300 K)(16). The shift of $E_{G,2}$ at 300 K, also obtained performing a line shape analysis, is shown in Fig.7. Comparing the PL data with room temperature absorption measurements, again, a significantly smaller band gap reduction is found by absorption than

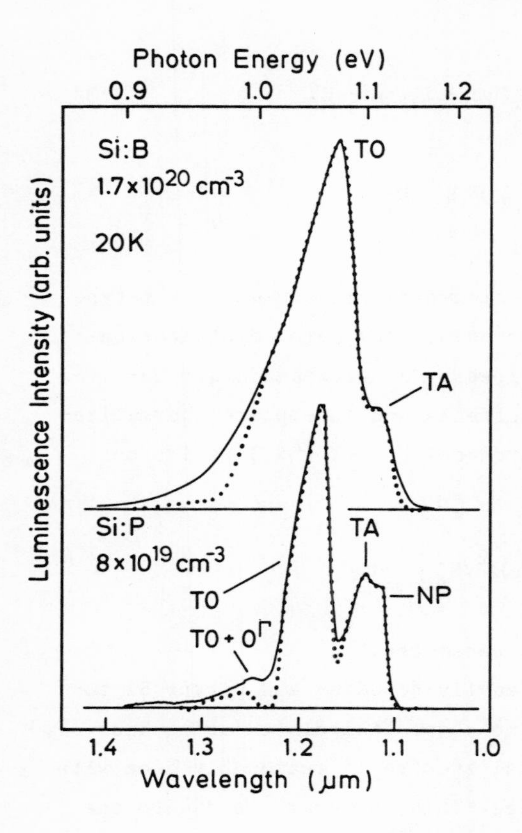

Fig.5 PL spectra (drawn line) for Si:P ($8 \times 10^{19} cm^{-3}$) and Si:B ($1.7 \times 10^{20} cm^{-3}$). The dotted curves show calculated emission line shapes fitted to the experimental spectra (Refs. 15 and 16).

by PL spectroscopy. Recent theoretical work by Thuselt and Rösler[28] is in good agreement with the gap reduction ΔE_G measured by PL spectroscopy. The theoretical data reported by Saunderson[40], in contrast, show systematically a too large gap reduction. Comparing the room temperature band gap reduction for n- and p-type material with the low temperature data, the same reduction is found within the experimental accuracy.

As discussed above, difficulties arise in the interpretation of absorption spectra due to the overlap of band-to-band and free carrier absorption. As photoluminescence excitation (PLE) spectroscopy is only sensitive to band-to-band transitions which create minority carriers, this technique looks through the free carrier absorption. As shown in Fig.8, PLE spectra show directly the onset of band-to-band absorption and allow a straight forward determination of the optical band gap $E_{G,1}$. In Fig.8, the square root of the luminescence intensity, which

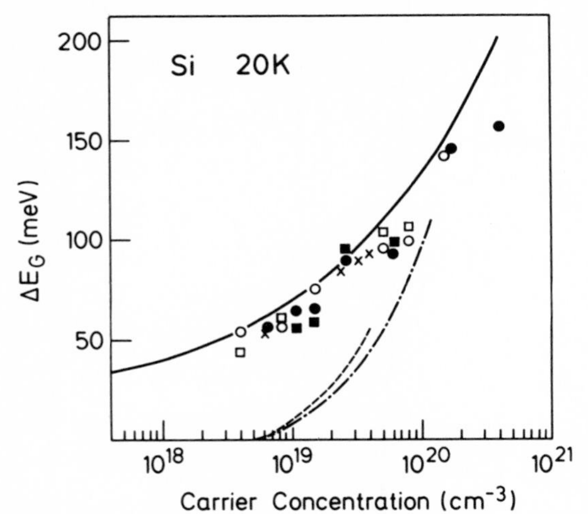

Fig.6 Shift of the reduced band gap $\Delta E_G = E_G$ (pure) - $E_{G,2}$ [E_G (pure) = 1.17 eV] versus carrier concentration at 20 K. The open (n-type) and filled (p-type) circles show PL data from Ref.16, the crosses data for n-type doping from Ref.11. The open (n-type) and filled (p-type) squares represent data from PLE. The solid line refers to calculations by Berggren and Sernelius (Ref.25). The dashed (n-type) and dash-dotted (p-type) curves give the results of low-temperature absorption measurements (Ref.7).

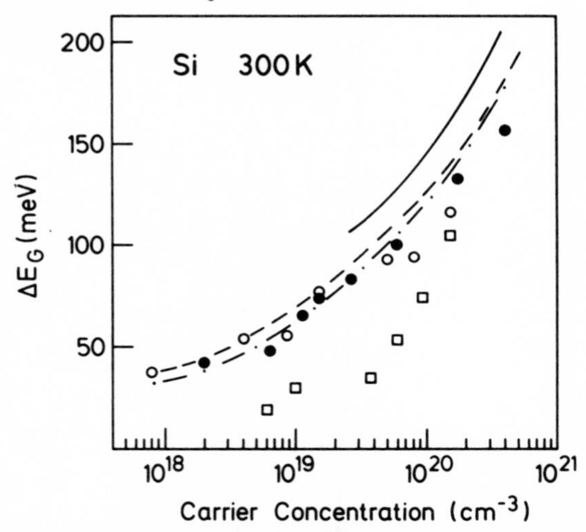

Fig.7 Same as Fig.6, but at 300 K [E_G (pure) = 1.11 eV]. The open (n-type) and filled (p-type) circles show PL data from Ref.16. The open squares represent absorption data from Ref.6. The solid line refers to calculations for n-type doping (Ref.40); the dashed (n-type) and dash-dotted (p-type) curves refer to theoretical data from Ref.28.

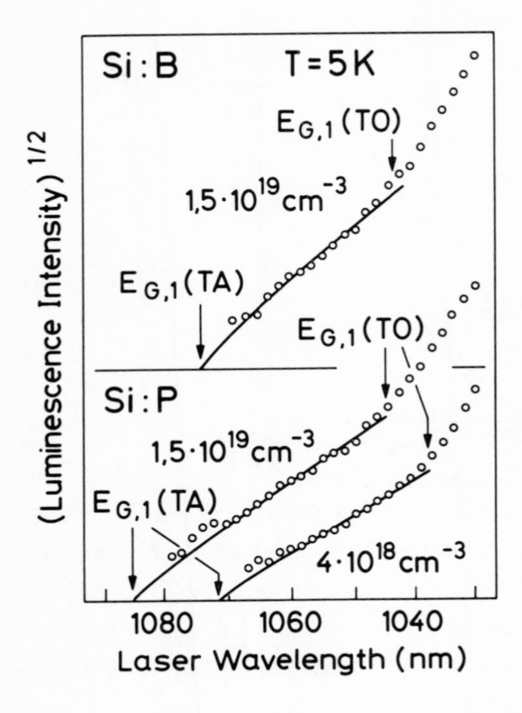

Fig.8 PLE spectra for p-type (top) and n-type (bottom) Si recorded at 5 K. $E_{G,1}$ (TO) and $E_{G,1}$ (TA) denote the onset of TO and TA phonon assisted band-to-band absorption. The drawn lines are fitted absorption curves (Ref.15).

is for low absorption proportional to the absorption coefficient, is plotted versus the exciting laser wavelength(12,15). The drawn curves are fitted absorption spectra calculated using Equs. 3.1 and 3.2. The corresponding shift ΔE_G of the reduced gap $E_{G,2}$, obtained from the PLE data ($E_{G,1}$) by subtracting the band filling E_F, are plotted in Fig.6. They coincide with the PL data within the experimental scatter of the data points. This shows that luminescence and selective absorption (PLE) spectra are consistent with each other(12,15). The too small gap reduction measured by conventional transmission spectroscopy arises most probably from the fact, that one has to rely on band-to-band transitions involving electronic states high above the band edge to extrapolate the band gap energy. This procedure, however, is very sensitive to a non-rigid shift of the valence and conduction band upon heavy doping as discussed by Dumke(11).

The experiments presented so far were carried out on bulk doped silicon. Recent work showed that photoluminescence spectroscopy is also useful to study heavily doped silicon layers prepared either by ion implantation and subsequent annealing(41) or by epitaxial growth techniques(42). Thereby it is important to use excitation light with

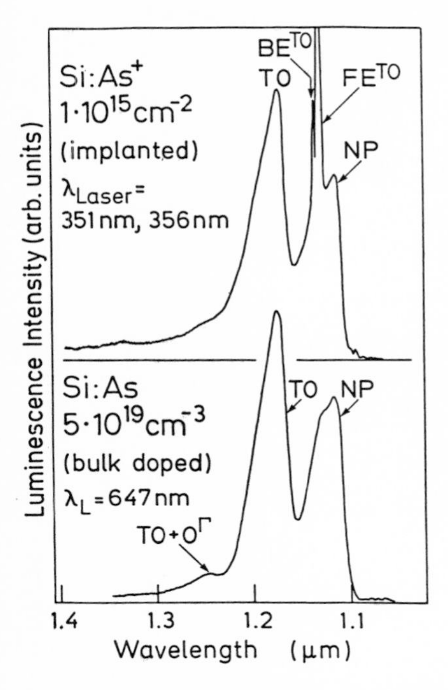

Fig.9 Low temperature PL spectra of rapid thermal annealed Si:As$^+$ implanted at a dose of 1×10^{15} ions/cm^2 (upper trace). FE^{TO} and BE^{TO} denote luminescence from the bulk. NP and TO are the no-phonon and TO phonon replica of the band-to-band luminescence in the implanted region. The lower trace shows for comparison the band-to-band PL spectrum of a bulk-doped Si:As sample (Ref.41).

a sufficiently small penetration depth to excite only the heavily doped region. Fig.9 depicts a low temperature PL spectrum from a ~ 100 nm thick As$^+$ implanted and rapid thermal annealed layer recorded with uv (~ 350 nm) excitation. The spectrum shows, besides some remaining luminescence features from the lightly doped substrate at ≃1.13 μm, typical band-to-band emission from the heavily doped layer. The comparison with the PL spectrum from a bulk doped Si:As sample with a carrier concentration of 5×10^{19} cm^{-3} also displayed in Fig.9 indicates a similar carrier concentration for the implanted and annealed layer(41).

In this Section we have focused so far on heavily doped silicon. Experimental results for the doping induced band gap narrowing in p-type germanium are shown in Fig.10. Using luminescence spectroscopy the shift of the indirect band gap $E_{G,2}$ as well as the shrinkage of the first direct gap E_0 have been measured(14). The comparison with theoretical data(14) based on the band structure type calculations by Viña and Cardona(23) shows a satisfactory agreement between experiment and theory. For heavily bulk doped as well as ion implanted and laser annealed n-type germanium similar luminescence work is reported in the literature(8,13).

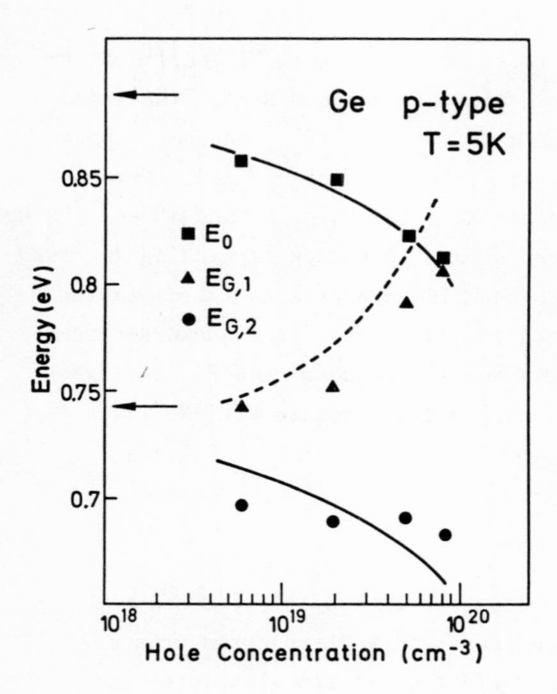

Fig.10 Shift of the direct band gap E_0, of the indirect gap $E_{G,2}$, and of the optical gap $E_{G,1}$ vs hole concentration in p-type germanium. The drawn curves show the calculated shrinkage of E_0 and $E_{G,2}$, the dashed line indicates the calculated shift of $E_{G,1}$. The arrows on the vertical scale show the gap energies E_0 and E_G in pure germanium (Ref.14).

Optical absorption and emission spectroscopy have also been used to study the doping induced band gap narrowing and band filling in III-V compound semiconductors such as GaAs. For a review of the work on GaAs see the paper by Casey and Stern(43) and some recent work by Olego and Cardona(44), by Titkov et al.(45), and by J. De-Sheng et al.(46). Heavily doped n-type InP has been studied very recently by Schwabe et al.(47) using photoluminescence and excitation spectroscopy.

4.
IMPLICATIONS FOR DEVICE PHYSICS

A basic relation to describe the electrical properties of doped semiconductors is the np product, which can be expressed by a mass-action type of equation

$$n \cdot p = n_i^2 = C \cdot T^3 \exp(-E_G/kT) \qquad (4.1)$$

with n and p as the concentrations of electrons and holes, n_i as the intrinsic carrier concentration, and C as a constant containing the electron and hole effective masses. This equation is valid for moderate doping levels and assumes Boltzmann statistics. For heavily doped material with carrier concentrations exceeding the 10^{18}-$10^{19}\,cm^{-3}$ range deviations from this law are observed[17]. Therefore Equ.4.1 has been reformulated as[17,48,49]

$$n \cdot p = n_{ie}^2 = n_i^2 \cdot \exp(\Delta E_G^{eff}/kT) \qquad (4.2)$$

with n_{ie} as the "effective" intrinsic carrier concentration and ΔE_G^{eff} as the "effective" band gap narrowing. Thereby ΔE_G^{eff} accounts for three different effects. One is the band gap reduction due to many body effects in the degenerate carrier system as discussed in Section 2. The other two effects are nonparabolicities in the density of states and band tailing effects as well as deviations from Boltzmann statistics. For degenerate semiconductors Fermi-Dirac statistics have to be used to describe the majority carrier system. For this statistics effect, ΔE_G^{eff} can be corrected by[18]

$$\Delta E_G' = \Delta E_G^{eff} + kT \cdot \ln(1 + .27\exp(E_F/kT)). \qquad (4.3)$$

The exact relationship between $\Delta E_G'$, which still contains both the many body and the density of states effects, and the band gap narrowing ΔE_G as discussed in Section 2 and 3 (Figs. 6 and 7), is still unclear[21]. Van Vliet and Marshak[50] formulated a relation between $\Delta E_G'$ and ΔE_G as

$$\Delta E_G' = \Delta E_G + \Theta_h + \Theta_p \qquad (4.4)$$

where Θ_n and Θ_p contain the nonparabolic density of states of the electrons and the holes, respectively.

The intrinsic carrier concentration n_i or n_{ie}, on the other hand, is an important parameter to describe the properties of semiconductor devices. The $I_c - V_{EB}$ characteristic of a npn transistor is given by

$$I_c = I_o \exp(q \cdot V_{EB}/kT)$$

$$\text{with } I_o = n_{ie}^2 \mu_n kT/p . \qquad (4.5)$$

I_c denotes the collector current density, V_{EB} the voltage between emitter and base, q the elementary charge, p the concentration of majority carriers (holes) in the base region, and μ_n the mobility of the minority carriers (electrons). From Equ. 4.5 it is clear that changes of n_{ie} induced by heavy doping of the base region (np^+n transistor) strongly affect the device properties.

The measurement of such device characteristics has been used to deduce "electrical" data for the dopant induced band gap shrinkage ΔE_G [17,18]. Using np^+n transistors Slotboom and de Graff[17] determined the product n_{ie}^2 by measuring the I_c-V_{EB} characteristic. The electron mobility μ_n is assumed to be the same as the one in n-type material with the same impurity concentration.

A somewhat different approach is to measure the temperature dependence of device properties, such as I_c [17,51]. For such an experiment, not all the parameters, which determine the measured quantity, have to be known, provided they are temperature independent in the range of temperatures investigated. One has further to assume that ΔE_G does not depend on temperature.

A review of the present understanding of heavy doping effects on the properties of devices is given in the paper by de Alamo et al.[21]. In Fig.11 is plotted what these authors call their best "guess" for ΔE_G^{eff} versus dopant concentration. Thereby ΔE_G^{eff} is not corrected for Fermi-Dirac statistics. Correcting for this statistics effect using Equ.4.3 a larger "electrical" band

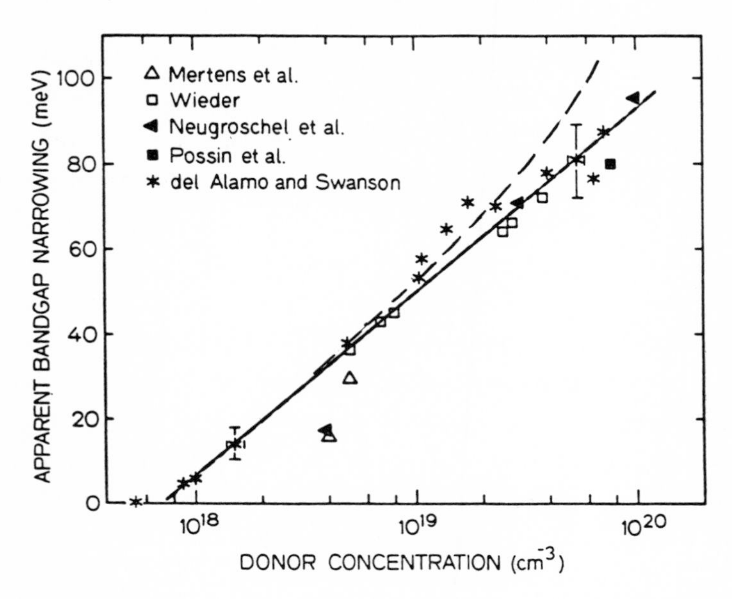

Fig.11 Best "guess" of the "apparent" band gap reduction (ΔE_G^{eff}) deduced from device measurements (Ref.21) assuming Boltzmann statistics. The drawn curve is fitted to the experimental data. Correcting for Fermi statistics the dashed curve is obtained.

gap reduction $\Delta E_G'$ is obtained as indicated in Fig.11 by the dashed curve. The agreement of these electrical data with the band gap narrowing ΔE_G obtained from PL measurements (Figs. 6 and 7) is surprisingly good taking into account the difficulties which are still present in the understanding of the relationship between ΔE_G and ΔE_G^{eff}.

5.
CONCLUSIONS

In the present paper the reduction of the fundamental band gap in heavily doped semiconductors has been discussed. The theoretical description of this band gap narrowing, caused by many body interaction of the majority carriers among themselves as well as by carrier-impurity interaction, has been reviewed. Special emphasis has been laid on the experimental measurement of the band gap reduction and on the implications on device physics.

It has been demonstrated that photoluminescence and selective absorption (photoluminescence excitation spectroscopy) measurements give a consistent set of data for the doping induced band gap shrinkage in heavily doped silicon. However a further improvement in the accuracy of these data should be possible using a refined description of the nonparabolic valence and conduction band density of states in the calculation of the luminescence lineshape. Also more work has to be done to establish the exact relation between the band gap reduction measured by optical spectroscopy and the "effective " gap shrinkage used in the description of semiconductor devices.

The heavy doping effects discussed in this paper are closely related to the phenomena found in highly photo-excited semiconductors(52). In this case the photoexcitation creates a dense gas of electrons and holes at equal concentrations with both positive and negative charges mobile. This is in contrast to heavily doped material where only the electrons (holes) form a gas whereas the opposite charge (ionized donors or acceptors) is fixed.

Band gap renormalization is also observed in quasi-two-dimensional systems such as modulation doped GaAs/(AlGa)As quantum-well heterostructures(53-57). In these structures a two-dimensional electron (n-type) or hole (p-type doping) gas is present in the quantum wells with the donor or acceptor ions built into the (AlGa)As barriers separating the GaAs wells. Therefore the electron or hole gas does not interact with the ionized dopant impurities. For p-type doping with a hole concentration of $5.3 \times 10^{10} cm^{-2}$ and a well width of 10.7 nm the reduction of the first electron-heavy hole subband transition energy amounts to 12.3 meV which is somewhat less than 1 % of the total transition energy(55). The energy spacing between the individual electron or hole subbands, on the other hand, is not affected indicating a rigid shift of the valence and conduction band(55).

ACKNOWLEDGEMENTS

I would like to thank M. Cardona, J. del Alamo, P.T. Landsberg, A. Neugroschel, R.M. Swanson, and L. Viña for many helpful discussions. Thanks are further due to S. Pagel for typing the manuscript.

REFERENCES

1. R.A. Abram, G.J. Rees, and B.L.H Wilson, Adv. Phys. 27, 799 (1978)
2. Proceedings of the "International Conference on Heavy Doping and the Metal-Insulator Transitions in Semiconductors", Solid State Electron 28, No.1/2 (1985).
3. C. Haas, Phys. Rev. 125, 1965 (1962).
4. J.I. Pankove and P. Aigrain, Phys. Rev. 126, 956 (1962).
5. A.A. Volfson and V.K. Subashiev, Fiz. Tekh. Poluprovodn.1, 397 (1967) Sov. Phys. Semicon. 1, 327 (1967).
6. M. Balkanski, A. Aziza, and E. Amzallag, Phys. Stat. Sol. 31, 323 (1969).
7. P.E. Schmid, Phys. Rev. B 23, 5531 (1981).
8. C. Benoit à la Guillaume and J. Cernogora, Phys. Stat. Sol. 35, 599 (1969).
9. R.R. Parsons, Solid State Commmun. 29, 763 (1979) and references therein.
10. P.E. Schmid, M.L.W. Thewalt, and W.P. Dumke, Solid State Commun. 38, 1091 (1981).
11. W.P. Dumke, Appl. Phys. Lett. 42, 196 (1983); J. Appl. Phys. 54, 3200 (1983).
12. J. Wagner, Phys. Rev. B 29, 2002 (1984).
13. J. Wagner, G. Contreras, A. Compaan, M. Cardona, and A. Axmann, Mat. Res. Soc. Proc. 23, 147 (1984).
14. J. Wagner and L. Viña, Phys. Rev. B 30, 7030 (1984).
15. J. Wagner, Ref. 2, p. 25.
16. J. Wagner, Phys. Rev. B 32, 1323 (1985).
17. J.W. Slotboom and H.C. de Graaff, Solid State Electron 19, 857 (1976).
18. R.P. Mertens, J.L. van Meerbergen, J.F. Nijs, and R.J. van Overstraeten, IEEE Trans. Electron. Dev. ED-27, 949 (1980).
19. H.S. Bennett, Ref. 2, p. 193.
20. R.W. Keyes, Comments Solid State Phys. 7, 149 (1977).
21. J. del Alamo, S. Swirhun, and R.M. Swanson, Ref. 2, p. 47; International Electron Device Meeting, Washington D.C. (USA), (1985), p. 290.

22. L. Viña and M. Cardona, Physica 117 B & 118 B, 356 (1983)
23. L. Viña and M. Cardona, Phys. Rev. B 29, 6739 (1984) and references therein.
24. G.D. Mahan, J. Appl. Phys. 51, 2634 (1980).
25. K.F. Berggren and B.E. Sernelius, Phys. Rev. B 24, 1971 (1981); Ref. 2, p. 11.
26. A. Selloni and S.T. Pantelides, Phys. Rev. Lett. 49, 586 (1982).
27. R.A. Abram, G.N. Childs, and P.A. Saunderson, J. Phys. C 17, 6105 (1984).
28. F. Thuselt and M. Rösler, Phys. Stat. Sol. (b) 130, 661 (1985); Phys. Stat. Sol. (b) 130, K 139 (1985).
29. P.T. Landsberg, A. Neugroschel, F.A. Lindholm, and C.T. Sah, Phys. Stat. Sol. (b) 130, 255 (1985).
30. H.J.J. de Man, IEEE Trans. Electr. Dev. ED-18, 833 (1971).
31. W.L. Kauffman and A.A. Bergh, IEEE Trans. Electr. Dev. ED-15, 732 (1968).
32. F.A. Lindholm, S.S. Li, and C.T. Sah, 11th IEEE Photovoltic Specialists Conf. 1975, p. 3.
33. R.A. Abram, Ref. 2, p. 203.
34. P.B. Allen, Phys. Rev. B 18, 5217 (1978).
35. M. Takeshima, Phys. Ref. B 27, 2387 (1983) and references therein.
36. E.O. Kane, Ref. 2, p. 3.
37. G. Dresselhaus, A.F. Kip, and C. Kittel, Phys. Rev. 98, 368 (1955).
38. J.C. Hensel and and G. Feher, Phys. Rev. 129, 1041 (1967).
39. H.D. Barber, Solid State Electron. 10, 1039 (1967).
40. P.A. Saunderson, Ph. D. thesis, University of Durham, 1983 (unpublished).
41. J. Wagner, J.C. Gelpey, and R.T. Hodgson, Appl. Phys. Lett. 45, 47 (1984).
42. J. Wagner, W. Appel, and M. Warth, J. Appl. Phys. 59, 1305 (1986).
43. H.C. Casey and F. Stern, J. Appl. Phys. 47, 631 (1976).
44. D. Olego and M. Cardona, Phys. Rev. B 22, 886 (1980).

45. A.N. Titkov, E.I. Chaikina, E.M. Komova, and N.G. Ermakova, Fiz. Tekh. Poluprovodn. 15, 345 (1981) [Sov. Phys. Semicond. 15, 198 (1981)].
46. J. De-Sheng, Y. Makita, K. Ploog, and H.J. Queisser, J. Appl. Phys. 53, 999 (1982).
47. R. Schwabe, A. Haufe, V. Gottschalch, and K. Unger, Solid State Commun. 58, 485 (1986).
48. D.D. Tang, IEEE Trans. Electron. Dev. ED-27, 563 (1980).
49. A.W. Wieder, IEEE Trans. Electron. Dev. ED-27, 1402 (1980).
50. C.M. van Vliet and A.H. Marshak, Ref. 2, p. 214.
51. A. Neugroschel, S.C. Pao, and F.A. Lindholm, IEEE Trans. Electron. Dev. ED-29, 894 (1982).
52. For a review see: T.M. Rice and J.C. Hensel, T.G. Phillips, and G.A. Thomas in "Solid State Physics," Vol 32, edited by H. Ehrenreich, F. Seitz, and D. Turnbull (Academic Press, New York, 1977).
53. A. Pinczuk, J. Shah, H.L. Störmer, R.C. Miller, A.C. Gossard, and W. Wiegmann, Surf. Sci. 142, 492 (1984).
54. A. Pinczuk, J. Shah, R.C. Miller, A.C. Gossard, and W. Wiegmann, Solid State Commun. 50, 735 (1984).
55. D.A. Kleinman and R.C. Miller, Phys. Rev. B 32, 2266 (1985).
56. D.A. Kleinman, Phys. Rev. B 32, 3766 (1985).
57. G.E.W. Bauer and T. Ando, J. Phys. C 19, 1537 (1986).

SEMINARS GIVEN AT THE SCHOOL

Dr. C. Klingshirn Physikalisches Institut, Universitat, Robert-Mayer-str 2-4,D 6000 Frankfurt am Main W. Germany	Localisation in Semiconductors
J.C. Licini, MIT Room 13-2150 77 Massachusetts Ave. Cambridge MA02139 USA	Quasi-1 dimensional Mosfets
Dr. V Vieira Centro Fisica Materia Condensada Av Prof Gama Pinto 2 1699 Lisboa Portugal	Non-Equilibrium Methods for Quantum Disordered Systems
Y. Meir School of Physics Tel Aviv University Ramat Aviv 69978 Tel Aviv Israel	Localization and Quantum Percolatior
U. Sivan School of Physics Tel Aviv University Ramat Aviv 69978 Tel Aviv Israel	Level Correlation Function and AC Conductivity in a Finite Disordered Sample
V. Chandrasekhar Becton Center Yale University New Haven CT06520 USA	Quantum Transport in Small Metal Structures
Dr. J-L Pichard DPh G/PSRM Orme des Merisiers 91191 Gif sur Yvette Cedex France	Large Volume Limit of the Distribution of the Localisation Lengths and One Parameter Sealing

C. Gros
Inst Fur Theoretische Physik
ETH
CH 8093 Zurich
Switzerland

An Introduction to Heavy Fermion Systems

M. Lakrimi
Physics Division
Sussex University
Brighton
BN1 9QH
England

Magnetoresistance Measurement in 1-d ALGaAs Heterostructures

F.I.B. Williams
DPH SPSRM
Cen Saclay
91191 Gif-sur-Yvette Cedex
France

Electrons on Liquid Helium

Dr. H. Yurtseven
Department of Physics
University of Ankara
Besevler
Ankara
Turkey

General Thermodynamic Treatment of Crystal Systems close to Phase Transitions

Dr. P.C.W. Holdsworth
Physics Department
6224 Agriculture Road
Vancouver
BC V6H 1H9
Canada

Theory of Conductivity and Thermopower in Localized Electronic States

POSTERS PRESENTED AT THE SCHOOL

The Soap Froth: a prototypical system with topological disorder.

C.W.J. Beenakker
Philips Research Laboratories, Eindhoven, The Netherlands.

Localization and Interactions in One-Dimensional Metals

T. Giamarchi and H.J. Schulz
Laboratoire de Physique des Solides UPS bat.S10 91405
Orsay, France.

Non-metal Metal Transition in Semimagnetic Semiconductors

T. Wojtowicz, T. Dietl, M. Sawicki, W. Plesiewicz and
J. Jaroszyinski
Institute of Physics, Polish Academy of Sciences, Warsaw,
AL Lotnikow 32/46, Poland.

Zero Current Voltage Oscillations in Al_xGa_{1-x}-GaAs Heterostructures.

A.D.C. Grassie, J.M. Hutchings, M. Lakrimi
Physics Division, University of Sussex, Brighton, UK.

Magnetoresistance Measurements on Very Narrow $Al_xGa_{1-x}As$-GaAs
Heterostructures.

A.D.C. Grassie, K.M. Hutchings, M. Lakrimi
Physics Division, University of Sussex, Brighton, UK.
J.J. Harris and C.T. Foxon
Philips Research Laboratories, Redhill, UK.

Hopping Conductivity for Localized Electronic States

R. Barrie, P.C.W. Holdsworth, M.R.A. Shegelski
Department of Physics, UBC, Canada.

Normalized Resistance and Magnetoresistance Scaling in $Bi_{1-x}Sn_x$.

H. White and D.S. McLachlan

University of The Witwatersrand, Johannesburg, South Africa.

Quantum Transport in Small Metal Structures.

V. Chandrasekhar, W. Wind, M.J. Rooks and D.E. Prober
Yale University.

Mr Sayed Abboudy	Dept. of Physics, Royal Holloway and Bedford New College, Egham, Surrey TW20 0EX, UK. (Alexandria, Egypt).
Dr. Osman Adiguzel	Dept. of Materials Science, Surrey University, Guildford, Surrey GU2 5XH UK. (Elazig, Turkey).
Dr. David V. Baxter	Dept. of Physics, McGill University, Montreal, Quebec, H2X3R4, Canada.
Dr. Carlo W.J. Beenakker	Philips Research Labs, PO Box 80.000, 5600 JA Eindhoven, The Netherlands.
Mr Ward P. Beyermann	Dept. of Physics, California University, Los Angeles, CA90024, USA.
Mr John H. Burnett	Dept. of Physics, Harvard University, Cambridge, MA 02138, USA.
Mr Thierry Capra	D.P.M. Bat 203, 43 bd du 11 Novembre 1918, 69622 Villeurbanne, France.
Dr. Eleuterio Castano	Blackett Laboratory, Imperial College, Prince Consort Road, London, SW7 2BZ, UK.
Mr Venkat Chandrasekhar	Becton Center, Yale University, New Haven, CT 06520, USA.
Mr Rafael Chicon	Dpto de Fisica, Facultad de Ciencias, Universidad de Murcia, Murcia 30001, Spain.
Nadine de Courtenay	17 rue Gramme, 75015 Paris, France.
Mr Pascal Dellouve	Lab de Physique des solides, Universite de Paris-Sud, Bat 510, 91405 Orsay, France.
Dr. George Dousselin	Lab Physique des Solides, INSA Rennes, 20 Av des Buttes de Coesmes, 35043 Rennes Cedex, France.
Mr Driss El-Khatouri	Lab de Phys des Solides, Universite P et M Curie, 4 Place Jussieu Tour 13, 75230 Paris Cedex 05, France.
Mr Mauricio Esguerra	Landwehrstr 52, D 8000 Munchen 2, West Germany.

Mr Didier Gauthier	Dept. Physique, INSA Toulouse, Av de Rangueil, 31077 Toulouse, France
Mr Robert S. Germain	Dept. of Physics, Cornell Univeristy, Ithaca, NY 14853-2501, USA
Dr. Gérard Ghibaudo	LPCS-ENSERG, 23 rue des Martyrs, 38031 Grenoble, France.
Mr Thierry Giamarchi	Lab de Physique des Solides, UPS Batiment 510, 91405 Orsay, France.
Dr. Andrzej Gorski	Inst. of Nuclear Physics, Radzikowsk-iego 152, 31-342 Krakow, Poland.
Mr Claudius Gros	Inst. fur Theoretische Physik, ETH, CH 8093 Zurich, Switzerland.
Mr Selman Hershfield	Clark Hall, Cornell University, Ithaca, NY 14853-2501, USA.
Mr Nir Hess	Dept. of Physics, Tel Aviv University, 69978 Ramat Aviv, Tel Aviv, Israel.
Dr. Peter C.W. Holdsworth	Physics Dept., 6224 Agriculture Road, Vancouver, BC V6H 1H9, Canada.
Mr Peter F. Hopkins	Conant Hall 12, Harvard University, Cambridge, MA 02138, USA.
Mr Andrew Kent	1053 Cardinal Way, Palo Alto, CA 94303, USA.
Dr. Claus Klingshirn	Physikalisches Institut, Universitat, Robert-Mayer-str 2-4, D 6000 Frankfurt am Main, West Germany.
Dr. Fumio Komori	Physics Dept., Tokyo University, Hongo Bunkyo-Ku, Tokyo, Japan.
Dr. Tadeusz K. Kopec	Inst. of Low Temperature Research, Polish Acad. of Sciences, Prochnika 95, 53-529 Wroclaw, Poland.
Mr Hans-Rainer Kraus	Max-Planck Institut, Heisenbergstr 1 7000 Stuttgart 80, West Germany.
Mr M'hamed Lakrimi	Physics Division, Sussex University, Brighton BN1 9QH, UK.
Mr M.L. Leadbeater	Physics Dept. Nottingham University, Nottingham, NG7 2RD, UK.

Mr James Leo	Physics Dept. Imperial College, Prince Consort Road, London, SW7 2BZ, UK.
Mr Jerome C. Licini	MIT Room 13-2150, 77 Massachusetts Ave, Cambridge, MA 02139, USA.
Mr Peter Lindqvist	Dept. of Solid State Physics, Royal Inst. of Technology, S-10044 Stockholm, Sweden.
Mr Hans-Peter Loebl	Physik Dept. Inst. E13, James Franck Str, D8046 Garching, West Germany.
Mr Kenneth D. Mackay	Cavendish Lab. Cambridge University, Madingley Road, Cambridge CB3 OHE, UK.
Mr Michael Maliepaard	Cavendish Lab. Cambridge University, Madingley Road, Cambridge CB3 OHE, UK.
Mr Vladimir Matijasevic	Ginzton Lab. Stanford Univeristy, Stanford, CA 94305, USA.
Mr Yigal Meir	School of Physics, Tel Aviv University, Ramat Aviv 69978, Tel Aviv, Israel.
Mr Timothy L. Meisenheimer	Physics Dept. Purdue University, W Lafayette, Indiana 47907, USA.
Mr Michele Migliuolo	Physics Dept. University of Rochester, Rochester, NY 14627, USA.
Nancy A. Missert	Ginzton Lab. Stanford University, Stanford, CA 94305, USA.
Mr Moustafa A-K Mohamed	Dept. of Physics, University of Alberta, Edmonton, Alberta T6G 2J1, Canada (Cairo, Egypt).
Dr. Yutaka Nishio	Faculty of Science, Toho University, 2-1, Miyama, 2 chome Funabashhi-shi, Chiba, 274, Japan.
Mr Tomi Ohtsuki	Physics Dept. Univeristy of Tokyo, Hongo 7-3-1 Bunkyo ku, Tokyo, Japan.
Dr. M. Nilgün Ozer	Physics Dept. Bogazici University, Bebek, Istanbul, Turkey.
Dr. Jean-Louis Pichard	Dph G/PSRM, Orme de Merisiers, 91191 Gif sur Yvette Cedex, France.

Mr Jean Piquet
Ecole Nationale Supérieure de Techniques Avancées, 32 Boulevard Victor, 75015, Paris, France.

Mr Daniel H. Reich
Dept. of Physics, University of Chicago, Chicago, Illinois 60637, USA.

Mr Reinhart Richter
Dept. of Physics, McGill University, 3600 University Street, Montreal PQ H3A 2T8, Canada.

Mr Pedro D.S. Sacramento
Centro fisica Materia Condensada, Av Prof Gama Pinto 2, 1699 Lisboa Codex, Portugal.

Dr. Michel Saint Jean
GPS de l'Ecole Normale Superieure, 24 rue Lhomond, 75231 Paris, France.

Dr. Bernadette Sas
Central Research Inst. for Physics, Hungarian Academy of Sciences, H 1525 Budapest 114, POB 49, Hungary.

Mr Mattias Severin
Dept. of Physics, Universitetet i Linkoping IFM, 581 83 Linkoping, Sweden.

Mr Christopher Shearwood
Dept. of Physics, University of Leeds, Leeds LS2 9JT, UK.

Mr Uri Sivan
School of Physics, Tel Aviv University, Ramat Aviv 69978, Tel Aviv, Israel.

Mr Walter F. Smith
McKay Lab., Harvard University, 9 Oxford Street, Cambridge, MA02138, USA

Mr Roger Sollie
Inst. of Theoretical Physics, Universitetet I Trondheim, N-7034 Trondheim-NTH, Norway.

Dr Sandro Sorella
ISAS Trieste, Strada Costiera 11, 34013 Trieste, Italy.

Dr Giancarlo Strinati
Dipartimento di Fisica, Universita degli Studi di Roma, Piazzale A Moro 2 00185 Roma, Italia.

Mr Richard P. Taylor
Physics Dept. Nottingham University, Nottingham, NG72RD, UK.

Mr Leandro R. Tessler
Dept. of Physics, Tel Aviv University, Ramat Aviv 69978, Tel Aviv, Israel.

Mr Michel Trudeau	Dept. of Physics, Montreal Univeristy, CP 6128 Succ A, Montreal, Quebec H3C 3J7, Canada.
Dr. Witold Trzeciakowski	Unipress, Polish Academy of Sciences, Sokolowska 29, 01 142 Warsaw, Poland.
Dr. Vitor Vieira	Centro Fisica Materia Condensada, Av Prof Gama Pinto 2, 1699 Lisboa, Portugal.
Mr Juan C. Villagran	Dept. of Physics, University of Texas at Austin, Texas, 78712, USA.
Mr Jürgen Vogel	Inst. Theoretical Physics II,Gluckstr 6, 8520 Erlangen, West Germany.
Mr Joseph Vranken	Vastestof Fysika en Magnetisme, K U Leuven, B 3030 Leuven, Belgium.
Mr Rodolfo Wehrhahn	Inst. Theoretische Physik, Universitat Hamburg, Jungiusstrasse 9, 2000 Hamburg 36, West Germany.
Mr David Wharam	Cavendish Laboratory, Cambridge University, Madingley Road, Cambridge, CB3 OHE, UK.
Mr Hylton White	Dept. of Physics, University Witwatersrand, 1 Jan Smuts Ave., Johannesburg, South Africa.
Mr Geoffrey Whittington	Dept. of Physics, Nottingham University, University Park, Nottingham, NG7 2RD, UK.
Dr F.I.B. Williams	DPH SPSRM, Cen Saclay, 91191 Gif-sur-Yvette Cedex, France.
Mr Tomasz Wojtowicz	Inst. of Physics, Polish Academy of Sciences, Al Lotnikow 32/46, 02-668 Warszawa, Poland.
Mr Süleyman Yalcin	Metallurgical Eng. Dept., METU, Ankara, Turkey.
Dr. Hamit Yurtseven	Dept. of Physics, University of Ankara, Besevler, Ankara, Turkey.
Prof. Gerd Bergmann	Physics Dept., University of Southern California, Los Angeles, CA90089-0484, USA.

Prof. Hidetoshi Fukuyama	Inst. for Solid State Physics, University of Tokyo, Roppongi Minato ku, Tokyo 106, Japan.
Prof. Ja'nos Hajdu	Inst. fur Theoretische Physik, Universitat Koln, Universitatstrasse 14, D 5000 Koln 41, West Germany.
Prof. Donald F. Holcomb	Dept. of Physics, Cornell University, Ithaca, NY 14853, USA.
Prof. Sir Nevill Mott	Cavendish Laboratory, Cambridge University, Madingley Road, Cambridge, CB3 OHE, UK.
Dr. Angus MacKinnon	Blackett Laboratory, Imperial College, Prince Consort Road, London, SW7 2BZ, U
Dr. Robin Nicholas	Clarendon Laboratory, Oxford University Parks Road, Oxford, UK.
Dr. Michael Pepper	Cavendish Laboratory, Cambridge University, Madingley Road, Cambridge, CB3 OHE, UK.
Prof. Philip J. Stiles	Physics Dept. Brown University, Rhode Island 02912, USA.
Dr. Gordon A. Thomas	AT&T Bell Laboratories, 600 Mountain Avenue, Murray Hill, N.J.07974, USA.
Prof. Klaus von Klitzing	Max-Planck-Inst. Heisenbergstr 1, Postfach 80 06 65, 7000 Stuttgart 80, West Germany.
Dr. Joachim Wagner	Fraunhofer-Inst. fur Angewandte Festkorperphysik, Eckerstrasse 4, D 7800 Freiburg, West Germany.

AUTHOR INDEX

Numbers in parentheses are the reference numbers appearing at the end of the particular article